THE
OXFORD BOOK OF
HISTORICAL
STORIES

Michael Cox is a senior commissioning editor with Oxford University Press. He has edited *The Oxford Book of English Ghost Stories* (1986) and *Victorian Ghost Stories* (1991) (both with R. A. Gilbert), and *Victorian Detective Stories* (1992). He is the author of *M. R. James: An Informal Portrait* (1983).

Jack Adrian is a leading authority on popular fiction in the twentieth century and the editor of two collections of short stories from *The Strand* magazine (*Detective Stories* and *Strange Tales*), two collections of stories by E. F. Benson (*Desirable Residences* and *Fine Feathers*), a volume of stories by Rafael Sabatini, *The Fortunes of Casanova*, and together with Bill Pronzini he has edited *Hard-boiled: An Anthology of American Crime Stories*.

THE
OXFORD BOOK OF
HISTORICAL
STORIES

ϾϽ

Edited by

MICHAEL COX

AND

JACK ADRIÀN

Oxford   New York

OXFORD UNIVERSITY PRESS

1995

Oxford University Press, Walton Street, Oxford OX2 6DP

Oxford   New York
Athens   Auckland   Bangkok   Bombay
Calcutta   Cape Town   Dar es Salaam   Delhi
Florence   Hong Kong   Istanbul   Karachi
Kuala Lumpur   Madras   Madrid   Melbourne
Mexico City   Nairobi   Paris   Singapore
Taipei   Tokyo   Toronto

and associated companies in
Berlin   Ibadan

Oxford is a trade mark of Oxford University Press

First published 1994
First issued as an Oxford University Press paperback 1995

British Library Cataloguing in Publication Data

Data available

Library of Congress Cataloging in Publication Data
The Oxford book of historical stories / edited by
Michael Cox and Jack Adrian.
p.   cm.
1. Historical fiction, English.   2. Short stories, English.
I. Cox, Michael, 1948–   . II. Adrian, Jack.
PR1309.H4509   1994   823'.08108—dc20   94–647
ISBN 0-19-283208-5

1 3 5 7 9 10 8 6 4 2

Printed in Great Britain
Biddles Ltd
Guildford and King's Lynn

CONTENTS

ACKNOWLEDGEMENTS

We should like to thank the following individuals and institutions for help and advice received during the preparation of this anthology: Robert Adey, Mike Ashley, Dr Glen Cavaliero, Richard Dalby, Owen Dudley Edwards, Professor Douglas G. Greene, W. O. G. Lofts, Barry Pike, the late Robert Sampson, Professor Tom Shippey; the staff of the Bodleian Library, Oxford (especially those in the Old Library Reading Rooms, the Copying Department, the Stack, and, at Nuneham Courtenay, John Slatter), and in particular Richard Bell, Head of Reader Services; the University Library, Cambridge; the National Library of Scotland.

INTRODUCTION

History is too important to be left solely to historians. Or, to put it less provocatively, historical fiction shows that there are parts—important parts—of the experience of history that the professional historian cannot be expected to reach. Such is the thrust of the 'little word of warning' Conan Doyle puts into the mouth of Brigadier Gerard:

When you hear me speak, you must always bear in mind that you are listening to one who has seen history from the inside. I am talking about what my ears have heard and my eyes have seen, so you must not try to confute me by quoting the opinions of some student or man of the pen, who has written a book of history or memoirs. There is much which is unknown by such people, and much which will never be known by the world.[1]

In the hands of its most accomplished practicners, historical fiction finds a way into the internal life of the past that is denied to the scholarly method. For the activities of the historian and of the writer of historical fiction are not fundamentally inimical. Each seeks to recover the past, as R. G. Collingwood noted:

Each of them makes it his business to construct a picture which is partly a narrative of events, partly a description of situations, exhibition of motives, analysis of characters. Each aims at making his picture a coherent whole, where every character and every situation is so bound up with the rest that this character in this situation cannot but act in this way, and we cannot imagine him as acting otherwise. The novel and the history must both of them make sense; nothing is admissible in either except what is necessary, and the judge of this necessity is in both cases the imagination.[2]

History writing and historical fiction have this much common ground: they are both imaginative constructs since all historical narrative is, to some extent, history re-made and re-fashioned. The writing of history and historical fiction are—or can be—complementary means of understanding the past.

The future Professor Sir Herbert Butterfield brought his sympathetic and seminal little study, *The Historical Novel* (1924), to its close with one idea of the contrast between these activities:

---

[1] Arthur Conan Doyle, 'How the Brigadier Slew the Brothers of Ajaccio', *Strand* (June 1895); reprinted in *The Exploits of Brigadier Gerard* (1896).
[2] R. G. Collingwood, *The Idea of History* (Oxford, 1946), 245–6.

With a set of facts about the social condition of England in the Middle Ages the historian will seek to make a generalization, to find a formula; the novelist will seek a different sort of synthesis and will try to reconstruct a world, to particularize, to catch a glimpse of human nature. Each will notice different things and follow different clues; for to the historian the past is the whole process of development that leads up to the present; to the novelist it is a strange world to tell tales about.[3]

Thus considered, historical fiction is one of the oldest literary activities of which we know: the *Iliad* and the *Odyssey* contain fictional material thrown around real people and real events; the older stories in the Bible may have been conceived with comparable sentiments, whilst Dante can be seen as the most revolutionary historical fictionist of all time, with his use of historical figures and what he takes to be real events placed in the context of eternal justice. For many centuries no great distinction was necessarily made between historical factual writing and historical fiction. The need for national myths has always existed, from Livy's refurbishment of the legendary founding of Rome to Mason Weems's story of George Washington and his father's cherry-tree. It was partly in answer to the myth-making habit that William Robertson, the most dependable of the Enlightenment historians, preached the austere gospel of 'the dignity of history'. One effect was to drive social history into the hands of the historical fictionist, specifically of Robertson's pupil, Walter Scott.

Scott, the fountain-head of English historical fiction in prose, brought a historian's professionalism to the skills of the novelist. In the late eighteenth century, history in Gothic fiction was frequently—and literally —incredible. But in Scott (though he was an admirer of Ann Radcliffe) credulity is seldom outraged. The ability to write easily and fluently of the past, without any appearance of contrivance or pedantry, was the transforming fact of Scott's fiction. As T. B. Macaulay saw, Scott showed historians how much history they were losing:

At Lincoln Cathedral there is a beautiful painted window, which was made by an apprentice out of the pieces of glass which had been rejected by his master. It is so far superior to every other in the church, that, according to the tradition, the vanquished artist killed himself from mortification. Sir Walter Scott, in the same manner, has used those fragments of truth which historians have scornfully thrown behind them in a manner which may well excite their envy. He has constructed out of their gleanings works which, even considered as histories, are scarcely less valuable than theirs.[4]

---

[3] H. Butterfield, *The Historical Novel—an Essay* (1924), 113.

[4] Macaulay, 'History', *Miscellaneous Writings*, ed. Thomas Flower Ellis (1860); reprinted from the *Edinburgh Review* (May 1828).

To Macaulay, in 1828, historical fiction gave future historians their agenda:

> To make the past present, to bring the distant near, to place us in the society of a great man or on the eminence which overlooks the field of a mighty battle, to invest with the reality of human flesh and blood beings whom we are too much inclined to consider as personified qualities in an allegory, to call up our ancestors before us with all their peculiarities of language, manners, and garb, to show us their houses, to seat us at their tables, to rummage their old-fashioned wardrobes, to explain the uses of their ponderous furniture, these parts of the duty which properly belongs to the historian have been appropriated by the historical novelist.[5]

In the 1840s and 1850s the historical novelist was in direct competition for readers with the historian. Macaulay declared that in writing his *History of England* he would not be satisfied unless it 'shall for a few days supersede the last fashionable novel on the tables of young ladies'.[6] It was no coincidence that Thackeray's *The History of Henry Esmond* (1852) appeared only three years after the publication of the first two volumes of Macaulay's *History*. And just as Thackeray desired to assume something of the historian's aura, so Macaulay defined the perfect historian in terms of literary technique, as one 'in whose work the character and spirit of an age is exhibited in miniature': 'by judicious selection, reflection, and arrangement, he gives to truth those attractions which have been usurped by fiction.'[7]

The vogue for historical fiction after Scott spanned European and American culture. In the USA, James Fenimore Cooper became one of Scott's most devoted admirers, his doomed Amerindians paralleling Scott's doomed Highlanders, whilst Nathaniel Hawthorne more directly confronted the shame and glory of his own New England Puritan ancestry; European writers, from Dumas to those like Victor Hugo, Alessandro Manzoni, and Benito Pérez Galdós who used historical fiction in the cause of political liberty, extended the genre's boundaries, culminating in one of the greatest of all historical novels, Tolstoy's *War and Peace* (1865–72); in England, W. Harrison Ainsworth, Bulwer Lytton, Dickens, Charles Reade, and Charles Kingsley produced best-selling historical novels.

History in the English-speaking world was animated by what seemed the mutually antagonistic impulses of progress and retrospection, and the fictionists followed suit. History and its satellite fictions might be written to show how much superior was the present to times past, or how much

---

[5] Macaulay, 'Hallam's Constitutional History', *Critical and Historical Essays* (1844), reprinted from the *Edinburgh Review* (Sept. 1828).

[6] Macaulay to Macvey Napier, 5 Nov. 1841, in Thomas Pinney (ed.), *The Letters of Thomas Babington Macaulay*, iv (1977), 15.

[7] Thackeray to Mrs Brookfield, 13 Sept. 1849; quoted in Pinney, *Letters of Thomas Babington Macaulay*, v (1981), 72n.

we had lost in moving onward from the past, and in the hands of the best authors it often did both. Ghost stories—another popular nineteenth-century genre—dramatized an active relationship between the two states of time present and time past: the ghost as a metaphor for the ever-living past. But historical fiction, a more ubiquitous and varied literary form, perhaps spoke more profoundly for the times—not just by its colonization of the past to serve present needs, but, more particularly, by its relation to the wider search for origins—whether geological, biological, national, or cultural. The Victorian age—modern, expansionist, technologically and socially advancing—was an age obsessed by its past. In its own way, historical fiction—like Pre-Raphaelite art or the pseudo-medievalism of Tennyson—typified this urge to turn away from the insistent clamour of the present; yet, like Macaulay's exotic narrative in its voluptuous prose, the moral was so often a comforting assurance of the superiority of the present, whether in human compassion or modern convenience.

Though the prevailing fictional temper of the Victorian period was one which gave precedence to realistic depictions of contemporary life, fiction that located itself in a past that could only be imagined struck a responsive chord in many readers. History also offered a virtually endless supply of incidents, settings, and characters that could be used by those whose principal aim was to tell a good story. Here, indeed, lay the source of its mass appeal: history *as* story; history as drama; history as action.

History, too, as propaganda, whether blatant or insinuating, appealed to young readers when presented in fictional form. As the century advanced, it was used to inculcate the values of Empire in the young: excitement, danger, and suspense coated the expository pill. In book after book the aggressive appropriation of other lands and other peoples was disguised as adventure and romance, with the imperial idea sometimes legitimized by settings in a remote past—amongst the followers of Alexander the Great or the legions of Rome. George Alfred Henty was to bring this approach to a (not too) fine art, using Scott's audience-identification technique by which the story is seen through the eyes of a marginal youth (in Henty a boy). R. L. Stevenson may have generously identified with his predecessors in boys' fiction in the dedicatory poem to *Treasure Island* (1883), but he was a far greater writer and lacked the moral bullying with which purveyors of juvenile historical fiction—W. H. G. Kingston, R. M. Ballantyne, Frederick Marryat, George Manville Fenn, and others—often invaded the past.

The didactic potential of historical fiction was keenly perceived by many Victorian writers. Charlotte Yonge, author of *The Book of Golden Deeds* (1864), saw historical tales as 'enabling the young to realize history vividly—and, what is still more desirable, requiring an effort of the mind which to read of modern days does not'.[8] As late as 1912 J. A. Buckley and W. T. Williams (schoolmasters both) opened their *Guide to British Historical Fiction* with the words: 'No attempt need be made to demonstrate the value of historical fiction as a handmaiden to history proper.' The symbols of servitude as a fixed station in life, propriety as synonymous with history, value as implicitly moral no less than economic, helped make these teachers of what would be the Great War generation appropriate heirs to Kingsley, who wrote *Westward Ho!* (1855) during the Crimean War 'to make others fight' and to 'put into people's heads some brave thoughts'.[9]

Throughout the nineteenth century, self-contained historical short stories had remained somewhat in the shadow of the historical novel. Scott's short historical tales—such as 'The Two Drovers' and 'Wandering Willie's Tale' (a historical ghost story set into *Redgauntlet* (1824))—were few in number and it was as a novelist that Scott exerted his influence over Victorian historical fiction. As Conan Doyle remarked:

there are far fewer supremely good short stories than there are supremely good long books. It takes more exquisite skill to carve the cameo than the statue. But the strangest thing is that the two excellences seem to be separate and even antagonistic. Skill in the one by no means ensures skill in the other. The great masters of our literature, Fielding, Scott, Dickens, Thackeray, Reade, have left no single short story of outstanding merit behind them . . . On the other hand, men who have been very great in the short story . . . have written no great book. The champion sprinter is seldom a five-miler as well.[10]

Historical short stories can be found in the early Victorian magazines (*Bentley's Miscellany*, edited by Dickens from 1837 to 1839, is a particularly rich source), but they are uneven in quality and often suffer from the prolixity to which the period was prone. It was the technological advances that led to the appearance in the 1890s of the lavishly illustrated general interest monthly magazine that heralded the golden age of short historical fiction, and of popular short fiction in general.

The great exemplar was George Newnes' *Strand Magazine*—launched in January 1891 with a mix of serialized novels, short stories, and entertaining, rather than improving, articles, and profusely illustrated. The Sherlock Holmes stories established the template of the short story series; and short historical stories—including Conan Doyle's Brigadier Gerard series—followed in their wake. By the end of the 1890s the *Strand* had become a national institution, with a circulation approaching half a million

---

[8] Charlotte M. Yonge, Preface to *The Chaplet of Pearls; or, The White and Black Ribaumont* (1868).

[9] Charles Kingsley to the publishers Macmillan's, 22 July 1854; quoted in J. A. Sutherland, *Victorian Novelists and Publishers* (1976), 118.

[10] Arthur Conan Doyle, *Through the Magic Door* (1907), 113.

copies a month. Rivals rapidly proliferated—*Pearson's*, the *Royal*, the *Windsor*, *Harmsworth's* (later the *London*), *Nash's*, *Pall Mall*, the *Ludgate*, *Cassell's*; and though none quite matched the *Strand*'s success, they still sold in the hundreds of thousands. At the same time, fiction-only magazines—no articles and few illustrations—began to appear, amongst them the *Red Magazine*, the *Storyteller*, the *Grand Magazine*, the *Novel Magazine*, and the *Premier Magazine*. From 1910 onwards, in any given month, there might be roughly a dozen all-fiction magazines and perhaps fifteen or twenty fiction-carrying periodicals on sale, representing something like two million words of fiction every month, of which a significant proportion was historical in character. In book publishing the statistics are even more striking. In 1914 the librarian and bibliographer Ernest Baker published his *Guide to Historical Fiction*, an exhaustive descriptive catalogue of all historical novels and short-story collections published since the early nineteenth century on both sides of the Atlantic. Baker seems to have missed very little: from 1830 to 1914, over 1300 historical novels had been published. Even more significant, however, is that in the dozen or so years from 1901 to 1914 over 1200 works of historical fiction were issued, of which nearly a quarter were short-story collections.

In its second issue the *Strand*, following a policy of translating established European authors, ran a short story by Balzac, 'A Passion in the Desert', set in the Napoleonic era. Two issues later it presented Maupassant's 'Two Fishers', commending the author as 'one of the best writers of short tales now living . . . his remarkably artistic style . . . tells a story in its full effect without a word too much or too little'.[11] Within the same April number the *Strand* introduced a contemporary author swashbuckling his way through the France of Henri IV: Stanley J. Weyman's 'The King's Stratagem'. It was a foretaste of things to come. As yet largely unknown, despite his *The House of the Wolf* (1890), Weyman was to prove the great historical fiction bestseller of his time; indeed, much of the pre-Great War popularity of historical fiction can be attributed to Weyman's success. A second boom in historical fiction after the First World War owed much to Weyman's follower Rafael Sabatini, who first came to mass prominence with *Scaramouche*, his engrossing tale of the French Revolution, in 1921. *Scaramouche* was a world bestseller, but its success was swiftly eclipsed by *Captain Blood* in 1922. Both books had an enormous impact, not only on historical fiction but also, through the new storytelling medium of film, on popular cultural consciousness.

Since the late nineteenth century, some historical periods have held more attraction for writers than others, to the extent that a few have almost become sub-genres in their own right: the Regency Romance (linking

Conan Doyle's *Rodney Stone* (1896), Jeffery Farnol's *The Broad Highway* (1910), and, later, the best-known work of Georgette Heyer); tales set in the English seventeenth century (for example, J. H. Shorthouse's *John Inglesant* (1881) or Rose Tremain's *Restoration* (1989)); stories of Ancient Rome (a still vigorous strain, from Robert Graves's fictional autobiography of Claudius to Allan Massie's of Augustus); above all, the French Revolution and the literary children of Dickens's *A Tale of Two Cities* (1859), the most successful of all being Baroness Orczy's clever transformation of the languid English fop into fearless hero in *The Scarlet Pimpernel* (1905). In America, the defeated South has been the focus for a tradition of historical fiction that encompasses Margaret Mitchell's *Gone With the Wind* (1937), the stories of William Faulkner, and the novels of Robert Penn Warren. The grim potentialities of historical fiction were revealed in the Negrophobic Thomas Dixon's *The Clansman* (1906), which portrayed the struggle for equality in post-war Reconstruction as the black rape of white virginity, inspiring outbreaks of lynching when brought to the screen with Lilian Gish as the suicide victim in D. W. Griffith's *Birth of a Nation* (1915). However loathsome Dixon's ethos and its fictional expression, it kept pace with history proper, which in the hands of austere academics up to 1950 was as supportive of white Southern supremacy as any fiction.

If history can be fictionalized it can also be re-imagined in other ways, producing fictions of alternative history: in Keith Roberts's *Pavane* (1968) and Kingsley Amis's *The Alteration* (1976), the Reformation has never happened; in L. Sprague de Camp's *Lest Darkness Fall* (1941) the barbarians never over-ran Rome; in Philip Dick's *The Man in the High Castle* (1962), Len Deighton's *SS-GB* (1978), and Robert Harris's *Fatherland* (1992) the Allies have lost the Second World War. Speculation of another sort has attempted to understand the sexual history of the past, as in Margaret Yourcenar's exploration of male homosexuality in *Mémoires d'Hadrien* (1951) or the novels of Mary Renault.

A particularly fertile cross-breed has been the 'Mystorical'—the detective story set in a past era. Scott's Paulus Pleydell (in *Guy Mannering*) and Dumas's D'Artagnan (in *Louise de la Vallière*) might be considered as detectives in a broad sense; but the historical detective first appeared fully formed in Melville Davisson Post's *Uncle Abner; Master of Mysteries* (1918), set in Jeffersonian Virginia. The vogue was given concerted impetus by another American writer, Lillian de la Torre, who, in 1944, made use of the similarity in the relationships of Sherlock Holmes and Dr Watson and of Dr Johnson and James Boswell (who indeed had contributed to the creation of Conan Doyle's famous partnership). In 1945 Agatha Christie set *Death Comes as the End* in ancient Egypt, though her sure hand as a writer of detective fiction deserted her in the anxiety to defer to the demands of authenticity. A contemporary variation of Holmes and Watson, and one with high claims of literary status, appears in Umberto Eco's medieval *tour de force*, *The Name of the Rose* (first published in English 1983); on a more modest level, the Middle Ages have also been used successfully by Edith Pargeter (writing as Ellis Peters) in her Brother Cadfael mystery novels.

The manipulation of history, whether in the naval novels of Patrick O'Brien or George MacDonald Fraser's Flashman series (one of the great charms of which is to have Flashman, a character borrowed from classical Victorian fiction, invade other works of fiction, such as *The Prisoner of Zenda*), continues to find a mass readership. We are, it seems, as receptive to the imagined past as our Victorian predecessors. Yet there has been no anthology of historical short stories since Sabatini's *A Century of Historical Stories* (1936), though this also contained novel extracts (Mike Ashley's *The Mammoth Book of Historical Whodunnits* (1993) is limited to the 'mystorical' sub-genre). The reason for this has much to do with the continuing dominance of the historical novel, both in the market-place and in terms of critical interest; but it also reflects the fact that, like short stories in general, the genre's natural habitat—the fiction-carrying monthly magazine aimed at a mass readership—has virtually disappeared (the *Strand* died in 1950, its rivals having long predeceased it). To some extent, therefore, our aim in this collection has been rehabilitatory: to show the range and fecundity of short historical fiction over the last 150 years. We have included only genuine short stories (i.e. no extracts from longer works), and our definition of historical is a fairly broad one. By it we mean a piece of fiction set in a period (usually at least fifty years) prior to the author's birth. Within this definition we have tried to include a diversity of stories: swashbuckling adventure; stories that set out seriously to engage with historical issues or personalities; ghost stories with historical settings; stories that evoke great events; historical detective stories; nostalgic or empathetic re-creations; pastiches; stories that reassemble the ordinary experiences of past humanity.

We hope the result will be a collection with a high entertainment value for anyone with a taste for fiction set in the past. A more serious, though subsidiary, aim has been to test the view that historical fiction—in the most accomplished hands—can sharpen our sense of the past by reminding us that history is about human beings and about the universality of human experience. Historical fiction can be narrowly escapist in its appeal; but it can also convey, in a uniquely powerful way, the awareness that we are all part of the unravelling fabric of history. Thomas Carlyle said that Scott's historical novels taught a truth that was

as good as unknown to writers of history and others, till so taught: that the bygone ages of the world were actually filled by living men, not by protocols, state-papers, controversies and abstractions of men. Not abstractions were they, not diagrams and theorems; but men, in buff or other coats and breeches, with colour in their cheeks, with passions in their stomach, and the idioms, features and vitalities of very men.[12]

The words hold good for many of the writers represented in this anthology.

We have tried to select examples in which the pure narrative skills of the storyteller are combined with a demonstrable sense of responsibility towards the historical record. The effective deployment of necessary historical fact is a perpetual technical problem for the writer of historical short stories, for, unlike the historical novelist, the form affords no leisure to develop background authenticity, no scope to show off the fruits of historical research, and requires an almost instantaneous creation of ambience, so that the reader is drawn quickly and effortlessly into an age that is not his own. One of the most important criteria for inclusion in this collection, therefore, has been the ability to bring off what Hugh Walpole called the 'double art'—of 'eliminating Time and in the same moment re-creating Period'.[13] This is impressive in a novel; in a short story it is doubly so.

Sabatini's introduction to the *Century of Historical Stories* asserted that historical fiction was 'still regarded with a measure of disfavour in certain pseudo-intellectual quarters'. Perhaps it still is. But he went on to maintain, robustly, that 'the achievement of a vivid narrative of events that might have taken place a century or so ago is one that places its author midway between the novelist and the historian; and this, for a storyteller, seems to me to be a position of some dignity'. It seems so to us also.

We have arranged the stories in a rather unconventional way—not chronologically by publication date, but according to the historical period in which each story is set, beginning in prehistory and continuing as far as the early years of the present century. We hope this has produced some interesting juxtapositions; but for those who prefer to read the stories sequentially according to date of publication, we have provided a separate contents list at the back of the book. The source notes contain bibliographical and biographical details, as well as brief comments on the historical settings where this has seemed appropriate.

M. A. C.
J. A.

January 1994

---

[12] Carlyle, 'Lockhart: Life of Scott', *London and Westminster Review*, vi (1837), 337; reprinted in *Critical and Miscellaneous Essays* (1839) as 'Sir Walter Scott'.

[13] Hugh Walpole, 'The Historical Novel in England [sic] Since Sir Walter Scott' in *Sir Walter Scott Today: Some Retrospective Essays and Studies*, ed. H. J. C. Grierson (Edinburgh, 1932), 170.

CHARLES G. D. ROBERTS

The World Without Man

∽∞∾

It lay apparently afloat on the sluggish, faintly discoloured tide—a placid, horse-faced, shovel-nosed head, with bumpy holes for ears and immense round eyes of a somewhat anxious mildness.

The anxiety in the great eyes was not without reason, for their owner had just arrived in the tepid and teeming waters of this estuary, and the creatures which he had already seen about him were both unknown and menacing. But the inshore shallows were full of water-weeds of a rankness and succulence far beyond anything he had enjoyed in his old habitat, and he was determined to secure himself a place here.

From time to time, as some new monster came in sight, the ungainly head would shoot up amazingly to a distance of five or ten, or even fifteen feet, on a swaying pillar of a neck, in order to get a better view of the stranger. Then it would slowly sink back again to its repose on the water.

The water at this point was almost fresh, because the estuary, though fully two miles wide, was filled with the tide of the great river rolling slowly down from the heart of the continent. The further shore was so flat that nothing could be seen of it but an endless, pale-green forest of giant reeds. But the nearer shore was skirted, at a distance of perhaps half a mile from the water, by a rampart of abrupt, bright, rust-red cliffs. The flat land between the waterside and the cliffs, except for the wide strip of beach, was clothed with an enormous and riotous growth of calamaries, tree-ferns, cane and palm, which rocked and crashed in places as if some colossal wayfarers were pushing through them. Here and there along the edge of the cliffs sat tall beings with prodigious, saw-toothed beaks, like some species of bird conceived in a nightmare.

Far out across the water one of these creatures was flapping slowly in from the sea. Its wings—eighteen feet across from tip to tip—were not the wings of a bird, but of a bat or a hobgoblin. It had dreadful, hand-like claws on its wing-elbows; and its feet were those of a lizard.

As this startling shape came flapping shoreward, the head afloat upon the water eyed it with interest, but not, as it seemed, with any great apprehension. Yet it certainly looked formidable enough to excite misgivings in most creatures. Its flight was not the steady, even winging of a bird, but spasmodic and violent. It came on at a height of perhaps twenty feet above the sluggish tide, and its immense, circular eyes appeared to take no notice of the strange head that watched it from the water's surface. It seemed about to pass a little to one side, when suddenly, with a hoarse, hooting cry, it swerved and swooped, and struck at the floating head with open jaws.

Swift as was that unexpected attack, the assailant struck nothing but a spot of foam where the head had disappeared. Simultaneously with the lightning disappearance, there was a sudden boiling of the water some eighty-odd feet away. But the great bird-lizard was either too furious to notice this phenomenon or not sagacious enough to interpret it. Flopping into the air again, and gnashing his beaklike jaws with rage, he kept circling about the spot in heavy zigzags, expecting the harmless-looking head to reappear.

All at once his expectations were more than realized. The head not only reappeared, but on a towering, leather-coloured column of a neck it shot straight into the air to a height of twenty feet. The big, placid eyes were now sparkling with anger. The flat, shovel jaws were gaping open. They seized the swooping foe by the root of the tail, and, in spite of screeches and wild flappings, plucked him down backwards. At the surface of the water there was a convulsive struggle, and the wide wings were drawn clean under.

For several minutes the water seethed and foamed, and little waves ran clattering up the beach, while the owner of the harmless-looking head trod his assailant down and crushed him among the weeds of the bottom. Then the foam slowly crimsoned, and the mauled, battered body of the great bird-lizard came up again; for the owner of the mysterious head was a feeder on delicate weeds and succulent greenstuff only, and would eat no blood-bearing food. The body was still struggling, and the vast, dark, broken wings spread themselves in feeble spasms on the surface. But they were not left to struggle long.

The water, in the distance, had been full of eager spectators of the fight, and now it boiled as they rushed in upon the disabled prey. Ravenous, cavern-jawed, fishlike beasts, half-porpoise, half-alligator, swarmed upon the victim, tearing at it and at each other. Some bore off trailing mouthfuls of dark wing-membrane, others more substantial booty, while the rest fought madly in the vortex of discoloured foam.

At the beginning of the fray the grim figures perched along the red ramparts of the cliff had shown signs of excitement, lifting their high shoulders and half unfolding the stiff drapery of their wings. As they saw their fellow overwhelmed they launched themselves from their perch and came hooting hoarsely over the rank, green tops of the palms and feathery calamaries. Swooping and circling they gathered over the hideous final struggle, and from time to time one or another would drop perpendicularly downward to stab the crown or the face of one of the preoccupied fish-beasts with his trenchant beak. Such of the fish-beasts as were thus disabled were promptly torn to pieces and devoured by their companions.

Some fifty feet away, nearer shore, the harmless-looking head which had been the source and inspirer of all this bloody turmoil lay watching the scene with discontent in its round, wondering eyes. Slowly it reared itself once more to a height of eight or ten feet above the water, as if for better inspection of the combat. Then, as if not relishing the neighbourhood of the fish-beasts, it slowly sank again and disappeared.

Immediately a heavy swirling, a disturbance that stretched over a distance of nearly a hundred feet, began to travel shoreward. It grew heavier and heavier as the water grew shallower. Then a leather-coloured mountain of a back heaved itself up through the smother, and a colossal form, that would make the hugest elephant a pigmy, came ponderously forth upon the beach.

The body of this amazing being was thrice or four times the bulk of the mightiest elephant. It stood highest—a good thirteen feet—over the haunches (which were supported on legs like columns), and sloped abruptly to the lower and lighter-built fore-shoulders. The neck was like a giraffe's, but over twenty feet in length to its juncture with the mild little head, which looked as if Nature had set it there as a pleasantry at the expense of the titanic body. The tail, enormous at the base and tapering gradually to a whip-lash, trailed out to a distance of nearly fifty feet. As its owner came ashore, this tremendous tail was gathered and curled in a semicircle at his side—perhaps lest the delicate tip, if left too distant, might fall a prey to some insignficant but agile marauder.

For some minutes the colossus (he was one of the Dinosaurs, or Terrible Lizards, and known as a Diplodocus) remained on all-fours, darting his sinuous neck enquiringly in all directions, and snatching here and there a mouthful of the rank, tender herbage which grew among the trunks of fern and palm. Apparently the spot was to his liking. Here was a wide beach, sunlit and ample, whereon to bask at leisure. There were the warm and weed-choked shallows wherein to pasture, to wallow at will, to hide his giant bulk from his enemies if there should be found any formidable enough to make hiding advisable. Swarms of savage insects, to be sure, were giving him a hot reception—mosquitoes of unimaginable size, and enormous stinging flies which sought to deposit their eggs in his smooth hide, but with his giraffe-like neck he could bite himself where he would, and the lithe lash of his tail could flick off tormentors from any corner of his anatomy.

Meanwhile, the excitement off-shore had died down. The harsh hootings of the bird-lizards had ceased to rend the air as the dark wings hurtled away to seek some remoter or less disturbed hunting-ground. Then across the silence came suddenly a terrific crashing of branches, mixed with gasping cries. Startled, the diplodocus hoisted himself upon his hind-quarters, till he sat up like a kangaroo, supported and steadied by the base of his huge tail. In this position his head, forty feet above the earth, overlooked the tops of all but the tallest trees. And what he saw brought the look of anxiety once more into his round, saucer-eyes.

Hurling itself with desperate, plunging leaps through the rank growths, and snapping the trunks of the brittle tree-ferns in its path as if they had been cauliflowers, came a creature not unlike himself, but of less than half the size, and with neck and tail of only moderate length. This creature was fleeing in frantic terror from another and much smaller being, which came leaping after it like a giant kangaroo. Both were plainly dinosaurs, with the lizard tail and hind-legs; but the lesser of the two, with its square, powerful head and tiger-fanged jaws, and the tremendous, rending claws on its short forearms, was plainly of a different species from the great herb-eaters of the dinosaurian family. It was one of the smaller members of that terrible family of carnivorous dinosaurians which ruled the ancient cycad forests as the black-maned lion rules the Rhodesian jungles today. The massive iguanodon which fled before it so madly, though of fully thrice its bulk, had reason to fear it as the fat cow fears a wolf.

A moment more, and the dreadful chase, with a noise of raucous groans and pantings, burst forth into the open, not fifty feet from where the colossus stood watching. Almost at the watcher's feet the fugitive was overtaken. With a horrid leap and a hoot of triumph, the pursuer sprang upon its neck and bore it to the ground, where it lay bellowing hoarsely and striking out blunderingly with the massive, horn-tipped spur which armed its clumsy wrist. The victor tore madly at its throat with tooth and claw, and presently its bellowing subsided to a hideous, sobbing gurgle.

The diplodocus, meanwhile, had been looking down upon the scene with half-bewildered apprehension. These creatures were insignificant in size, to be sure, as compared with his own colossal stature, but the smaller one had a swift ferocity which struck terror to his dull heart.

Suddenly a red wrath mounted to his small and sluggish brain. His tail, as we have seen, was curled in a half-circle at his side. Now he bent his body with it. For an instant his whole bulk quivered with the extraordinary tension. Then, like a bow released, the bent body sprang back. The tail (and it weighed at least a ton) struck the victor and the victim together with an annihilating shock, and swept them clean around beneath the visitor's feet.

Down he came upon them at once, with the crushing effect of a hundred steam pile-drivers; and for the next few minutes his panicky rage expended itself in treading the two bodies into a shapeless mass. Then he slowly backed off down into the water where the weedy growths were thickest, till once more his whole form was concealed except the insignificant head. This he reared among the swaying tufts of the 'mares' tails', and waited to see what strange thing would happen next.

He had not long to wait. That hideous, mangled heap there, sweating blood in the noon sun, seemed to have some way of making its presence known. Crashing sounds arose in different parts of the forest, and presently some half-dozen of the leaping, kangaroo-like flesh-eaters appeared.

They were of varying sizes, from ten or twelve feet in length to eighteen or twenty, and they eyed each other with jealous hostility. But one glance at the weltering heap showed them that here was feasting abundant for them all. With a chorus of hoarse cries they came hopping forward and fell upon it.

Presently two vast shadows came overhead, hovering a moment, and a pair of the great bird-lizards dropped upon the middle of the heap. Hooting savagely, with wings half uplifted, they struck about them with their terrible beaks till they had secured room for themselves at the banquet. Other unbidden guests came leaping from among the thickets; and in a short time there was nothing left of the carcases except two naked skeletons, dragged apart and half dismembered by mighty teeth. In the final mêlée one of the smaller revellers was himself pounced upon and devoured.

Then, as if by consent of a mutual distrust, the throng drew quickly apart, each eyeing his neighbour warily, and scattered into the woods. Only the two grim bird-lizards remained, seeming to have a sort of understanding or partnership, or possibly being a mated pair. They pried into the cartilages and between the joints of the skeletons with the iron wedges of their beaks, till there was not another titbit to be enjoyed. Then, hooting once more with satisfaction, they spread their batlike vanes and flapped darkly off again to their red watch-tower on the cliff.

When all was once more quiet the giant visitor fell to pasturing among the crisp and tender water-weeds. It took a long time to fill his cavernous paunch by way of that slender neck of his, and when he was satisfied he went composedly to sleep, his body perfectly concealed under the water, his head resting on a little islet of matted reeds in a thicket of 'mares' tails'. When he woke up again the sun was half-way down to the west, and the beach glowed hotly in the afternoon light. Everything was drenched in heavy stillness. The visitor made up his drowsy mind that he must leave his hiding-place and go and bask in that delicious warmth.

He was just bestirring himself to carry out his purpose, when once more a swaying in the rank foliage of the cycads caught his vigilant eye. Discreetly he drew back into hiding, the place being, as he had found it, so full of violent surprises.

Suddenly there emerged upon the beach a monster even more extraordinary in appearance than himself. It was about thirty-five feet in length, and its ponderous bulk was supported on legs so short and bowed that it crawled with its belly almost dragging the ground. Its small head, which it carried close to the earth, was lizard-like, shallow-skulled, feeble-looking, and its jaws cleft back past the stupid eyes. In fact, it was an inoffensive-looking head for such an imposing body. At the base of the head began a system of defensive armour that looked as if it might be proof against artillery. Up over the shoulders, over the mighty arch of the back, and down over the haunches as far as the middle of the ponderous tail, ran a series of immense flat plates of horn, with pointed tips and sharpened edges. The largest of these plates, those that covered the centre of the back, were each three feet in height, and almost of an equal breadth. Where the diminished plates came to an end at the middle of the tail, their place was taken by eight immense, needle-pointed spines, set in pairs, of which the chief pair had a length of over two feet. The monster's hide was set thick with scales and knobs of horn, brilliantly coloured in black, yellow, and green, that his grotesque bulk might be less noticeable to his foes among the sharp shadows and patchy lights of the fern jungles where he fed.

The sluggish giant moved nervously, glancing backwards as he came, and seemed intent upon reaching the water. In a few moments his anxiety was explained. Leaping in splendid bounds along his broad trail came two of those same ferocious flesh-eaters whom the great watcher among the reeds so disliked. They ranged up one on each side of the stegosaur, who had halted at their approach, stiffened himself, and drawn his head so far back into the loose skin of his neck that only the sharp, chopping beak projected from under the first armour-plate. One of the pair threatened him from the front, as if to engross his attention, while the other pounced upon one of his massive, bowed hind-legs, as if seeking to drag it from beneath him and roll him over on his side.

But at this instant there was a clattering of the plated hide, and that armed tail lashed out with lightning swiftness, like a porcupine's. There was a tearing screech from the rash flesh-eater, and he was plucked back sidewise, all four feet in air, deeply impaled on three of those gigantic spines. While he clawed and writhed, struggling to twist himself free, his companion sprang hardily to the rescue. She hurled herself with all her weight and strength full upon the stegosaur's now unprotected flank. So tremendous was the impact that, with a frightened grunt, he was rolled clean over on his side. But at the same time his sturdy fore-arms clutched his assailant, and so crushed, mauled and tore her that she was glad to wrench herself away.

Coughing and gasping, she bounded backwards out of reach; and then she saw that her mate, having wriggled off the spines, was dragging himself up the beach toward the forest, leaving a trail of blood behind him. She followed sullenly, having had more than enough of the venture. The triumphant stegosaur rolled himself heavily back upon his feet, grunted angrily, clattered his armoured plates, jerked his terrible tail from side to side as if to see that it was still in working order, and went lumbering off to another portion of the wood, having apparently forgotten his purpose of taking to the water. As he went, one of the grim bird-lizards from the cliff swooped down and hovered, hooting, over his path, apparently disappointed at his triumph.

The watcher in the reeds, on the other hand, was encouraged by the result of the combat. He began to feel a certain dangerous contempt for those leaping flesh-eaters, in spite of their swiftness and ferocity. He himself, though but an eater of weeds, had trodden one into nothingness, and now he had seen two together overthrown and put to flight. With growing confidence he came forth from his hiding, stalked up the beach, coiled his interminable tail beside him, and lay down to bask his dripping sides in the full blaze of the sun.

The colossus was at last beginning to feel at home in his new surroundings. In spite of the fact that this bit of open beach, overlooked by the deep green belt of jungle and the rampart of red cliffs, appeared to be a sort of arena for titanic combats, he began to have confidence in his own astounding bulk as a defence against all foes. What matter his slim neck, small head and feeble teeth, when that awful engine of his tail could sweep his enemies off their feet, and he could crush them by falling upon them like a mountain! A pair of the great bird-lizards flapped over him, hooting malignantly and staring down upon him with their immense, cold eyes, but he hardly took the trouble to look up at them.

Warmed and well fed, his eyes half-sheathed in their membraneous lids, he gazed out vacantly across the waving herbage of the shallows, across the slow, pale tides whose surface boiled from time to time above the rush of some unseen giant of a shark or ichthyosaur.

In the heavy heat of the afternoon the young world had become very still. The bird-lizards, all folded in their wings, sat stiff and motionless along the ramparts of red cliff. The only sounds were the hiss of those seething rushes far out on the tide, the sudden droning hum of some great insect darting overhead, or the occasional soft clatter of the long, crisp cycad leaves as a faint puff of hot air lifted them.

At the back of the beach, where the tree-ferns and the calamaries grew rankest, the foliage parted noiselessly at a height of perhaps twenty feet from the ground, and a dreadful head looked forth. Its jaws were both long and massive, and armed with immense, curved teeth like scimitars. Its glaring eyes were overhung by eaves of bony plate, and from the front of its broad snout rose a single horn, long and sharp. For some minutes this hideous apparition eyed the unconscious colossus by the waterside. Then it came forth from the foliage and crept noiselessly down the beach.

Except for its horned snout and armoured eyes, this monster was not unlike in general type to those other predatory dinosaurs which had already appeared upon the scene. But it was far larger, approaching thirty-five feet in length, and more powerfully built in proportion to its size; and the armoury of its jaws was more appalling. With a stealthy but clumsy-looking waddle, which was nevertheless soundless as a shadow, and his huge tail curled upwards that it might not drag and rattle the stones, he crept down until he was within some fifty feet or more of the drowsing colossus.

Some premonition of peril, at this moment, began to stir in the heavy brain of the colossus, and he lifted his head apprehensively. In the same instant the horned giant gathered himself, and hurled himself forward. In two prodigious leaps he covered the distance that separated him from his intended prey. The coiled tail of the colossus lashed out irresistibly, but the assailant cleared it in his spring, fell upon the victim's shoulders, and buried his fangs in the base of that columnar neck.

The colossus, for the first time, was overwhelmed with terror. He gave vent to a shrill, bleating bellow—an absurdly inadequate utterance to issue from this mountainous frame—writhed his neck in snaky folds, and lashed out convulsively with the stupendous coils of his tail. But he could not loosen that deep grip, or the clutch of those iron claws.

In spite of the many tons weight throttling his neck, he reared himself aloft, and strove to throw himself over upon his assailant. But the marauder was agile, and eluded the crushing fall without loosing his grip. Then, bleating frightfully, till the sounds re-echoed from the red cliffs and set all the drowsing bird-lizards lifting their wings, he plunged down into the tide and bore his dreadful adversary out of sight beneath a smother of ensanguined foam.

Now, the horned giant was himself a powerful swimmer and quite at home in the water, but in this respect he was no match for his quarry. Refusing to relinquish his hold, he was borne out into deep water; and there the colossus, becoming all at once agile and swift, succeeded in rolling over upon him. Forced thus to loose his grip, he gave one long, ripping lunge with his horn, deep into the victim's flank, and then writhed himself from under. The breath quite crushed out of him, he was forced to rise to the surface for air. There he rested, recovering his self-possession, reluctant to give up the combat, but even more reluctant to expose himself to another such mauling in the depths. As he hesitated, about a hundred feet away he saw the mild little head of the colossus, apparently floating on the tide, and regarding him anxiously. That decided him. With a crashing bellow of rage and a sweep of his powerful tail he darted at the inoffensive head. But it vanished instantly, and a sudden tremendous turmoil, developing into a wake that lengthened out with the speed of a torpedo-boat, showed him the hopelessness of pursuit. Turning abruptly, he swam back to the shore and sulkily withdrew into the thickets to seek some less unmanageable quarry.

The colossus, so deeply wounded that his trail threw up great clots and bubbles of red foam, swam onward several miles up the estuary. He realized now that that patch of sunny beach was just a death-trap. But in the middle of the estuary, far out from either shore, far removed from the unseen, lurking horrors of the fern forests, spread acre upon acre of drowned marsh, overgrown with tall green reeds and feathery 'mares' tails'. Through these stretches of marsh he ploughed his way, half-swimming, half-wading, and felt that here he might find a safe refuge as well as an unfailing pasturage. But the anguish of his wounds urged him still onwards.

Beyond the reed-beds he came to a long, narrow islet of wet sand, naked to the sun. This appeared to him the very refuge he was craving, a spot where he could lie secure and lick his hurts. He dragged himself out upon it eagerly. Not until he had gained the very centre of it did he notice how his ponderous feet sank in it at every stride. As soon as he halted he felt the treacherous sands sucking him down. In terror he struggled to free himself, to regain the water. But now the sands had a grip upon him, and his efforts only engulfed him the more swiftly. He reared upon his hind legs, and immediately found himself swallowed to the haunches. He fell forward again, and sank to his shoulder-blades. And then, the convulsive thrashings of his tail hurling the sands in every direction, he lifted his head and bleated piteously.

The struggle had already drawn the dreadful eyes of those grim, folded figures perched along the cliff-tops miles away; and now, as if in answer to his cry, they came fluttering darkly over him. Seeing his helplessness, they flapped down upon him with hoots of exultation. Their vast beaks tore at his helpless back, and stabbed at the swiftly writhing convolutions of his neck. One, more heedless than his fellows, came within reach of the thrashing tail, and was dashed, half stunned, to earth, where the sands got him in their hold before he could recover himself. With dreadful screeches, he was sucked down, but his fellows paid no attention to his fate. And meanwhile, in a ring about the islet, not daring to come nearer for terror of the quicksand, crocodiles and alligators and ichthyosaurs, with upturned, gaping snouts, watched the struggle greedily.

As the lower part of his neck was drawn down into the quicksand, the colossus lost the power to move his head quickly enough to evade the attacks of his horrid assailants. A moment more, and he was blinded. Then he felt his head enfolded in the strangling membranes of wings and borne downwards. Once or twice the convulsions of his neck threw his enemies off, and the bleeding, sightless head re-emerged to view.

But not only his force, but his will to struggle, was fast ebbing away. Presently, with a thunderous, gasping sob, the last breath left his mighty lungs, and his head dropped on the sand. It was trodden under in an instant; and then, afraid of being engulfed themselves, the hooting revellers abandoned it, to crowd struggling upon the arched hump of the back. Here they tore and gorged and quarrelled till, some fifteen minutes later, their last foothold sank beneath them. Then, with dripping beaks and talons, they all flapped back to their cliffs; and slowly the fluent sand smoothed itself to shining complacency over the tomb of the diplodocus, hiding and sealing away the stupendous skeleton for half a million years.

EUGENE O'SULLIVAN

Overture

ɔ๐ɔ

I *detest* trade missions.

I mean, when you've seen one of them, you've seen a reasonable approximation of them all. The same stupid conversations, the same elephantine stone-walling, the same bovine diplomatic let-outs, the same greasily wearying convivialities.

And the miserable insincerity of it. Pledges of undying peace and prosperity to us both, undying friendship and understanding, growth of exchange of ideas born of the cross-fertilization of cultures . . . for God's sake! They know, and we know that they know, and they know that we know that they know, and so on till hell freezes over (and then they will offer it to us at cut rates)—that we're spies. We know . . . and all the way round . . . the same thing, with all of its loathsome permutations and combinations. I honestly believe that the whole thing would work much better if we said it to each other at the outset. We don't think you've got much, but we're sure as hell going to find out what you've got. You, ditto.

Instead—the Olympian fart, doubled.

Deepest respect for your technological achievements. Overwhelming admiration for your conquest of the world trade. Deepest respect for your technological underpinning of the whole schmeer. Wonderful delight at your contribution to the betterment of the human race by your latest naval counter-turbulence invention and, incidentally, how does it work? Era of peace and prosperity in sight.

And, in reply, deeper respect for your outstanding preservation of ancient industries. Magnificent admiration for your custody of commodities world trade hasn't been able to get at. Deeper than deepest respect for your antedeluvian methods of staying outside the world market. Marvellous ecstasy at your contribution to the betterment of the human race by your latest troglodyte army siege-warfare device, as if anyone would ever bother themselves having a war with you in which you could lay siege. There would be rather more interest if you could lay an egg. At least, we could market the egg.

And the frightful dinners. I like a little wine, provided it is good. My genial hosts appear to like a lot of wine, no doubt because it is bad. And the entertainment! When the fifth or sixth refill has been around, surely there is an excuse for going to bed. Not on your eternal life. There will be dancing. There will be nuts. There will be instruments. I seriously thought of asking them whether they wouldn't save a great deal of expense by getting a few shepherds and goats to provide the song and dance. I could remember far better evenings when they did, but, then, I'm not supposed to remember that kind of thing.

Anyhow, just when I think I really am going to be able to get up, and say it's been ever so nice a party, what does my oaf of a so-called host do but scream for his wife? She, not having been an object of barter, I haven't seen her to date, and indeed I've been wondering to whom I owed my bread-and-butter letter. I wasn't altogether sure that he was married. No doubt, if I checked it out on my briefing data I'd get all the dope, sordid official mythology about courtship, deepest family enthusiasm for nuptials, hopes or realizations for infants, and so forth, but I've been around many of these towns in this miserable promontory, and I've had my glandful. Some are married, some are not. From the way he rubbed up to my thigh till I could get away from him, right from the moment we met, I'd have said he was not. It turns out he is.

And out she came. And my God, what an act! I've not expected very much from these dives so far, but she is the limit. One second she's the receptacle of the sexual offerings of the entire universe—no doubt her fellow-countrymen have a word for it, but I'll survive without the information—and the next, nobody ever loved a husband like she's all over hers. A good looker, but tarted up. I suppose with jaded appetites like his, and the grossness of the court, she has a natural inferiority complex, and on come the daubs and smears. And her loathsome husband! 'Don't you like my wife? Aren't you really excited by my wife? Did anything ever get you leaping out like my wife?' And the old grab at the thigh, and something more, if he's any luck, which if I can help it, he hasn't.

At last she goes off. There's the usual corn-ball caterwaul first about the gods prospering our associations, meaning the two cities, hers and mine, but somehow suggested in her song as being symbolized by me and her. If the future prospects arising from the mission were summed up by the relations between me and her, they could call the whole thing off right now. Anyhow, for my money, that would be a darn good idea right there. We'll never get anything out of this collection of obscurantist jocks, nothing we'd want to market anyway.

Then she goes, and here, I say to myself, is where I go too. Do I? Do I, hell. He gets me aside, and begins selling her to me. And I mean, selling. I mutter something about a useful basis of import–export tariffs with relation to some God-damn commodity of theirs—not that we want it, but it passes the time to talk about it—but he sweeps it aside. It's her, nothing but her. You'd swear she was the only thing he wanted to sell me. He has a hope, the market being overcrowded where I come from. He even insists she is a stimulus to his naval policy. I choke, rather than saying what navy, which is what I'd like to say.

And I can't get rid of him. Father can't have expected this kind of thing in getting me to beef up the exports. It was bad enough when I was seeing this guy's brother down the way, with the squint-eyed wife, and clinging brats, one looking as if she expected him to murder her, and one as if she wanted to murder her mother—not that I blame her—and the baby boy, an infant epileptic if ever I saw one, biting away at his mother's breasts as if he meant to make lunch and dinner of it. But at least they left me alone. I guess they had enough to deal with, looking at themselves. Here, I was the big number.

He gets rid of the guests and attendants. Usual delicacy. Get the hell out of here, see you tomorrow, can't you see I want to talk business. If they couldn't, I could. And then, the empty room, the sloshing of wine, the hand moving drunkenly in, me dodging as politely as I could, the big grab. Oh Father, dear Father, come home to me now and tell me what in blazes you expect from me. Surely the export market doesn't need sacrifices like this?

Well, eventually, we're up against the wall Mother Reilly, and I say no. I mean, I do it nicely. I don't tell him that there was a time when I was fair meat for shepherds of any given sex that was going, provided it was relatively attractive. But with a fat like that, with his bad breath and his worse breadth? I explain I am not that way inclined, it isn't the custom of my country, maybe they do things differently in his place. Evidently they do. He begins by telling me who his conquests have been. It turns out they include his brother, they include that devious bastard on the island, they include the old man on the sand lots, they include the kid in drag, and heaven knows how many others I haven't heard of. More, he says they're all leagued together to stand by him. I ask if this means they're on a common customs agreement. He bursts into tears and I say I'm going to bed.

That starts it off again. To bed, yes, but not alone, surely? With him, please. Don't I realize, he's making me an offer thousands would be delighted to get. Have I not noticed his so beautiful wife? Yes, I have noticed his so beautiful wife. Don't I realize what he's giving up to go to bed with me? I tell him I'm not worth the sacrifice. He assures me I am. He decides I need more wine. I decide I need a hole in the head.

I don't know what happened, and I don't want to know what happened. I have vague recollections of a grunting, sweating fat body sticking to me, and me wishing I was talking to anybody, anywhere, even my mad sister. I do remember, probably the worst horror of all, that stinking breath whispering to me that we'd do it again, and better, tomorrow night, when we hadn't drunk so much. And that's it. I'm getting out of here, father or no father, trade mission or no trade mission, and I'm going to find some way of hamstringing that stupid bastard if it's the last thing I do, whether by kicking his fat backside down his throne steps, or getting his brother to disown him, or kidnapping his wife who must be about ready for something.

Say, that's an idea.

She's dumb, but not so dumb that she can't want a little diversion.

What the hell is her name. Yeah, I have a note of it in the briefing I didn't bother to read.

He's hunting today, he told me. Must be an exciting exercise. I daresay the scent of him sends any self-respecting animal to ground for about fifty miles about the place. But she won't be likely to accompany him. Who'd blame her?

I'd better get the name right. Yes, that's it. Diplomatically sounding, now. I'm on a trade mission. '*Helen!*'

ARTHUR D. HOWDEN SMITH

A Trooper of the Thessalians

ᕁᕁ

You are a stranger, eh? I thought so! I always talk with the strangers who come to Antioch; some of them are from the countries I helped to conquer in my day. A long time since—a long, long time! But I can still sit a horse with any youngster, and handle a drungos on the citadel parade or in battle. Messala they call me. I command the garrison for Seleucus. . . . Ho, varlet! Another cup for this stranger, and a jar of your Cyprian—a mellow wine that, as good as we used to get from Tempe's vineyards before I marched from Macedonia, little thinking—empty-head that I was!—that I should never see the sunset fade on the Troad's peaks again.

Well, well, it is all by and done with. . . . Where did you say you were from? Rome? Humph! Rome! Ha, I remember! A city in Italy, midway of the peninsula. It grows in power, men say. No, I never saw it. Alexander turned east, not west, or I doubt not I should have tried the height of your walls. . . . I? Why, yes, I rode with him from the beginning. The crossing of the Granicus, Issus, Tyre, Egypt, Arbela and afterwards, I was through all. Up and down the world we marched, and back and across, and everywhere we conquered. Men said there was magic in it, and worshipped Alexander as a God, but I know better. It was discipline and valour and consummate strategy. But mostly discipline.

Ha, this wine is good after a hot day in saddle. By your leave, I'll slip off my armour. What said the poet? 'A cuirass to an old man's back is like a peasant's heaping sack.' Ha, ha, ha! There was a time I recked no more of my mail than that varlet does of his tunic. . . . This sword? You like it? I see you have a soldier's eye, Roman. Yes, it is not as other swords. There is a saying—and that I am alive to repeat it may go for proof of it—that he who wields it can not be slain by any other blade. And there is this further proof: that he who owned it last was not slain as I have ever seen a man slain in the four score and seven fights and sieges I have known, as also, that the man who owned it before *him* died of an arrow. Before that I have only legend to go by. You may see the marks upon the blade. Those at the top are Egyptian; below it is Greek. He who died of the arrow was my shieldmate, a true friend, Roman, one in ten thousand. May the Gods remember him! Agathocles was his name—he used to tell us that the sword had been in his family since Marathon, when an ancestor of his who was a slave, took it from a Persian and won great renown.

You think it strange that a slave should fight for his city? Well, so do I, Roman. But we Greeks ordered it so in the old days, and if the histories go for anything the slaves fought better than many freemen. At Marathon they beat the Persians, four to one against them. After all, a Greek slave is a better man than any Persian or—Humph! Hrrrrrrumph! Yes, as I was saying, it is a strange sword Those who use it without fear it will protect, but let a coward touch it! . . . The story? Yes, there is one, if you can tolerate an old soldier's rambling tongue; but we shall have to leap far over the past, back to Arbela, where we Greeks crushed the Persians once and for all, and made Asia a satrapy for Macedon. Ha, lad, that was a fight!

I'll pass over all that came first. For two years we had been tramping about from the shores of the Euxine to the Nile, besieging cities and beating armies. We beat King Darius, himself, at Issus, but he got away, and Alexander was too busy consolidating our conquests to pursue the Persians then. But after we had taken Tyre, Gaza, Pelusium, and Memphis and made Egypt safe, we turned eastward again, and collected an army the like of which no man has ever seen. Forty-seven thousand Greeks, Roman; forty-seven thousand veterans. Gods, what men! There was the Phalanx, six drungoi, each of three thousand men, eighteen thousand altogether, the pick of Macedon, who dug a way with their sarissas through every enemy that dared to meet them. There were the shield-bearers, heavy infantry that could stand against anything short of the Phalanx. There were archers and javelin-troops and slingers, who were not afraid to face the Persian elephants. There were light horse used to charging without thought the Persian cataphracti. And last, there were us of the cuirassiers, cataphracti who—but of what use are words?

Roman, did you ever see two drungoi of armoured-men on armoured horses, sheathed in mail like metal statues, thundering to the attack? The Gods, themselves, never witnessed a more sublime spectacle. A gigantic lance-head forged out of three thousand men, striking with the impact of a Hercules! And in truth, we were the lance-head of Alexander's army, as he, himself, would say: 'The Phalanx to thrust with, and the Cataphracti to hew the way for them.' But I am wandering from my story.

My shieldmate Agathocles and I were hekatonarchs in the first regiment of the Thessalian drungos. . . . No, I never served with the Macedonians. They were the King's Guard, but for that very reason we of the Thessalians made it a point of honour to outride and outfight them. It was well known in the army. Alexander had one of his sayings about it: 'My Macedonians at my back, and the Thessalians to make the flank good.' And it is true that we were always put to it to resist the mass of the enemy the while the Macedonians overrode the line elsewhere. Our replacements were twice theirs. . . . Indeed, it was by reason of our constant drain of replacements that my story had its origin.

The *strategos* of the Thessalians was a Pharsalian named Philip, a good soldier, but a courtier—I say no more of him, since he was my general. He had a son, Dion, for whom he secured the appointment of chiliarch of our regiment soon after Issus, where Tigranes, our first chiliarch, was slain; and I will not deny that this left a bad taste in the mouths of all of us squadron commanders. We had thought Agathocles should have it, for you must know that Dion had served his father on the staff and never ridden in the charge with the drungos. But we swallowed twice and kept our mouths shut, thankful when Philip sent Dion home to Macedonia to recruit for the drungos, leaving it to the hekatonarchs to command the regiment by turns.

We went on after Issus, and stood by aiding the infantry as best we could at the sieges which followed. The next year found us in Egypt, with more work to do, and Dion still lingered in Macedon, sending us recruits and stores—horses we obtained from Asia, where the finest were ours for the seizing. When we turned back at last, with Egypt a Greek province, and the word trickled through the ranks, as it will in any army, that the King was determined to force conclusions with Darius, and either break the Persian power or else wreck his own army in final defeat, we of the Thessalians were overjoyed, and we hoped, too, that Philip would give our regiment a new chiliarch in recognition of all that we had accomplished in his son's absence.

But our hopes were disappointed. At Damascus Dion rejoined us with a draft of recruits, and took over the regiment. Did I say that he was as vain as he was cowardly? For coward he was. Such are to be found amongst all races, Roman, although I thank Hercules that we Greeks have fewer than most. . . . Yes, he was as vain as a Mede, who will tell you how great his people once were, forgetting that he is now little better than a slave. His first thought was to muster the regiment for inspection outside Damascus, and try to find fault with the discipline and equipment. But his father, who was no fool, put a quick stop to that, and smoothed over the ill-feeling that was aroused by complimenting us officers on the efficiency we had shown and serving out an extra wine ration to the men. I made no doubt that he dropped a word of advice in Dion's ear, too, as that was the last we heard of inspection. But the fool's vanity was incurable, and when he was denied the opportunity of exploiting it by humiliating those who were better men than he and who had done his work for him, he must needs vent it in peacock parading and thrusting the regiment into all manner of additional service, in which he never took part, but for which he always claimed credit.

It was this cursed vanity of his which brought about my shieldmate's death. From Damascus we marched north by Hamath to the Euphrates, which we crossed at Thapsacus towards the end of summer. There was a small force of Persian horse watching the crossing, and they would have retired without any pressure in face of such an array as ours; but Dion must rush to his father and Alexander and offer to send a squadron of ours to teach the Persians what was in store for them. Alexander said nothing, and Philip gave the boy leave—I think, because he was always hoping that Dion, himself, would lead one of these dashes and quickly win honour thereby. The result was that Agathocles' squadron was ordered out to brush back the Persians. And mark you, Roman, it was foolish to waste armoured horsemen on a venture which could have been performed by light horse. So said many of the generals afterwards, although none of us could find fault with Alexander for not checking Dion, for we loved the young King because he would never refuse a man who desired to attempt some dashing exploit.

I can see him now—he was sitting his horse close by where we were grouped above the river crossing—watching Agathocles lead his men down against the Persians. Only a boy like the rest of us, with a boy's eager, restless manners, forever twitching at his weapons or rubbing his beardless chin, but with a brain that never slept and as keen an eye for battle strategy as old Parmenio, who was his right hand and the one man who dared to talk to him like a father. Fever-hot he was in all he did, fierce in battle, tireless in carouse, unbounded in imagination. Who but he would have thought of invading India and dreamed of conquering the yellow men beyond? He lived in a day an ordinary man's life, and to prove that he was no god the Gods smote him while his boyhood was unblemished. If he had lived! Give thanks that he did not, Roman, for your turn would have come had he been spared us a few years more. And then, instead of a score of kingdoms and principalities carved out of the wreckage of what he won we should have had one land, all Greek, from the Pillars of Hercules to the east where the world ends.

Be sure, though, that no such thoughts as these were in my head as I sat behind Alexander on the Euphrates' bank and watched Dion in his silvered mail, pompously erect and trying to achieve the look of the eagle that Alexander assumed by nature. . . . You know the eagle, Roman? He is your emblem? That is well for your people. The eagle is a conqueror; a people who take him for emblem should carry their arms far, yes, as far as the eagle flies—and wherever I have travelled, there the eagle has assailed his prey . . . But Dion was no eagle. Not he! He looked like a plump partridge hoisted into saddle and hung about with a housewife's pans from the kitchen.

The Persians had already commenced to retire when they saw Agathocles coming down to the ford, and because they were brave men and were five times the number of his squadron they turned back and attacked him savagely as he rode up from the water, smiting his ranks with the arrow-hail of their Scythian archers and flinging other companies in headlong charges upon the lances of his troopers. But Agathocles rode right through them, split his squadron in halves, and then stormed from two sides upon the wreckage he had made. The Persians were beaten, and fled, raining arrows behind them, with Agathocles pelting at their heels, his sword flickering like a blown flame. We on the opposite bank set up the shout of victory, and Alexander waved the light troops forward to the crossing. And it was that moment, Roman, that Agathocles fell from his horse. An arrow from a wounded Scythian on the ground had pierced his cheek and driven upward into the brain. Gods! What an ill hour. He was—But of what use words now? A man has but one shieldmate of his youth. Other friends, perhaps. Others to share his shield. But none like the shieldmate of youth. . . . I pour this libation to him, Roman. Ha, you do well! The Gods will look favourably upon one who honours the dead, even a barbarian . . .

I was first to reach his side, and it was I who stripped off his armour and loosened his fingers from the hilt of his sword. My thought was to make the sword my own. I knew that would have been his wish. But that fop Dion must ride up as we were burying Agathocles and cry out in his squeaky voice:

'Where are the hekatonarch's equipments? They must be sold according to camp law.'

I pointed to the little heap of armour and weapons beside Agathocles' horse that a trooper held by the bridle.

'But the sword?' persisted Dion. 'Is the sword there? I have heard much of that sword. Men say he who carried it cannot be slain by another blade—and all of us saw that Agathocles died by the arrow.'

'I have taken the sword,' I answered briefly. 'Mine can be sold in its place.'

'No, no,' squeaked Dion. 'Camp law is camp law.'

At this several of my brother officers spoke up and exclaimed that Agathocles had been my shieldmate, and that they were willing for me to have the sword to remember him by. But the chiliarch shook his head stubbornly.

'Let Messala buy it in, then,' he said. 'It is camp law that a dead man's effects be sold at auction.'

And off he rode without another look at the man he had sent to his death.

He was chiliarch. There was nothing to do but obey him. And I had this reassurance in doing so: that I knew no other man in the regiment, yes, no other in the drungos, would bid against me for the sword, whatever magical qualities it possessed. What is camp law to soldiers in face of the love betwixt shieldmates?

That night the camp marshal offered Agathocles' equipment at auction before the headquarters of the chiliarch. The horse was bought in for the reserve of the drungos; the horse-armour went with it. A man in another regiment bought the body-armour; a dekarch of ours bought the shield and lance. Last, the sword was offered. There was a dead silence while the circle of officers looked at me in sign that none would compete with me, and my sadness began to lift with this evidence of their affection.

'Ten staters,' I called. For I did not wish to set a cheap price upon what had been my shieldmate's or to take advantage of the consideration my brother officers showed me. But the words were scarce out of my lips when Dion's mincing voice squeaked from the shadows by his tent.

'Eleven staters, marshal.'

Roman, I could have slain the swine where he sat. My heart boiled with rage.

'Fifteen staters,' I shouted.

The marshal looked uncomfortable. He would have liked to declare the bidding ended, for no sword—as a sword—was worth fifteen staters. But he dared not, and there was a grating edge to Dion's squeal as the chiliarch called:

'Sixteen staters.'

I had no more money, but men behind me in the circle thrust upon me whatever was in their pouches, and so I was able to carry on my bidding to twenty-two staters.

'Twenty-three staters,' shrilled Dion.

The marshal looked towards me enquiringly.

'Your bid, Messala?'

'I have offered my last bid,' I growled.

An answering growl echoed from the circle of officers. The marshal hesitated a moment, and Dion stepped out and lifted the sword from the cloak on which the equipment had been spread.

'If there is no more bidding the sword is mine,' he declared with the abrupt insolence we hated.

After all, as I have said, he was chiliarch. What could the marshal do?

Dion drew the long grey blade and flashed it in the firelight. And as I sit here, Roman, it hissed as a man does in derision—or better still, as a woman does, vibrant with the contempt that scorches and sears. Not one of us stirred or breathed, for that note we had never heard before. Not so had the sword sung when Agathocles wrought with it in battle! Then it had whistled and purred. This—this was unspeakable, horrible—like a chained beast that waits to strike—or perhaps a snake—yes, perhaps a snake.

Men who were closest to Dion moved away from him, uneasy for the sword's evil voice. But there was no misgiving in his face, only puzzlement, as he held the blade before him—thus—lengthwise, and studied the clean sweep of edge and point.

'Man's voice or woman's voice,' he squeaked, 'it has a woman's shape.'

And therein he was right, Roman, as I have ever maintained. Look closely, and you will perceive that it suggests the gracious slimness and strength of some green maid, most wildly perfect and thirsty of life. Yes, very thirsty! . . .

'It seems to cry out against you,' returned the marshal. 'There is no luck for you in that sword, Dion.'

But Dion frowned upon him reprovingly, striving for the expression with which Alexander received a captive satrap.

'Not lucky!' he squealed. 'You babble. The sword hisses at the enemies it will guard me against. It serves notice that the Gods have given me into its protection. Yes, yes, it is a good servant, this sword. I have heard Agathocles tell how it guarded all who possessed it.'

'Yet Agathocles is dead,' I said, the wrath throbbing in my voice.

'Slain by his own foolishness,' he twittered. 'Rash one that he was! And he should have had his shield up to guard his cheek below the helmet-rim. So much the Gods do for us—no more!'

No more! Two men caught me by the arms and dragged me back from the fireside lest I should smite him. And that would have meant one thing in any army Alexander led—death! Right, too. Discipline overrides personal hatred. Remember that, Roman. Yes, yes, discipline before all . . .

Where was I? Ah, yes! They dragged me away from the fireside, and I lay out under the stars that night and prayed and cursed by turns until exhaustion drugged me. But I was up with the trumpets at dawn, and my squadron was first standing to horse. Sleep is a great chastener, and an hour dulls bitter memories—dulls where it does not erase . . .

We left the Euphrates behind us, and plunged boldly into the heart of the Persian dominions, marching east towards the Tigris. South of us was the rich plain which lies betwixt the two rivers, full of fat, prosperous cities and all manner of wealth. Mark, Roman, how our King's craft prevented him from falling into the trap which would have caught most invaders. Southward was loot to repay ten campaigns; eastward was the broad Tigris—and we already had the broad Euphrates at our backs. Men snarled and grumbled as we continued eastward. 'Why put our necks in the noose?' 'Why run headlong into space?' 'Why pass two rivers when we have only to turn right, and find all the plunder we can handle?'

Ah, but Alexander thought further than they! He knew that if we had turned southward between the rivers Darius would have fallen upon our rear and straddled his mighty army across the plain from river bank to river bank, and we should have been sealed up there with no chance to exploit the strategy and discipline which were our only means of overcoming the myriads of our enemies. He knew, too, that booty and plunder did not spell empire, any more than the victories we had been winning for three years. What he must have was the defeat and humiliation of Darius, the dispersal of the Persian armies. It was that he sought—or destruction for himself.

So we crossed the Tigris as we had the Euphrates, forty-seven thousand men, with an immense column of transport, a column that stretched for parasang after parasang, two days' march from van to rearguard. Four days we marched south after we passed the Tigris, and then our scouts reported the enemy in force ahead; and Alexander, himself, rode forward with some of his Macedonians, and shattered a considerable body of Persian horse, who fled away across the low, rolling hills that clutter this country. There was no sign that more of the enemy were within striking distance, but from prisoners we learned that the whole bulk of their army, with Darius in command, was lying a few parasangs south, awaiting us in a prepared position.

We had been marching steadily for many weeks, and were become footsore and weary, besides which we had several thousand sick and wounded men who were unfit for duty; and Alexander decided to halt where we were for four days to rest the army and fortify a camp to contain the invalids and our train of military stores and engines. A drungos of Thracians was appointed to garrison it, and in the darkness of the early morning of the fifth day we resumed our advance, planning to attack the Persians at sunrise. But the Gods decreed otherwise.

We climbed several ridges of hills, maintaining the battle order at every step, and as we were ascending the last we heard the trumpeting of the Persian war-elephants and the clamour of their cymbals, growing louder and louder, then fainter and fainter as the alarm spread to the distant wings. When we reached the summit the sun was just rising, and in the crimson light we could see the hosts of Darius spread out before us like a forest, a great, thick hedge of men and horses, war-chariots and elephants. Their shouting as they sighted us was like—was like—Roman, did you ever hear the sea pounding the Euxine's beach? It was like that.

It was plain now that we had failed to surprise the enemy, yet we still thought to be sent forward, and were disgruntled at the command to halt in our tracks. But officers of the staff soon brought us an explanation of Alexander's caution. In the growing light we had observed a wide area of fresh-turned earth in front of the Persian centre, and suspected that Darius had constructed a system of pitfalls to disorder our cavalry. And while the rest of us squatted on the grass and studied the far-off Persian lines, the young King rode close up to it and determined the strength of its several parts, as likewise, that what he had believed pitfalls were no more than a levelling of the ground to improve the speed of the Persian war-chariots.

All day the two armies faced each other, without exchanging a blow. We of the Thessalians simply sat beside our horses or gathered in groups to talk over the latest bit of gossip that had come down the line; Dion stood apart by himself, nursing Agathocles' sword. I watched him whenever I could, and I knew his coward's soul was shrivelling inside him. I could tell by the way his throat contracted and expanded, and the feverishness with which he clasped and unclasped the sword's hilt. And once when his father rode up to inspect us I heard Dion ask if there was to be a night-attack in a tone that convinced me he hoped to make use of the darkness to conceal himself out of harm's way. But old Philip gave him no encouragement. Any one could have told the fool that it was not Alexander's way to risk the disorder that fighting in darkness brings.

On the verge of dusk the strategoi drew up the drungoi in ranks and announced to us that the attack was postponed to the morrow. In the meantime we were to eat heartily and sleep in confidence of the result. Later Alexander rode through the ranks and shouted encouragement to every regiment, singling out officers and men he knew. How we cheered him! It was the first cheer that had come from us since we topped the hill, and you should have seen the Persians running this way and that, tightening their formations and preparing to meet an immediate attack. It put us in a good humour, the men forgot the tension of the day and we slept on our arms as easily as though we had been in winter-barracks.

Morning brought another story. Gods! My blood quickens with the memory, Roman. Remember, we were in open country, with no natural flank defences, and the enemy were five times our number. His horse alone were equal to our entire strength. It was a certainty that he would lap around us, flank us at the least, probably come at us from the rear as well as the front. So we marched in a great hollow square, arranged so that each part could be supported by the other parts. But our principal strength was massed in the front line, and mainly in the right wing of that line, which Alexander led. Do you see his plan? It was simple. He could not beat the Persians all at once with the army he had, but what he could do was to beat them so badly at one point, holding them in check elsewhere with inferior numbers, that their resistance would crumble away.

And it was the old story: Alexander and the Macedonians were to deliver the whirlwind attack, while we of the Thessalians resisted the better part of the Persians to give them their opportunity. Not that we minded, any of us. No, no! It was a point of pride with us to do the heavy thrusting. And we knew that the King knew what we did.

We were on the far left of the front line, and old Parmenio, who commanded the left wing, rode with us in the midst of Philip's staff. Our regiment was first in the column, and Dion led, his cheeks tallowy, his hands moist with sweat. Next to us was a drungos of light cavalry of the Greek allies. Then two drungoi of heavy Greek infantry. Then the Phalanx in an oblong bristle of pikes. Beyond them the shield-bearing infantry, two drungoi. And on the right Alexander and the Macedonian cataphracti. In advance of the right wing was a swarm of light troops.

The second line was one drungos of the Phalanx and three or four bodies of Greek and Macedonian infantry, mostly light-armed troops. It was linked with the first line at either extremity by regiments of light-armed infantry and horse; they were the sides of the square. And it was for the two drungoi of the cataphracti, we of the Thessalians and the Macedonians, to make sure that those sides were not burst in. By Hercules, every face of the square, except only the right of the front line, was slender enough. Our whole front was no longer than the Persian centre, and to avoid driving full into the midst of the Persians Alexander gave the word as we descended the hillside to incline to the right, with the result that the Phalanx and the Macedonians were directed at the juncture of the Persian centre and left wing.

This did not suit King Darius, who had hoped to engage us with his centre and then fold his two wings around us, and he launched two attacks to herd us back. First, he loosed a cloud of Scythian and Bactrian horse to strike our right flank. Alexander stopped them with the light troops of the right side of the square. And while this fight was going on, the Persians sent their scythe-armed chariots at our front. They came rattling and bumping across the plain in a cloud of dust, and Dion turned chalk-white as the sunlight was reflected from the long blades that projected from their sides. But they never even reached us. Our light troops met them in the open, and slew horse and driver with arrows and javelins.

So far, you see, we of the left wing had not lifted a weapon. We marched along quietly, keeping our alignment with the front of the square, and watched the cavalry combats that swirled in and out of the dust-clouds away over on the right. But our turn was coming. I saw a gap open between the Persian left and centre as a huge column of Susian horse abandoned their position to go to the assistance of the Scythians and Bactrians who were waging a losing fight with our light horse on Alexander's right. And quick as a flash Alexander struck. The Macedonians darted out from our line like a spearhead of galloping horses and men, and flung themselves into the gap, shearing deep into the unguarded flank of the Persian centre. After them charged the shield-bearing infantry and the Phalanx.

There was a roar of exultation all along our front, but the Persians answered it; and their right wing and half of their centre charged us of the left at a run. One moment we had a clear view of the battlefield from end to end. The next we were swallowed up in a sea of enemies. They engulfed our front; they boiled around the left side of the square and smote the second line in the rear; they even started to wedge in between our two drungoi of heavy infantry and the Phalanx, who were striding forward like a machine, crushing all the opposition before them. But Simmias, who commanded the left drungos of the Phalanx saw what the Persians were up to, and he wheeled his men out of position and dropped back to make good our right. Thanks to the Gods for that! But for him, I think, we should have been torn apart in that first mad rush of half the races of Asia. Wave after wave of man beat upon our spears, and the clashing of weapons and armour, the neighing and screaming of horses, the shouts and the thudding of feet were deafening. The elephants lumbered at us, waving their trunks and bellowing with rage, and it seemed as though nothing might stay them from crushing a bloody path through our ranks. But, Roman, know this: disciplined troops need never fear those beasts. A few well-directed arrows, and they turned about despite their drivers' prodding and wreaked amongst the rearmost Persians the very havoc their masters had sought to accomplish against us.

And what of the Thessalians you ask? Well you may! For if Simmias and his drungos of the Phalanx made good the right of our half of the front it was we of the cataphracti who must resist the greater assault upon the far left. We were like a shield a warrior interposes between his body and an enemy's blows—yes, and at times we ceased to be a shield, and became a weapon, a lance, a sword! Ah, yes, a sword! And that reminds me of what happened at the beginning, when the Persians stormed down upon us, and old Parmenio shouted to Philip to hurl Dion's regiment forward to break the force of their rush.

My squadron was leading the column, Dion was not two spear-lengths from me and I heard every word that passed as Philip trotted up to give the order.

'Honour to you, boy,' said the strategos. 'Yours shall be the first charge.'

Dion fumbled with the hilt of the sword that he held naked in his hand, and his voice quavered in reply.

'Only one regiment to—'

'One is enough,' Philip cut him short. 'Be off! Time presses.'

I have often wondered whether Philip knew his son was a coward. Some suspicion he must have had, but I think he hoped the boy would gradually accustom himself to danger; many men do.

'Do not go too far,' he added. 'Swing around and cut your way back to our rear. The second regiment will fall to as soon as you have driven home your blow. We must check the enemy before they reach us, else they will overrun the foot.'

'It is foolish not to strike with the whole drungos,' whined Dion, gathering his bridle shakily. 'One regiment! We shall be devoured—'

His father gave him a frown.

'I hear you have a new sword,' said Philip. 'Prove it!'

Dion snarled the order to the trumpeter, and Philip rode aside as we urged our horses to a gallop. For myself, I admit I paid little attention to the enemy. My thought was for Dion, and I made up my mind that he should stand the shock if I had to hold him in his saddle.

We crashed into the Persian line where Babylonian and Uxian infantry were formed a hundred men deep. They yielded to us, and receded; but troops flocked to their aid and no matter how deeply we penetrated the mass there were always ranks in front of us. Very brave men, these, who tried to grapple our spears when we pierced them or to leap up and drag us from our saddles. Sometimes they succeeded.

Through all this turmoil Dion rode without raising his sword, and presently I understood that he was dazed by fear. Persians struck at him in passing, their blades scraping on his armour. But he ignored both friend and enemy, galloping on like a man in a dream.

I judged now that we had thrust as far as was safe into the Persian array, and so I called to him.

'Back, chiliarch!'

He did not even hear me, and I ordered the trumpeter to sound the call. The brazen notes rang clear above the uproar, but Dion would have continued to ride forward if I had not seized his bridle-rein. His eyes were glassy with terror.

I was of two minds what to do. First, of course, my duty was to extricate the regiment; but I hungered to lesson Dion as he deserved, and as we turned I peered across the field in search of an opportunity. The Gods favoured me.

The Babylonians and Uxians had fled, and a column of Persian cavalry rode out to intercept us, led by a very tall warrior in a high, gold headdress.

I bade the trumpeter sound the charge again, and contrived our course so that Dion should come face to face with the Persian, dropping my hold on his bridle-rein a moment before they met.

The Persian hewed at him, and Dion squatted low in the saddle, ducking his head, as though that would save him. He was no weakling, that Persian, and sparks flew from his sword as it clattered on Dion's helm. Yet the stout mail turned the blade, and Dion involuntarily raised his own sword. I thought he was going to strike back. But no. His bemused face stared right and left, stark with the agony of utter fear. Armed men hemmed him in. He could not fly.

The Persian struck again. Gods, the marvel of it! His blade fell upon Agathocles' sword, wavering over Dion's helm, and the grey steel that my shieldmate had loved bit through the crest of the helm and deep into the coward's skull. No hostile blade had smitten him, Roman. The sword, itself, had turned upon him. The marvel of it! And some men deny that the Gods deal justly!

I wrenched the sword free as he slipped from his saddle, and cut the Persian to the lungs with it. An ill-requital for a good deed. But in battle a man may not choose his course. Smite your enemies or they will smite you. . . . The rest? We broke through the enemy, and regained the rear of the drungos, where Philip and Parmenio received us.

'Well done, Thessalians!' cried old Parmenio, his beard standing out from his face as it always did in the excitement of battle.

Philip's mouth was set in a straight line.

'You will take the chiliarch's place, Messala,' he said.

Not a question. Not a comment. Perhaps he knew it was better to say nothing. I don't know. I have often wondered. . . .

The battle? Yes, I was forgetting it. The sword meant more to me than any battle, just as Dion's death meant more than all the Persians I slew with the blade that had come to me in so strange a manner. For I wielded it diligently. That first charge was one of many, at first by regiments, but as the Persians heaped more men around us we used the drungos as a whole, hammering away wherever our infantry were most sorely tried.

By afternoon we decided the battle was lost. An immense column of Indian horse from the Persian centre had slashed through the square—not on our side of it, of course—and ridden back over the hills to the camp and captured that. We were too busy with our own troubles to observe what Alexander was doing, so we were as surprised as the Persians when the cry started in their ranks:

'Darius flees! Run, brothers, run!'

They commenced to scatter, and we headed our tired horses into their midst again, lest they should regain self-confidence—our infantry were worn out, finished—when what should we see but a brazen streak of armour, and Alexander galloped up to meet us with his Macedonians. Lucky fellows! *They* had been chasing beaten Persians ever since their first charge, with the Phalanx at their backs to keep the Persians beaten. But we were glad to see them, for all that, and we croaked an answer to their cheers. I remember I waved my sword around my head—thus! Do you hear it? As if a man spoke low to a friend. It is always so; I have never heard the hiss since Dion held it. Many times it has kept my head for me. Men say I shall die in my bed, but I do not like to think of that. If the years carry me to feebleness I will divorce the Grey Maid—so I call it— and give her to some youth of promise—Eh? Well, it might be, Roman. Prove your worth, and I'll take thought of it. Why not? She has drunk the blood of every race in the East. Let her try the West. But first prove your worth. Agathocles' sword is not for any chance passer-by. Remember Dion, Roman. If there is luck in the blade there is peril, too.

JACK LINDSAY

Greek Meets Greek

ιοοι

The usual stir and hubbub in the village did not greet the arrival of the ass-train, and in a few moments Lampon discovered the reason for this neglect. A rival ass-train already stood in the centre of the village, surrounded by the natives. Lampon saw indignantly that the other asses, like his own, were laden with wine-jars, and he felt as if he had found a thief in his pantry at home in Lissos on the Adriatic Sea. He could not have defended his indignation, for there were numberless customers among the Dardanian mountaineers and no means of establishing a monopoly. Still his first emotion was that he was being criminally crowded out of the enormous territory over which he was used to trek.

Then his anger passed, and his heart warmed. Obviously the other ass-train belonged to a fellow-Greek, and it would be good to hear Greek talked once more.

Only the more courageous traders dared to enter these uplands. The roads were bad, seldom more than goat-tracks; the distances were vast; the returns uncertain; the dangers of robbery or murder extreme. But Lampon had always been an obstinate man, though his favourite remark was: 'My handicap has always been that I can see the other fellow's point of view'; and he had made his first expedition as the result of a wager. Having once begun, he kept on repeating the exploit, attracted by the difficulties.

But how had the stranger reached the valley of the Margos, a tributary of the Danube? He couldn't have come from the same coast as Lampon. He must have come from the east, along the Danube, and branched off down the Oiscos—or up the Axios from Macedonia in the south. Traders occasionally ventured along those routes, driven by greed and restlessness, to chaffer with the quarrelling Celts who were pressing down towards Greece.

Lampon made out his rival amid the disorderly crowd: a tall, wiry man with a lock of dark hair over his forehead, who was busy cuffing one of the ass-drivers. The gaping villagers opened to let the newcomer through, and Lampon strode forward, overcome by the burden of the situation. He felt as if he and the stranger were ambassadors from Hellas charged with momentous responsibilities.

'Greetings,' he said with off-hand affability.

The stranger stared at him, screwed up his eyes, then smiled. His smile seemed to come painfully, with effort; then fell away abruptly from the lined face, like a curtain twitched by a flaw of wind. He wasn't the kind of man that Lampon, a genial talker who never managed to discard all his fat despite his travels, would have chosen for a companion at home in Lissos; but probably there wasn't another Greek handy for hundreds of miles. That was a bond. Besides, they were both wine-merchants.

The two men distrustfully exchanged a few words and found that they were bound for the same market-town. But there was no need to fear each other's competition, since the thirst for Greek wine in these vineless districts was unquenchable. The marauding tribes found no time to cultivate more than a little spelt and millet for the cupboards of winter; and, though they brewed some fermented drinks, the true wine of the grape could always command a payment in valuable metals. Money the tribesmen did not employ.

The village where the meeting had taken place was too poor to buy wine. It consisted of a few huts of clay and wattle, one larger than the others where the chief lived. The window-holes were roughly closed by stones or bits of bark; there were no doors, only holes which could be barricaded in times of storm or assault; and the chimneys were smoke-holes over the flat hearthstones. Some dwellings were merely dug-outs under dung-heaps.

Both Lampon and the stranger, whose name was found to be Sphodrias, had goatskin tents, which they ordered to be set up close together. Their servants, who included a few armed slaves for bodyguard, ran energetically round, pushing back the curious villagers, preparing the evening meal of boiled cabbage and barley-cakes, unloading the asses and turning them out to feed, chattering at the top of their voices, and ranging the wine-jars in regular rows so that a careful watch might be kept upon them.

Lampon and Sphodrias, with side-glances at this activity and with various interruptions, continued their conversation, enquiring politely about each other's business, home-town, and lineage.

Night was closing down, and they warmed towards one another in the barbarian solitude, enclosed by rocky mountains and arid fields, representatives of a superior culture who deigned to enrich themselves at the expense of the excitable Celts, forgetting their own ancestors who had come out of the north with fire and sword, and recognizing no kinship with these later invaders. They grumbled together, talking of the incredible disadvantages of trading in Dardania, recounting how asses had fallen over precipices, browsed on poisonous herbs, been bitten by ferocious spiders as big as rats, or been stolen by villainous tattooed raiders, men who swore by the Sun and Thunder but had no respect for Themis or Just Dealing. The gains were trivial in comparison with the outlays; whenever a tribe was strong enough, it imposed a tax at some valley-head or river-ford and yet hadn't the decency to make good roads in return; the perils were incessant and nerve-racking, from landslides to wolf-packs; the mountain-tracks wore a man down into a premature grave. Things were particularly bad at the moment, as the tribes were fighting over the ownership of some salt-deposits found in a gulley. There were rumours of a coming border foray into Macedonia. And what was a civilized man away from the beloved city of his home, the source of all culture and comfort and spiritual fecundity? Certainly they must both be mad to wander into the desolation where the gods lacked graceful bodies.

Having thus disburdened themselves, they recognized in each other the adventurous urge that would never be satisfied with the city's sacred boundaries, preferring the terrible fate of an unburied death in the salt wastes of the sea to the benefits of literature and the rights of a free man in politics. They began to like one another, to ignore the antipathy of their characters; and when the village chief appeared in a battered bronze breastplate, an old helmet once shaped as a boar's head, and an odd pair of embroidered sandals, they felt bound together in racial solidarity.

They received the chief courteously and contemptuously, offered him a beakerful of unmixed wine from both supplies, and were pleased to see him grow fuddled at the unusually potent drink. A boy in Lampon's train did the interpreting, though both the masters knew enough of the various hill dialects to follow what was said. Tonight they felt too thoroughly Greek to condescend to the paltry speech of barbarians, the swallow-twitterings of the baby-mind.

Dusk had thickened into night, and the voluble servants sat round a fire of sticks and cow-dung. The natives peered from their doors. At last one of them grew brave and sidled up with a crudely made harp. He plucked at the strings, and grinned. Then, finding that no missiles were thrown, he began an interminable performance, joined after a while by a man with a melancholic reed-pipe. Others gathered round and started singing. The music was more tolerable then; and the two merchants, having obtained at the last moment an aged fowl (to be boiled in goat's milk) as an addition to their dinner, sat listening with complaisant pleasure.

Stars hung thickly in the moonless, unclouded sky, wavering like water-drops. The air had no chill, and the wide silence of the mountains hemmed the village in. At moments a breath of wind rifted the warmth of the huddled group round the bonfire. At such moments the smallness of human effort, its isolation in the elements of immensity, was felt. But the cheerful, inconsequent song brushed aside the weight of the starry unknown, and once again nothing lived except the human space cleared by fostering fire.

Lampon and Sphodrias drank. Each had broached, in honour of the occasion, an amphora of wine, though not expecting to drink more than a few beakers at most. But neither liked to leave to the other the act of hospitality. Lampon must drink of the wine of Thasos fetched by Sphodrias, and Sphodrias must drink of the wine of Corcura fetched by Lampon. Then honour would be satisfied. They touched cups and drank dutifully.

'I congratulate you on the Thasian vintage, my friend Sphodrias.'

'The Corcuran wine could not be bettered, my friend Lampon.'

'Help yourself to some more, I beg you.'

'Don't forget the beaker at your elbow.'

The chant in the foreign tongue gave the savour to the ceremony. The merchants felt that they were adequately upholding the name of Hellas among the uninitiated. The little band of Dardanian villagers seemed the whole world of barbarians, the outer darkness of countless enemies moving down on the small environed land of Hellas and threatening to stamp out the cultural light kindled by Hellenic pride.

Now things were as they should be. In the wilderness the two Hellenes met and embraced in brotherly understanding, while the despised and dangerous world sat in its place, contributing a hymn of gratitude. By drinking one another's wine—Hellenic wine of the West and of the East, wine from the Ionian Sea and from the Aegean—the two merchants acclaimed the land of their love, the harmonizing achievement of their people.

'I insist that you pour out some more, my good friend Sphodrias.'

'I am already a cup ahead of you, my good friend Lampon.'

As the wine glowed through their blood, sentiment strengthened. Each man told of his home. Lampon told of his wife and three children. Sphodrias told of his mother and unmarried sister. Lampon confessed that he hated wandering and wanted only to be a good husband superintending his children's education; but fate wouldn't let him. Sphodrias confessed that he travelled only to raise the money to buy the next-door house and have the property that he coveted. He wanted, above all things, to be merely a good son, but found the same trouble with fate as Lampon.

'My wife's a most remarkable woman. The perfect mother. If she wasn't, I couldn't have left her in charge of everything. I haven't seen her for two and a half years.'

'You see, I have to get that house next door. It spoils the appearance of my house. Only a trifle of alterations and I'll have a magnificent court-yard. The best in the town.'

Tears rolled down the cheeks of Lampon.

'Give me your hand, Sphodrias. I feel homesick. There's nothing in the world like a good woman.'

'I agree with you. This is my last expedition.'

'Same with me. I'm sorry we never met before.'

'So am I. Let's go on together. Always. On and on.'

'That's right. On and on.'

The music drifted pleasantly past. They forgot that barbarians sang. The song was the voice of humanity defying the forces of chaos, the blind fury and corruption of nature. It was the sweetness and sadness separating man from the elements of his begetting, giving him the earth as his plaything and yet making him a stranger on the earth. Lampon and Sphodrias, the two wanderers, felt themselves sanctified; for they were now becoming drunk.

A humble native approached, deputed by the villagers, to ask for a little wine in payment of the music. Only a very little, little wine, he said, and they'd sing till morning.

The merchants, softened now, poured out a few cupfuls of both Thasian and Corcuran wine, mixed the result with much water, and handed it over. The songs weren't wanted 'till morning', but meanwhile could go on. Listening to the brawls occasioned by the distribution of the wine, the merchants smiled at the uncontrol shown by the lesser breed. They themselves began discussing the vintages.

'The wine from Corcura always captures the market,' observed Lampon, confidently. Then, remembering, he added: 'Unless, of course, there's some Thasian about. That's almost as good, practically as good, one might say.'

'I was about to remark the same thing about the Thasian as you do about the Corcuran. I've never had a complaint once.'

'Neither have I. Not about Corcuran, though I must say I've heard complaints about other brands.' Lampon feared that his remarks would be misunderstood as offensive, and in his agitation he produced the very effect he wished to avoid. 'Not Thasian, of course.'

'I've heard complaints, too—about other people's wines.' Sphodrias spoke nastily, feeling that he was somehow being jeered at. 'I object to that kind of thing. It spoils the market.'

'What I meant to say. It lowers the dignity of the profession.'

'Of course, I'm not speaking about anyone present, but there are some wine-merchants——'

'I know. They water their stuff in the cask, and put turpentine and resin in to make up the strength.'

'Now, don't say I accused you of any such thing.'

'I didn't mean you. Why, I've never seen you before. I don't know anything about you.'

The merchants sat back on their pack-seats, glaring and clutching their cups. They felt suspicious and irritated, and couldn't recall which of them had been responsible for the note of antagonism. They wanted to break the emotion of debate and rivalry, but it gripped them tighter every moment. After all, they did know nothing of one another. Strangers in wild places couldn't be too careful.

'Well, I'll back Thasian wine against all others,' said Sphodrias, fiercely, smiting his thigh. 'I don't care who knows it.'

Lampon reddened and breathed heavily through his nose.

'Corcuran can't be beaten. There's something in the soil. It isn't only that the grapes are bigger than anywhere else, though no one can deny that. It's a richness in the soil.'

'At Thasos,' said Sphodrias, determined to lie rather than be worsted, 'the vines don't need supports. They stand up of their own accord.'

'I don't believe it,' shouted Lampon.

'Have you ever been to Thasos?'

'I don't need to go there to know a thing or two about it. That's the island where a wasp was once blown over from the mainland and it ate up the whole grape harvest.'

'That wasp must have been as big as the lies they tell in Corcura.'

The men were now frankly furious. All their doubts rushed back. Each saw the other as a scoundrelly competitor who threatened life and livelihood. Probably the meeting was a put-up job and the natives had been hired to commit murder as soon as bedtime arrived. Lampon remembered that the Aegean was filled with pirate ships, and Sphodrias that the Dalmatian coast was riddled with pirate haunts. There was a trick somewhere.

Both men drank strenuously, and the more they drank the surer they became of ambushes and stratagems.

The songs were throbbing louder. Neither Lampon nor Sphodrias listened, but the pulse of the music jangled against their own pulse-beats, like the encouraging voice of an ally at one moment, like the voice of a jeering enemy at the next.

Neither man knew which it was that first suggested gambling, but each eagerly seized the idea as an outlet for overcharged emotions. Dice? Neither had any. Lampon had left his behind, and Sphodrias had lost his in crossing a river a week ago. Knuckle-bones? It was no use asking the Dardanians. There would be an endless discussion, during which everything else but knuckle-bones would be produced, even if knuckle-bones were available. That was the Dardanian way of making people feel at home. The natives themselves played with pieces of coloured stone, but the merchants felt it derogatory to ask for such toys.

Lampon found the solution. 'Count-finger, then.'

Sphodrias nodded with sombre gravity, being afraid of toppling on to the ground, for they had departed from the Hellenic custom of drinking watered wine and were now gulping from undiluted cups. The finger game was played by two men suddenly thrusting out their hands with all, some, or none of the fingers extended. Each had to call at once the number of fingers shown by himself and his opponent. Though simple, the game needed quick observation. The player had to observe in a flash how many fingers his opponent was extending, and add this number to the number of his own extended fingers.

No worse game could have been devised for men bleared with wine.

The merchants faced one another.

'What's the stakes?' asked Lampon, snorting.

Sphodrias thought heavily, took another drink, and said, 'Wine, of course. Thasian wine will beat your squill-juice any day.'

'We'll see about that. But I don't want your filthy concoction. I wouldn't feed my donkeys with it. It would give them the mange.'

'As for me, I'm not a vinegar-merchant. When I win your Corcuran hemlock, I'll use it for killing weeds.'

They sat glowering at one another. The servants felt the tension and clustered discreetly round. The song of the natives died down, but the harpist still plucked at random notes.

'Are you going to play, or are you not?' snarled Sphodrias, lifting his hands.

Without replying, Lampon lifted his hands in turn, and the two men searched each other's eyes. Then with a common movement they lowered their hands rapidly.

'Eleven,' called Sphodrias.

'Ten,' called Lampon.

Immediately both claimed to have won. Sphodrias said that one of Lampon's fingers had been three-quarters sticking out and that it had therefore been rightly counted. Lampon denied the contention. He said the finger in question had been clearly tucked away. The men stamped their feet and roared.

'You're a cheat!' cried Sphodrias, leaping up. 'I claim my winnings against you.'

He ran to the line of Lampon's stock, snatched up an ass-staff and lashed out at the first earthenware jar. Lampon watched for a moment in agonized suspense, then dashed at the stock of Sphodrias.

'You're cheating by pretending I cheated,' he shouted, and aimed a blow at the nearest jar. Wine gushed out. Both men looked on their destructive handiwork and felt the exultation of seeing enemy blood pouring from the ruined ribs of a wine-jar.

'You cheated!' Sphodrias shouted again, and repeated his assault on a second jar. Lampon followed his example. Shouting, they worked along the lines of wine-jars, dominated by an insensate glee of destruction. Streams of wine dabbled their feet, and their arms were tired with the violent blows, but still they went on beating and shouting. The wines from the West and the East, the blood of Hellas, ran together along the grooves of thirsty alien soil.

The servants followed the progress of destruction, weeping, howling, imploring, quite unheard by their possessed masters. The villagers looked on aghast, and then awoke to the possibilities of the situation. Some scudded to the huts for bowls. Others were too impatient and threw themselves under the broken jars, careless of cuts or scratches, lapping up the trickles, scooping up handfuls of wine and mud and swallowing the mixture with coughs of appreciation. The mud was inclined to rasp the throat, but what did that matter?

The noise and gestures of the scavengers were felt as a stimulus by the wreckers, who, however, needed no stimulus. Shouting drunken insults, neither Lampon nor Sphodrias realized that his own stock was being demolished. The only emotion was a battle triumph, a sense of routing all competition for evermore. Finishing at last, they stood wearied out, leaning on the staffs which had dealt such damage.

Then Sphodrias, with a sneer, mumbled 'Cheat', made a deep mock bow, fell on his face, and was carried away by his moaning slaves. Lampon swayed on his feet, overjoyed, considering the collapse of Sphodrias the last tribute to his doughty staff. He passed a puzzled hand over his brow and began weeping—but not over spilt wine. He had forgotten such things. He was weeping over the astonishing beauty of the world.

'Stars,' he said, looking upwards and blinking. He held out his hand as if expecting the starlight to rain softly into it. 'Let's have another song.'

Sobbing, he allowed himself to be led back to his tent.

The natives continued to grovel, licking at the mud and inspecting the crannies of the broken jars for small deposits of wine. They were struck with awe, convinced by the mad destruction of the assault on the wine-stocks that they had witnessed a scene of god-inspired passion. They would tell the story for generations, till the merchants appeared as heroes, god-begotten, from beyond the North Wind, who fought a monster that bled fertilizing wine and shed its bones across the bouldered hills.

But the servants and slaves had no delusions. They knew that in the morning two liverish merchants would suffer the tortures of hell and would turn remorseful wrath on those nearest—the servants and slaves who must pay the penalty for the masters' shame. Muttering, they argued whether it would not be better to stick the masters in their sleep, and gained a slight comfort from the feeling that it was in their power to make that sleep eternal. But they knew very well that they lacked the courage and would merely mutter and complain till it was too late for action.

Meanwhile, mercifully, the masters snored.

NAOMI MITCHISON

Cottia Went to Bibracte

:∞:

You see, there was nowhere very much else for me to go; M. couldn't leave me in Rome, because of that dreadful man making love to me—and I couldn't do anything about it, could I, with father encouraging him, not to speak of his position?—and the Province wasn't much better; you've really no idea of all the tenth-rate people one meets there! And besides, I saw so much more of him this way; after all, we'd only been married six months, so we couldn't help being rather glad! But anyhow, there I was in that funny little town Noviodunum, and there I should stay till the end of the campaign. Of course we never expected anything to happen like it did; we thought there'd perhaps be a little fighting—those tribes in the north, who were always trying to revolt—but punishment, not war; if he'd had any idea of how things really were, M. would never have let me stay, and I should have missed it all.

It was dreadfully cold up there that winter, but I had my new furs, and besides M. used to get leave half the time. We had the dearest, silliest wooden house, all thatched into a point at the top; of course it was all very primitive to begin with—almost nothing but a table and some cooking-pots—but I set to work and bought carpets and curtains, and got some chairs made on a nice Italian model, and a bed; and then I'd brought a few things with me from home. So soon I had the sweetest little place of my own for M. to come to.

Oh, those winter evenings with a great fire of pine logs and M. coming in and the gleam of his armour! My dear, you can't think what it's like; I was just *too* happy! Then we'd have supper together, just nothing, a chicken and some little cakes, and a cup of wine between us. And then sometimes after that I'd let my hair down and sing or dance for him; we used to pretend I was a dancing-girl, and then he'd make love to me—that way! And oh, do you know what happened one time?—we found half the slaves standing in the doorway, looking on, and me with my hair over my face and my dress slipping off one shoulder! M. *was* so angry: I don't know what he didn't want to do to them all; but I wasn't really angry myself—not a bit; I expect I ought to have been, but somehow I wasn't! You know one can't be dignified all the time.

Besides, they were nice people; when all's said and done I like the Gauls. One has to deal with them the way they're accustomed to: they're not like Greeks. But they're very clean, they don't mind what they do for you once you've been kind to them, and if they do break your best glass, well, they always come and tell you instead of hiding it or saying it was one of the others. Of course they eat a great deal—I was horrified to find how much meat my household got through in a week, but M. laughed at me—and they're quarrelsome, and if you give them a chance they talk all day, and of course if you let them get drunk they're perfectly horrid; but they do get through their work like no one else, and if you're a woman alone in a house I should certainly advise you to have Gallic slaves.

I had a lot of time on my hands when M. wasn't there; there was no one to visit, and of course no theatres or amusements, and I couldn't go into the country by myself. There was always my lute, to be sure, and I didn't play so badly, and I had my sparrow, 'Sweet', and my little dog. But still I had to do something—I never could bear sitting still!—so I learnt to speak Gallic (it made my housekeeping much more satisfactory) and I used to weave linen on a loom like I remember my old nurse doing.

Then one day the baggage and hostages were sent into the town with a great escort. I watched from a funny little balcony in front of the house —there was a garden towards the river at the other side—and by and by M. came and joined me. That was the first time I'd seen him looking worried about things: we were just beginning to get rumours of the chiefs conspiring, and Caesar was in Rome. But of course I wouldn't think of leaving the town; we all believed the Haeduans were safe then.

M. had to see to the hostages and find quarters for them all; and he hadn't got it settled very satisfactorily when he came back. 'It's not at all easy,' I remember him saying. 'The townspeople here are old enemies to some of the hostages—it wouldn't be safe to leave them within murdering distance. And yet I don't quite see how I can keep the children in camp; I suppose we must, though.' 'What sort of children?' I asked. 'Oh, well, they're all sons of the chief men in some tribe or another; poor kids, they're as frightened as anything.' Then I had an idea: 'Oh, M.,' I said, 'couldn't you give me a couple to look after?' He looked at me then with the sort of smile that makes me pleased all over, and said, 'Do you think you really could, Cottia?' I almost wished I hadn't, but of course I said I'd love it, and he told me exactly how safe I was to keep them—it was rather like being trusted with someone's best pearls!—and how important they were. So I got the room next mine ready, and had a bolt put to it, and told them to cook some more supper; and then I fetched my cloak and went straight off to the camp with M. to find them.

And really I'd had so little to do with children! Being the youngest myself, and all the cousins older than me, and never seeing a child except Julia's two, who are dreadfully forward—well, I felt too silly saying I'd take charge of these ones! But of course I didn't say so to M. and we got to the camp. I do love soldiers! I can't help it, I always have: the Legate was so charming to me, too, and I do like being saluted: that's M. really, of course, not me, but still I always enjoy it. But I was much too nervous this time.

The hostages were all together, under guard; the big boys were in chains, and I did think it was rather a shame—they couldn't possibly have escaped. They all stared at me, and I felt so uncomfortable; M. had gone off to see the other officer in charge. I pretended not to look at them, but of course all the time I was, out of the corner of my eye. I wondered which I'd get: some of them looked such big rough things—the Belgians particularly—and I was sure I wouldn't be able to manage them. Then I heard such a miserable sniff from one side; I couldn't help looking round. There was a big pile of baggage on my right, and for a moment I didn't see the children crouching beside it; they were two little boys with long cloaks wrapped close round them. The smallest had his head turned away, but his shoulders were shaking wretchedly. The bigger one just sat looking at him; I think he was too nearly crying himself to be able to do anything. Suddenly he saw me; he stiffened all over and caught tight hold of his bundle. I smiled at him as nicely as I knew how, and after a minute he reached out a hand and touched the little one's shoulder, and he looked round too, with his face very white against the piled baggage.

Now, there's one accomplishment of mine that I'm rather ashamed of, but still I never forget it: and that's making faces. When I was small I used to practise hard wriggling my nose and eyebrows, so as to make the slaves laugh at dinner; and they always did, though they knew my father would be angry with them. So now I began doing it, at the children; first of all they just stared, and then, very slowly, the elder one let a huge grin waver across his face, and then the other gave a little, sudden laugh and whispered something to his brother. They obviously were brothers, when one began to think of it; they'd got the same curly, brown hair, and fine-cut faces, and they both had red woollen cloaks with yellow lines criss-cross all over them.

I wanted to speak to them, but I didn't quite know what to say. However, just then M. came back. 'I've settled it up,' he said. 'Oh, which?'—I was so afraid it was going to be a great, rough one! But he called aloud: 'Danorix!' and the bigger of my two brothers jumped up. I really was pleased: things do so seldom happen like that. I took his hand and he looked up at me; I suppose he thought I was going to make faces again. Then the little one scrambled to his feet and ran forward and caught hold of his brother, crying in Gallic, 'Oh, me too!' so I said 'Yes, both of you!' and M. nodded. They picked up their bundles and came along, one at each side of me.

Supper was ready, so we put them between us at the table; they sat up very straight and wouldn't drink anything but water, and when they thought we weren't looking, Danorix held his little brother's hand under the table.

What does one talk to children about? I wanted to start a conversation and make friends with them, but they only said yes and no; and at first I found their accent puzzling. I asked them all the obvious things: how old they were? seven and nine; were they tired? well, a little; could they ride? oh, yes; had they any brothers and sisters? 'I'm the eldest,' said Danorix. 'But we've got two sisters.' Iurca, the little one, added, 'Mother's dead: last year,' and choked. I tried to change the subject: 'What are you going to do when you're grown up?'—a silly question, but I couldn't think of anything else. Danorix threw his head back: 'We're going to be soldiers!' 'And fight the Germans?' asked M., amused. Danorix shook his head violently. 'No, the Hæduans!' Fortunately the slaves were all at the other end of the room, so they didn't hear.

I took them off to bed after that; they were very dirty after their journey, and I was so glad to see they wanted to wash without my telling them to. They were beginning to talk a little more now, and were very anxious to know what was happening to the others. Danorix had a funny mark on his chest in blue; I didn't like to say anything (you can't be too tactful with barbarians), but he saw me looking and said: 'That's the Owl; it means I'm going to be chief. If I get killed, Iurca can have it done on him.' I called one of the slaves to help us to undo the bundles which the children had fastened up with raw hide cords so tight that we couldn't get the knots undone. The man came in and he saw the owl mark too, and all at once he fell on his knees and began gabbling away in Gallic. Then Danorix started talking too, and put his hands on the man's head, while the man kissed the edge of his coat. I asked what it was. 'He's one of my men,' said Danorix. 'The Hæduans took him prisoner. Can he stay and help us?' 'Of course,' I said, and the man followed them about, worshipping them with his eyes, like a dog.

When they were rolled up in their blankets, both together like little dormice, I bent down to say good night to them. Iurca the kid, the little one, caught me round the neck and gave me a good hug. Danorix didn't— I'm sure he thought he was much too old!—but as I was going he caught me by the skirt, 'Do you know what I'd do if I was at home?' 'No, what?' 'I'd have a big, big sword, and I'd hang it up just there!'—he pointed to the wall by his head—'Has your man got a big sword?' 'Yes,' I said, thinking I might as well improve the occasion, 'in case little boys are naughty!' Both of them laughed at that, looking adorably round among the blankets. 'That's what father says,' said Iurca, 'but he never does!' 'Good night, you little villains!' And as I went out I heard them both laughing softly still.

They were very shy the next morning, and played a funny game by themselves with little white stones, marked with spots; they wouldn't tell me what it was, but the Gallic slave seemed to know. Such a queer little couple they were! Sometimes they'd be laughing and racing about and chattering away to me and every one else, as if the whole world belonged to them; and then half an hour later they'd both be down in the dumps, sitting in a corner holding hands and crying, not letting anyone come near them, just a pair of small, hard backs if I wanted to pet them! And then it would be gone as soon as it came, and they'd be as jolly as crickets. The only thing they always liked, however they were feeling, were my dog and bird; they were never tired of watching them; but I found them much more amusing to look after than any animals! They were very fond of music, too; they cut themselves reed pipes out of the river; Danorix feel in, doing that, and I was afraid he was going to catch cold—it was the middle of winter—but he didn't. The pipes weren't very good, though, because they ought to be made of fresh summer rushes, not old, dry ones.

That time M. went back to the legion looking rather worried; I couldn't make out quite what it was; he wasn't even very sure himself, only he had the feeling that something was going to happen. The last thing he said was, 'You will be careful of the hostages, Cottia? All our lives may depend on them!'

The next week we had the news of that terrible rising at Cenabum; they'd massacred every single man and woman—Roman, I mean—including an officer we knew. Perhaps you remember they caught the man who started it last year, and had him executed. Almost on the top of that came the first stories of trouble among the Arvernians. We weren't frightened at first, though we knew Caesar was away in Italy. The officer in command came to see me and asked whether he should take my two hostages to the camp, but I said they were quite safe—I was much too proud of their having been trusted to me ever to give them up like that! I asked him what he thought of the situation; he said he had just come from seeing the Hæduan magistrates, who all assured him of the loyalty of their state, and Noviodunum most of all, and that he thought we were quite safe. But one rumour came on the top of another. First we heard that Gergovia had stayed loyal to Rome, and the chief there had banished his nephew who was starting a rebellion; but the next day we heard that the nephew had come back at the head of a huge army, and it was the uncle and the loyalists who had been banished. Of course we didn't think very much of it, even so; every one said there was no need for me to be anxious: this young man and his followers couldn't do anything: if it had been some of the old chiefs from the north, Ambiorix, for instance, we might well be alarmed, but as it was it would all be put down in a few weeks. I remember being told the name and then forgetting it the next day; nobody'd heard of Vercingetorix then.

I'd had no news of M. for some time, when at last a message came through; he must have written it in a great hurry, and it had been some time on the way. He told me to send my hostages to the camp and ask for a guard as far as the Province for myself. It ended with an underlining, '*Noviodunum is not safe.*' I took the letter round to headquarters, but they said I couldn't possibly leave the town: it might have been safe when the letter was written, but now it would be most dangerous, and they couldn't let me; they were beginning to be uneasy that day. This was the last I heard of M. for months and months.

News came through to Noviodunum rather slowly, and often the townspeople got it before we Romans did. At first it was all Caesar's coming back from Italy and victory everywhere; we felt our friends at Cenabum were being well revenged. Vercingetorix could do nothing when he had Caesar against him—so we thought. Then when Avaricum was burnt we were sure the rebellion was really at an end. I found Danorix was going to have a birthday that same week, so I gave him a big box of dried fruits, and we had some of the hostage children in and played games on the grass by the river. One of them was Iurca's special friend; he was the son of a chief of the Bituriges, and his home was in Avaricum, so most likely he was an orphan and friendless by now; he was very sad, poor child, and wouldn't play with the others. The Hæduans were very much impressed too with our victory; several of the magistrates sent their wives and daughters to see me with all sorts of presents and compliments; I didn't believe in them though, and my two little boys used to hide when they came, and make faces from behind the curtains.

About now some more hostages were sent in, and a good deal of baggage too, and corn; we saw some lovely remounts for the cavalry being led along, and there were great chests full of official papers. Then we heard how Caesar was besieging Gergovia with Vercingetorix inside, and of course we all expected to hear of its fall in a week or two. But at first no news came, and then, when it did come, it was to say our army had been beaten and Caesar was marching north to join Labienus—and then? I was told I must prepare to move at an hour's notice. You ought to have seen how differently those horrid Hæduans behaved then! No more saying they were our brothers, no more visits from the chiefs' wives—not that

I cared!—no more running up to me in the streets and asking what I thought the fashions were going to be: oh, no! They'd tell me to my face that we were all going to be driven out of Gaul with our tails between our legs. I stayed in the house all day with the doors barred, and played with the children; they didn't like the Hæduans any more than I did, and though we didn't talk about that, it was something to know it was there.

One evening the children were in bed, and I was sitting at my loom working out a pattern in blue and yellow; it was quite quiet, and I could just hear the river outside. I had a lamp beside me as it was nearly dark, and I wasn't thinking of anything; only my mind was full of a picture of linen threads and the pattern that ought to be coming on the loom. Suddenly I woke up with a start and listened to the echoes of a scream somewhere quite close, and then I noticed that there'd been a noise growing for the last minute or two, men's voices and feet—and something else. I ran to the door and listened, and heard what I'd never heard before: the rattle of swords fighting in earnest.

Two of the slaves ran in: 'Mistress, they're coming! Down to the river!' For a moment I didn't understand; then I caught up my cloak that I had always ready, saying, 'Fetch the children!' But just then came a bang on the door and a clatter-clatter of metal hammered against it; the slaves stood gaping, and I called out myself as loud as I could, 'Who's there?' But it was a Roman voice I heard, a man asking desperately to be let in; I dropped the bar and opened, and he sprang through, a young centurion with his helmet off and a great bleeding cut on his head. I barred the door again; he held on to my loom with one hand and panted, 'I was sent to fetch you—my men—all killed—the devils!' and looked fearfully at the door. Another of the servants ran in: 'The river's no good—they're down there!' Then their own man came with Danorix and Iurca, still pink and yawning from their beds. The centurion told me to call all my slaves together; so I did, and bound up his head as well as I could.

They hurried in, ten or twelve of them, and my own maid, who could hardly stand for fright. I said, 'Who's going to stay by me?' and more than half shouted they would. I ordered the rest to go; they slunk off; then I promised the ones who were left their freedom if they saved us, and told them to see to the doors and windows. They ran off and we heard the noise outside getting louder, and with it more screams and a crackle of flames. Suddenly it all shifted round to the garden side, and there was a great crash, and then—they were in. The centurion jumped up, one hand at his head, the other at his sword. I picked up Iurca and ran behind the loom, calling to Danorix; I don't know what I hoped, only I never thought of doing anything heroic. I saw my groom come stagger-ing into the room with a spear right through his throat, and then a wild, horrible rush of the mob, and my centurion knocked over and trampled on. I remember certain things: seeing a jar falling and making a move to catch it, seeing my maid in the middle as white as chalk, with a dash of blood on her cheek, and seeing Danorix with all his teeth showing stab upwards into a man's side from behind. The loom checked them just for a moment and that saved me; I stood perfectly still, and Iurca was too frightened to move. Their leader came to a stand within a foot of me.

'Where is the other child?' I said. 'Bring him to me, please.' 'You're my prisoner!' yelped the man, but I only stared at him and said again, 'Bring me the child.' From somewhere in the middle of them a man hoisted over Danorix with his hands tied. 'Are they yours?' 'Yes,' I said, thinking that was best. 'Now take me to the magistrates.' They were yelling at me from the back of the crowd, but the ones in front seemed sobered, and no one even suggested binding me. I carried Iurca on one arm—he was a good weight!—and put the other round poor Danorix, who was badly bruised and crying a little. I bent down and whispered to him in Latin, 'You're my Marcus now,' and he understood and answered in Latin too, 'Yes, mother.' We had to step over the body of their man, who was lying dead in the doorway. As we went out I saw several of them glancing enviously at my necklaces (I used to go about with them all on in those days, just in case) and at last a boy snatched one from behind and it broke, but that was all the insult I had, except of course the things they shouted at me.

The streets were packed with people, but most of the killing was done indoors or in the camp outside; we saw little except a few flaming houses, though we heard quite enough that was horrible. Round one corner we ran into another armed band, who stopped us, shouting that they were after the hostages. One of them grabbed Danorix, but I held on to him, saying they were my children, and the man let him go. Somebody held up a torch to my face, and said, 'You aren't old enough!' but the others had gone on, and no one noticed.

Of course the magistrate I was taken to, who was a man M. and I had met several times, knew the children at once, but he let me keep them with me. We were all three locked up, and I untied Danorix; Iurca had never really woke up, so he went to sleep again at once on my cloak, but Danorix was far too wide awake; he'd killed a man and been in a battle and now he was a man himself! Then all at once he began crying again, and threw himself on to the floor: he went to sleep at last in my arms; I didn't get much sleep myself that night.

The next day a man unlocked the door and bade us follow him; I asked where we were going. 'Bibracte,' he said. I told him I must have food for the children, and after a little grumbling he produced it. We were all put into a cart, and set off for Bibracte. I'd time to wonder then; I questioned everybody I saw, but usually they wouldn't answer me, or else they'd say they were conquering the world, or some nonsense. I did find out that they'd murdered the officer in command and his nice wife, and in fact almost all the Romans at Noviodunum, and that they'd got hold of the baggage and Caesar's papers and all our hostages. There were hardly any Roman prisoners except me—my house was a little out of the way—but I saw some of the hostages. Of course a good many of them came from tribes friendly to the Hæduans, or too powerful to harm, like the Arvernians, and they'd been let go at once and were riding among the chiefs, but some of them were old enemies, or else they belonged to weak tribes that the Hæduans wanted to frighten; and these I didn't see. Horrible, horrible people to murder their own flesh and blood! I was afraid for my two, but as they'd not been killed in the riot that night they had a chance at least. It took us a good four days getting to Bibracte, and I can't say it was a comfortable journey. There was no news to be had of M.'s legion, so I didn't know if he was dead or alive, or what was happening at all. But we got there at last.

It's a steep climb up, but it's made easier for carts by the big ruts they've cut in the stone at each side of the main road. We were lodged in a queer little house, half underground, with one barred window at the level of the road, which had been an enamel-worker's forge. We got there in the evening, tired out and hungry, and were bundled in by the men. It was nearly dark and we looked about us, trying to see what sort of place it was. Then a woman came out of the inner room with a small clay lamp in her hand, and her hair in plaits over her shoulders. 'Who are you?' she asked. I answered, 'I'm a Roman prisoner.' She held up the lamp to look at me and the children and said, 'These aren't yours.' I didn't deny it to her. 'I'm Adianta,' she went on. 'My husband's Comm of the Atrebates, and he's here.' I said, 'My name's Cottia, and my husband's a Roman officer, but I don't even know if he's alive.' She took me by one hand and Danorix by the other, saying she would find us food, and we went into the inner room.

I'd been very lucky already, but I think my greatest luck was finding Adianta; she was kind to us, she stopped the soldiers from insulting us, and, what was almost more, she told us what was happening. Comm, her husband, had been our ally in the old days; Caesar had made him king and given him all kinds of honours; but now even he had joined with the Arvernian. They'd all met at Bibracte to elect Vercingetorix general again, but the Hæduans, who had only just come in, were jealous of him. They wanted to frighten and bribe the other tribes into choosing one of their chiefs; that accounted for their dealings with the hostages. We tried to find out if Celtigun, the children's father, was there, but there was a great crowd of chiefs and Adianta didn't know.

When Comm came to the house I kept out of the way in the little room where the enamelling tools used to be put, and told the children stories. But one day I heard him coming shouting along, and Adianta asked me to fasten the brooch on her shoulder. I did what she asked and ran back to my room. Comm came clattering in, and this time four or five others with him. They called for drink, which Adianta brought, and then they all began talking at the tops of their voices and singing. I didn't hear or heed what it was, till I heard Adianta protesting, and then suddenly Comm burst in on to me and pulled me out. Adianta ran between me and the chiefs, but her husband swung her back, laughing at her, and they all closed round me.

I can tell you I was frightened then! Those hairy great giants with their swords and wolf-skins, all shouting and roaring with laughter every way I looked! One of them threw a full wine-cup over me, and then the others, though I heard Adianta indignantly crying to them to stop. I stood there with my hair and clothes all drenched with the wine, and my hands over my face; then I was pushed from behind and one of them caught me and kissed me; I screamed, and he pushed me across to another. Then it was just a nightmare of a five minutes with all those drunk barbarians pulling me about one way and another; I was struggling all the time, trying to get away, and feeling an open mouth or a great mat of beard under my hand every time, and shrieking to Adianta to help me. I ended up in a man's arms with my hands held over my head, and he squeezing me up to him and kissing my neck; I could feel his great hot body against mine, and I didn't know what wasn't going to happen. But anyway I bit his ear as hard as I could, and he let go and swore.

I fell on to my knees on the floor, and Adianta knelt beside me and told Comm just what she thought of him. I looked up at the man who'd been holding me; he was the tallest of the lot, with coral brooches and great knobbly gold bracelets all up his arms; he wore bright red stuff sewn with gold, and long fringes to it, and his cloak was white fur. He was drunk and standing with one hand on the table and the other on his sword hilt; he'd blue eyes and ruddy-brown hair, and his ear was bleeding all down his neck, where I'd bitten it. He was half laughing and half scowling, and the others all had their eyes on him. Suddenly Comm gave a great shout, 'There's Rome kneeling to Vercingetorix!' Because it was Vercingetorix all the time, though I hadn't known it. Rather unsteadily he let go of the table and pointed to me: 'Gaul is merciful: Rome may rise.' I got to my feet and had a good look at him; suddenly he made a grab: 'Rome must pay tribute, though!' I ducked behind Adianta, and he caught one of her

plaits instead. Then I saw two of the chiefs holding Danorix and Iurca—
Adianta told me afterwards how they'd fought when I was screaming for
help—and I ran over to them.

I think Comm was ashamed of himself, and perhaps he didn't like
seeing his wife kissed, even by the general, so he interrupted: 'What's to
be done with these two?' Vercingetorix shook himself and frowned at the
children: 'Who are they?' 'Hostages from Noviodunum: for Celtigun.'
'Isn't he here, curse him?' 'No.' 'If he's not here by tomorrow, hang them
both. That'll teach him to be late!' 'Oh, you can't!' I cried, but Adianta
clapped her hand over my mouth and bade me be quiet. And Danorix
spoke for himself: 'Father will come! We hate the Hæduans, and they're
your enemies!' 'Listen to the Owlet, he's about right!' laughed one of the
others. But Vercingetorix was still angry: 'A pretty way he's shown his
friendship, that Celtigun! How did they get here? I thought those fools of
Hæduans would have killed them off.' 'She brought them.' Comm pointed
to me. 'Oh, did she! They're allies of Rome, then! I haven't done with
Rome!' He made a step towards me, and I didn't know what to do now,
but another man came in hurriedly, a man who wasn't drunk. He caught
Vercingetorix by the arm. 'They're waiting for you to speak!' 'You bird
of ill omen, Lucter! Tell them I'm busy!' 'Busy! Tell them you're in the
moon! Come out of this!' 'Oh, very well!' He stretched himself for a great
yawn and turned, then threw back over his shoulder: 'Remember, if
Celtigun's not here, they're to be hung tomorrow.'

Adianta looked very grimly after them. 'I shall see what I can do,' she
said. 'I'll talk to my husband; but I don't know . . .' She shook her head,
then, turning to me, 'They're all mad today because of the Arvernian
being elected general again in spite of everything; they were awake all last
night. What silly babies men are! and nobody to punish them when they
break things.' She began wiping up the wine: 'Your dress is all stained;
you must wear one of mine while we wash it.'

I don't think Iurca quite realized what the threat meant, but Danorix
did; his teeth were chattering, and he could hardly speak. Adianta did her
best, but all she could get was that they should be given till the latest
possible moment the next day. 'There's a good chance for you,' she said
to the children, but neither of us thought there was any. That was
another bad night, with Danorix trying hard to be brave; poor darling, he
was still afraid of the dark room, and they wanted to hang him till he was
dead!

The next morning seemed to rush away frightfully quickly. Adianta
and I tried hard to stop the children thinking about it, and gave them the
best dinner we could possibly cook, but it wasn't much good. I felt their
eyes on me all the time; I knew they must be thinking I could do something,

but of course I couldn't. The afternoon was worse. I asked Adianta if we couldn't possibly smuggle them away, but she said it was hopeless to try. Iurca wanted to look out of the window, so I went over with him and lifted him up. It was too low to see much in front—only men up to the waist and horses' legs. But further down the street one could see more. I tried to talk about everything that passed and amuse him that way, but he got tired of it, and I was just lifting him down when he said, 'Stop!' and then, 'That's our Owl!' 'What!' I cried, and Adianta and Danorix ran up too. Then Iurca began beating on the bars and crying, 'Father, father!' 'Is it?' I asked. 'Are you sure?' Danorix whispered yes, and then his knees shook and he fainted. But the Owl was drawing level with us; I shouted with all my might, 'Celtigun!' and Adianta ran to the door. 'Celtigun!' I shouted again, and saw all the legs running up and down. Then there was a great stream of people pouring in through the open door, and I picked Danorix up. A man snatched him away from me, and then Iurca, and every one behind yelled and clashed their swords together.

Adianta told Celtigun to take the children with him, as it might be safest. That was just as well, as they came for the children in the evening, and for a few minutes they wouldn't believe their father had come. That's the worst of a big tangled town like Bibracte.

I was sent for in the morning and begged Adianta to come too; I did my hair as well as I could, and put on the necklaces I had left, but my dress was badly stained. It was a glorious day, and all the streets were crammed with people, who stared at me. The chiefs were mostly in a big paved square in the centre of the town, talking and quarrelling and look-ing at one another's armour, all in those bright barbarian colours, with blazons painted on their shields. Celtigun and my two little hostages were standing together, watching it all; he kissed my hands and began thanking me, and the children jumped up and down, wanting to tell me everything! Then he broke off short and raised his open hand. It was Comm and Vercingetorix. They looked different in the open sunlight; or perhaps it was only their not being drunk.

Vercingetorix made me a bow, saying, 'Will Rome forgive?' 'No!' I said, but he was smiling at me so nicely that at last I changed my mind and said, 'Well, yes. But I wouldn't have if anything had happened to the children!' 'No,' he said, 'you're right,' and laid one hand on Iurca's shoulder. I saw Celtigun looking black at that, and so did Vercingetorix, for he said, 'I don't want to make enemies. If I forget how late you were in coming, will you forget what didn't happen to your sons?' He held out his right hand, which Celtigun looked at for a minute doubtfully, and then took with a great shake. After that he turned to me again: 'I'm glad you've forgiven me too, Rome. But your dress—how careless we were!

Adianta, you must get her a new one, a Gallic dress, the best you can find, because Rome must learn to admire us! And in the meanwhile'—he took out one of his coral shoulder brooches—'Rome must accept this.' I put my hands behind my back: 'No! I'm the wife of a Roman officer and you're a rebel!' The others looked terribly shocked, and Adianta made a movement to stop me; but he didn't mind, he only held it out, smiling still: 'Ah, but I'm not a rebel to this Rome!' and he pinned it on to my dress.

I wonder sometimes whether I ought to have taken it; but he's conquered and a prisoner, and the whole of Gaul is a province, so it can't matter. And M. thought it didn't matter when I told him; he said the brooch must go down as an heirloom. Anyhow, I did take it. It's a great thing for a man to have a smile like that!

Nothing much else happened, only after that I was allowed to do anything and go anywhere I liked. Adianta and I bought a dress, but of course it was only a great woollen thing with long sleeves in the Gallic fashion—M. hardly knew me in it. And then Celtigun said, would I marry him! I suppose he thought they were going to win and it was a great honour, but I told him very plainly that it wasn't. The children were disappointed, though, and couldn't make out why I wouldn't.

It was rather uncomfortable being free and respected and yet not being able to stop the things one saw going on. As a matter of fact there weren't many Roman prisoners, and when there were they'd usually let me help. But I did see some of the unfortunate Remi, who'd stayed loyal to us; they were treated worse than mad dogs, loaded with chains, knocked about, and beaten by anyone who'd a mind to; I don't know what happened to them in the end.

One never knew what was going on anywhere else—of course I didn't believe what they told me—and it was half a year since I'd heard from M. But everything happened unexpectedly; one day the chiefs were swarming all about the town, and their men rioting round the walls, as usual; and the next day they'd all cleared out and were gone. Adianta went too with her husband, but Celtigun left the children, and I took charge again. The hottest of summer was over, and we got cool winds sweeping over Bibracte; it was a healthy place, and the horse-flies didn't come up so high. We heard all sorts of rumours: battles and pursuits and stratagems; I didn't know what to believe; in the end I didn't believe anything, I didn't believe when I heard Vercingetorix was besieged and hard put to it in Alesia. But that story went on. All the men who were left marched down out of Bibracte, and I could see how every one was hating me more than usual for being a Roman.

One day the children were having dinner; I remember I'd promised to make them dolls afterwards with the crayfish claws (we used to have crayfish very often), and I was thinking what a funny mixture of man and baby they were. Suddenly we heard a clattering of horse-hoofs coming up the street and Danorix jumped up: 'That's father!' 'Nonsense!' I said. 'Sit down, Danorix.' But it was. He was blood and dust all over and wild-eyed like a beast. 'We're beaten!' he said, and snatched up a cup and drank. 'How?' I asked. 'Your people. Smashed! Broken in little bits! All that great army! I've come for the boys.' 'Oh, are they to go?' I cried. 'Yes, home, home. They must forget.' 'And me?' 'You forget too.' 'But I mean what's to happen to me?' 'You're right enough; the Romans will be here on our heels. They've beaten us, I tell you!' The children had been listening very quietly. 'Get the things, Iurca!' cried Danorix, and we bundled them up together, while Celtigun finished all the food there was on the table.

He had ridden ahead of his men, but some of them came up now, with their horses blowing and sweating. One who was badly wounded was lifted off and left in the house with me, and his horse was taken for the children. Iurca hung back, holding on to my hand: 'Oh, I do wish you were coming too!' But Danorix was flushed and excited, he ran up to me and kissed me hard, then scrambled on to the horse; he had their bundle, and I hoisted Iurca up in front of his father, who shouted to the men; they wheeled their horses and were off. I waved to the children as long as I could see them—goodbye, good luck—and they waved back.

For the next day or two I hardly went out of the house. I couldn't be sure how true it all was, but I couldn't help hoping; and that would mean I should hear about M. very soon—whatever had happened to him. I think I was more anxious in those few days than for months before. I looked after the wounded man as well as I could, and he seemed to be getting better. I was feeding him when I heard the first shouts outside, and I know I jumped up and left him to go on as best he could.

I stood in the doorway and watched them marching in, and heard the orders in crisp Latin, and the steady tramping, all in step, and saw the dark heads and the straight, firm ranks—Roman citizens and soldiers!— and I leant against the door-post crying with joy, though they'd all have thought I was a Gaulish woman. Then round the corner came the Eagles I knew, and I held my breath and waited. But he was there right in front, and I heard my voice all shrill and queer as I called to him. For a moment he didn't know me in my new dress with my hair plaited, and he and every one all stared at me. Then he cried out himself, and we ran into one another's arms in the middle of the street; and they all cheered.

I found out afterwards he'd thought I was killed at Noviodunum, and he'd made a vow to take no prisoners. But wasn't it extraordinary? The man I'd been looking after got well, so we took him back with us; you've seen him often enough; he's one of my litter-bearers. I've quite a lot of Gauls in the household now; it's so useful being able to talk their language. I don't know about Adianta or anyone else; we left Gaul as soon as we could.

But I wonder what happened to the hostages.

A Judgement of Tiberius

တတ

The bronze prow of the great galley rose and fell lazily to the long swell of the Mediterranean. Under a purple awning fringed with gold, Epaphroditus, Governor of Syria, lay stretched in deep thought. His narrow face, crowned with a head of red hair closely cut—the face of a young man aged with pleasure, and deeply seamed with the lines and the wrinkles of craft—was rested on a hand whose fingers were clogged with jewels. The nails of these fingers were bitten down to the quick. The man's face was turned towards a storied shore along which the galley was coasting. Had his cunning eyes seen anything of what they were fixed on, they would have recognized in distant shining columns smitten by an afternoon sun the great temples on the plains of Pœstum, and in a group of rock-bound islands against which the sea beat in one long murmur, the fabled home of the Sirens. It might have been well for Epaphroditus had he remembered how Ulysses, tiller in hand, and stern face set to the open sea, had fled for dear life from their fatal invitations. But the eyes of Epaphroditus saw none of these visible signs that he was within a few hours' sail of Capreæ, where the heart of the known world was revealed in a lonely rock, and Tiberius Caesar ruled over Solitude. The eyes of Epaphroditus were fixed on the Accounts of his Province. His stewardship of Syria had been a profitable one. He had not hesitated in the face of a golden opportunity. He had grasped the throat of his already half-strangled Province with both hands. He had got to the actual moment of squeezing the last breath of life from that exhausted body, when something which he least expected had happened. One of the Imperial galleys had come swiftly to her moorings in the sunset-lighted harbour, and a message of three words had told him that Tiberius wished to see him. The message was conveyed in three words: 'Come, old friend.' Who could fail to respond to so familiar a call?

The man's eyes removed themselves from that classic coastline, of which they had seen nothing, and looked upwards. Thus seen, Epaphroditus had the appearance of a devil struck with a sudden idea. But a fixity in his glance showed that this time he really noticed what he was looking at. Blazoned in gold on the purple awning over his head he saw the image and superscription of Tiberius. Lowering his glance and following the curved lines of the galley's sides he saw, on plaques set in them at intervals, the image and superscription of Tiberius. The image and superscription of Tiberius shone on the great brazen prow as it rose slowly against the skyline. Raising himself slightly from his couch and looking behind him, he saw, on stern-sheets and tiller, blazoned in jewels, and beaten in on gold, the image and superscription of Tiberius. The austere Emperor seemed everywhere. An uncomfortable feeling came upon Epaphroditus of being surrounded.

He bit his nails. Noticing then that one of the two boys who stood on each side of him had stopped from his duty of fanning him, to watch a shoal of porpoises, he struck him across the temple with a rod made out of rhinoceros hide which he always kept at hand, and smiled as he saw the blood come. Then he raised and lowered his eyebrows three times. This was a pre-arranged summons for his secretary to wait on him. Epaphroditus was too emasculated to talk. Leisurely once more he reposed himself.

A man came forward with a large scroll, his thick lips wreathed in a practised smile. The secretary's name was Hyginus. His swarthy complexion and unembarrassed manner revealed him as a son of Alexandria or Spain. A conscious look of assumed inspiration suggested that he was an author. He had indeed fathered on a protesting world the following foundlings of a fertile brain. A work called 'Poeticon Astronomicon' and a mythological history under the title of 'Fables'. His most mythological history, however, and his most imaginative fable were contained in the documents which he held under his arm. They were written in a finished hand, and showed, in form of a fraudulent balance-sheet, the finances of Syria during Epaphroditus's seven years' governorship of the Province.

Still under the influence of the uncomfortable feeling that had come upon him so lately, the master let fall upon the servant a cold look from which no meaning was intended to be extracted, and asked a question in one word: 'Tiberius?'

For answer, Hyginus gave a spontaneous imitation of a thirsty man drinking. Epaphroditus continued to look at this *improvisatore* fixedly. Presently he said:

'The accounts.'

'No eye,' said Hyginus, 'is practised enough to penetrate the art with which I have attributed withdrawals which were imperative to your Excellency's welfare, and to the proper adornment of your great hall of justice, to the devastating effects of successive famines. The financial history of your Syrian province is, lo! as a darkened room, into which he who would pry for peccadilloes enters confidently, to see, Nothing!'

With an easy gesture he dismissed to the vaulted sky whole columns of falsified figures.

'Tiberius sees in the dark,' said Epaphroditus.

'For a short time only, and immediately after awakening from sleep.'

'And then?'

'His eyes grow dim again.'

Epaphroditus said nothing. He continued to look at Hyginus with the same dull glance. Presently a viscous froth forced itself from between his teeth. This was the signal that this snake was satisfied. Hyginus, recognizing the fact, withdrew.

The Governor of Syria raised his right hand slowly. At the signal a slave on the watch always for his least movement offered on bonded knee a cup of wine. Epaphroditus drained it slowly. When he had done so, he let the cup fall from his nerveless hand and sank into a reverie.

He saw as in a conscious sleep ministered to by the monotonous lapping of the waves against the galley's sides, and by the measured beat of the oars at which slaves, stripped to the waist, were breaking their hearts, a certain great Hall of Justice, which his secretary had spoken of, and on which, for its real purpose of a lordly Pleasure House, a drained Province had given its last drop of blood. The roof sixty feet high, supported by pillars of the purest Parian marble; the ceilings and walls covered by the most expensive paintings that were procurable from the then degraded arts of Greece; the twelve golden statues of the Gods of Olympus, each capable of performing the office of torch-bearer, when the labours of the day were over and Night stole upon the marble palace, leading Orgy by the hand! The recollections of certain midnight moments soothed the Governor of Syria's overtaxed nerves. Still in a half-dream, he recalled riotous nights, and in a mental vision saw on his visionary palace walls mementoes of past triumphs ranged as if in some museum. It had been his whim that each episode should leave a token behind. That pearl necklace recalled the presence of Lalage. Neœra, in a fateful moment, had worn those emeralds in her hair; the diaphanous pale green robe had proved no armour of proof to the dancer from Syracuse. A pair of gold anklets recorded a further triumph. The silver armlets of some beauty whose name he had forgotten shone in mute rebuke. Suddenly the eyes of the visionary became fixed. In his mental review he had come to a certain bare place on the palace walls from which a certain token had been stripped. Epaphroditus recalled the momentary annoyance which had followed the disappearance, mysterious, and never accounted for, of three large trumpets made of ram's horn and profusely studded with jewels of great price. Strangely enough, too, these vanished memorials had not been associated, as were the others, with the victories of Venus. A simpler form of robbery had fixed them on the palace walls. Relying on Tiberius's intolerance of the Jews, the Governor of Syria had ravished them with his own hands from a place held particularly sacred, and a veritable Holy of Holies had witnessed, with an agonized dismay, a desecration which had brought no lightning from Heaven, but had, on the contrary, shown a robber walking off in broad daylight with the proceeds of his burglary under his arm. Jerusalem had shuddered through a long summer night while these three trumpets made of ram's horn were being fixed on the Governor's palace walls. How had they come to be so mysteriously removed? A Jewish girl had been suspected of the theft, huddling the sacred emblems of her faith under the loose robes which streamed behind her as she left that Hall of Justice in headlong and shamefaced flight. Their disappearances coincided, too, with the visit of a stout and genial Epicurean philosopher—one Sabinus, coming direct from the Court of Capreæ itself, with mulberry face and rotund belly, and returning there after a week's stay, with face more purple than ever, and with belly rounder still. This Sabinus the Epicurean, Epaphroditus remembered now, had examined these three ram's horn trumpets closely; had remarked upon them as being interesting relics; had pressed indiscreet questions as to how they had come into their present position as ornaments; had asked from what vendor of curios they had been obtained.

Irritated at softer memories being interrupted by recollection of so trivial a loss, Epaphroditus said aloud—'Chut! These three accursed trumpets were, after all, the least pleasing mementoes that I possessed. Why their loss should haunt me at the moment the Gods of Hell only can say! Nothing hangs on them!'

He started, surprised by a sudden gloom which darkened the sunshine as if night had fallen out of a midday summer sky; started, looked up, and saw towering above the galley, four hundred feet in the air, Capreæ, like some ominous lion crouched, frowning over a wine-dark sea.

It took the labour of fifty slaves, under the directing hand of Hyginus, to carry the Governor of Syria's litter, from the only landing-place in the island, up the steep path to the Villa of Jove. For though the ascent was in some places almost precipitous, Epaphroditus, reclined at full length on rose-coloured cushions, insisted on remaining horizontal. The laws of nature had to be set at defiance to please this Satrap, whose litter, as the hill grew steeper, was kept on an even plane, by two poles fixed in the back of it, and borne aloft by the combined efforts of the strongest porters Syria could supply. The sweat ran off these brawny slaves like water, as under Hyginus' directions they poised, raised, or lowered, according to the rise or fall of the ground, their exquisite and recumbent load. When the top of the hill had been reached their laboured breathing was interrupted by the following drawl omitted from the litter by an emasculate voice with a lisp in it.

'It seems a most extraordinary thing, that you beasts can never carry my litter up the slightest incline without shaking it so accursedly as to give me a brow-ache. It seems a most extraordinary thing.'

A more extraordinary thing had not escaped the speaker's notice, who thrust his head out of the litter's windows as he spoke, and who for all his pretence of lying in a faint of exhaustion, had, in fact, watched for any sign of life and movement in the island as narrowly as a lynx. He had seen none. He and his gilded litter and his gorgeously dressed retinue seemed to have invaded the Island of Silence. Epaphroditus had not seen a living thing stir during that long march up from the sea. Once he had caught sight of the glint of steel in a thicket by the side of the road. It was probably an ambushed guard. A solitary trumpet-call had also been heard. Beyond these two things nothing had been heard or seen. Even here, at the gates of the villa of Neptune, no guard was posted.

Epaphroditus told his trumpeter to sound a summons. Being a vain man, he was irritated at this seeming lack of respect. He was about to tell his trumpeter to sound a summons again, when a great door of the villa opened noiselessly, and a man with a painted face was seen bowing and smiling and beckoning him to enter. Epaphroditus recognized this man as one Priscus. He had known him in earlier days in Rome. His smile showed his character. A gold wand which he held in his right hand revealed his office. He was Tiberius's Master of the Ceremonies. He said not a word, but stood there smiling and bowing and beckoning Epaphroditus to enter.

The Governor of Syria raised himself in his litter, and looked at this silent embodiment of welcome steadily. 'Good-morrow, Priscus,' he said. 'Is everybody dead in this island save yourself? And are you stricken dumb?' Still smiling and making the same mechanical gesture for the other to enter, the man said three words:

'All is prepared.'

He turned his back, pointed down a long corridor, indicated by a gesture the fact that he was ready to usher the visitor to the host who was expecting him.

Epaphroditus looked at this gorgeously dressed automaton who was thus proposing himself as a guide to unknown hospitality, for three minutes. During this inspection his face remained impassive. But something seemed to be passing through his mind. He said nothing, however. Presently he signalled to Hyginus to marshal his retinue, and descended languidly from his litter.

A long procession was soon seen entering that villa perched on the highest cliff's edge of the rock-girt island, and in whose mysterious recesses the Lord of the known world was awaiting an unjust steward's approach. The unjust steward let no outward signs of uneasiness escape him. Leaning on his secretary's arm he walked with mincing steps down corridors whose marble walls bore so high a polish that they reflected passers-by as if in some mirror. Strange things had been pictured on this prolonged looking-glass, things which, if they had remained fixed there, instead of effacing themselves after an instantaneous impression, would have made visitors like Epaphroditus turn and fly for very life from a house whose polished walls had looked on such dreadful events. But the polished walls kept their secrets well, and Epaphroditus passed on to what was awaiting him.

What was awaiting him seemingly took the form of a large and empty room whose walls were covered with paintings reaching from the ceiling to the ground, and whose floor was decorated with a marble throne surmounted by an awning of the imperial purple, but under which no Caesar sat. After gazing at this hollow welcome for some moments with an astonishment which was fast merging into anger, Epaphroditus whispered an order. Seven boys bearing costly offerings advanced slowly and with deepest of prostrations laid them before that significant yet empty judgement seat. Then Epaphroditus turned to the Master of the Ceremonies, who still stood smiling with the gold rod of office in his hand.

'What is the meaning of this, Priscus?' he asked. 'I see no Caesar awaiting his faithful servant. No preparations for my advent. No guards. No retinue. No single sign of such respect as befits the reception of a Governor of one of the Emperor's chief provinces.'

He paused. He had suddenly become aware of the presence of an old man, bent with age, cloaked and hooded, standing with his back turned, in a distant part of the room, and intent on studying a picture. For some reason the sensation again came over Epaphroditus that he was being surrounded. But he continued sneeringly:

'Perhaps the old gentleman staring at the picture yonder, whom I have noticed now for the first time, can tell us where Caesar lurks! where this decrepit dotard airs his strenuous silence!'

He spoke with indignation, but amazement took indignation's place at the sound of a voice which he knew well, saying slowly, 'Good-morrow, Epaphroditus.'

The old man bent over the picture turned round slowly, and raised himself to the stature of six feet. Epaphroditus had recognized the voice. That unjust steward knew himself to be in the presence of a relentless master. From under the cowl two eyes, large, luminous, magnetic, read his very heart. Tiberius had turned upon him.

With neck stiff and upright, and with a frowning countenance which never relaxed itself, the Emperor moved slowly to the splendid throne and took his seat upon it. In his old cloak and hood he looked like some gigantic beggar enthroned, talking to a servant dressed like a Satrap. His luminous eyes—those eyes that could see in the dark—remained fixed on the Governor of Syria, who under their slow gaze felt as if he was in the power of some all-seeing operator bent on laying his very heart bare. The Emperor's great fingers, those fingers capable of boring a fresh sound apple through, and wounding a man's head with a fillip, remained clasped on the arms of the marble throne. On Tiberius's left wrist shone a massive gold bracelet studded with emeralds. The massive shoulders of this giant of seventy years of age told that his strength still remained Titanic. Epaphroditus felt like a fly entangled in the web of some monstrous spider. He heard his heart beating.

'Epaphroditus, our Governor of Syria,' said Tiberius, speaking deliberately and accentuating his words with a slow movement of the left hand, 'you are accused of stripping our good Syrian subjects, over whom we have deputed you our Governor, of many thousands of sesterces.'

'I indignantly deny the charge, Illustrious! Suspecting some such accusation at the hands of designing people who were themselves anxious to appropriate what they accuse me of appropriating, I have done myself the trouble of bringing my Secretary here with the full accounts of the province during my seven years' governorship of it. Hyginus, show Caesar the tablets.'

Hyginus crossed to the Emperor, prostrated himself, and put into his hand the immense bulk of falsified figures, of whose immunity, from all possibility of detection, he had assured his patron during the voyage from Syria.

Tiberius did not look at them.

'I am satisfied with your assurance,' he said.

The other, seeing that he had the advantage already, in what he had feared might be a difficult game, tried to meet the Emperor's unalterably fixed glance for the first time.

'I am glad,' he began, 'that Caesar is satisfied of my integrity, since from the cordial form of his summons to Capreæ, I had looked not for accusations, but for hospitality.'

'Hospitality will come in due course,' said Tiberius. 'Meanwhile be it known to you, my most excellent and esteemed Epaphroditus, that sesterces being matters of small account, a further charge of peculation has been levelled against you, from which you will doubtless free yourself with the ease that you have already shown. Briefly (since the hour for rest approaches), you are accused of stealing three trumpets made of ram's-horn, covered with jewels, from a certain shrine, temple, or theatre, we forget which. It may seem a small matter. But peculation is after all but peculation. And these said trumpets being my subjects' trumpets, they are also *Mine*.'

An instantaneous picture of that blank place on the walls of his Hall of Justice in Jerusalem came back to the Governor of Syria's mind, a picture of that blank place in those walls from which the three trumpets of ram's horn had been so mysteriously taken away, and round which the emeralds of poor Neœra, the diaphanous green robe of the Syrian dancer, and the silver armlets of the beauty whose name he had forgotten, showed as trophies of one long line of victories, in which that blank place in the palace walls recorded the one solitary defeat. He had a vision of that certain Jewish girl, suspected of the theft, hurrying from that Hall of Justice in headlong flight. His mind misgave him. A certain disagreeable feeling as of a net closing made itself felt. But with one hand on thigh he answered impudently:

'I indignantly deny, Caesar, that, consciously or unconsciously, I have appropriated your trumpets or anybody else's trumpets. Nor have I any recollection in truth of ever having seen them. In view of our past relation, Caesar, I should further like to say——'

The Emperor stopped this stream of hysterical excuse with a gesture. 'Summon the accuser,' he said.

In the accuser, who had evidently been waiting just outside the door, and who Priscus brought upon the scene with a wave of his gold wand, Epaphroditus recognized that same mulberry-faced Epicurean philosopher named Sabinus, who had visited Jerusalem from the Court at Capreae, who had expressed so curiously an interest in the three trumpets made of ram's horn, and had asked from what vendor of curios they had been obtained. Epaphroditus looked at the Epicurean. The Epicurean took no notice of him, but stood at the Emperor's right hand. Two slaves brought in an immense chest, which seemed so heavy that they were scarcely able to stagger under its weight. They put this chest before the Epicurean. From the weight of the chest the sudden thought flashed across Epaphroditus' mind that he was about to be confronted with some of the missing sesterces, whose disappearance his secretary had assured him that he had completely veiled. His mouth became dry; but once more with jewelled hand on hip he spoke impudently.

'I should like to tell Caesar,' he said, 'that as to these three trumpets, fashioned of ram's horn, and studded with jewels, which are reported to have been in my possession, I thought it desirable, before leaving Syria, to have some enquiries made as to the whereabouts of these said three trumpets. I could find no trace anywhere as to the whereabouts of these three trumpets: and so far as my recollection goes, cannot recall ever seeing them. Though I remember a high priest speaking to me about a trumpet'—here he turned to his secretary and added—'Or was it about a sacrificial ram?'

From the puzzled expression on his face Hyginus did not seem to be certain. He assumed an attitude of thought and said nothing.

It was left for the Epicurean Sabinus to produce a key to the mystery. He also produced something else. During the last declaration of Epaphroditus, he had been fumbling in the large chest. Remarking in unctuous tones, 'Behold an aid to defective memory,' he now slowly drew from its recesses, not some of those purloined sesterces which Epaphroditus had half expected to see, but the three identical trumpets made of ram's horn and thickly studded with jewels which had disappeared so mysteriously from the walls of his palace of pleasure in Jerusalem, and whose loss had given him those discomforting reflections on the long voyage from Syria.

The sensation came over Epaphroditus of a cold hand being slowly laid over his heart. He tried to look at those three trumpets, insignificant factors which had nevertheless exposed a whole career of fraud: and at the Epicurean who had sat at his table as a guest, seemingly an irresponsible agent, yet destined to undo him—he tried to look, but his eyes refused their office. His ears, however, came to his assistance. And these were the words he heard spoken in unctuous tones by Sabinus, who, to direct attention to the subject of his discourse, tapped the three trumpets three times significantly.

'These three aids to defective memory were found hanging in the Governor of Syria's private Hall of Justice in Jerusalem. The said private Hall of Justice is sixty feet high. At night when injustice is administered in it, it is lighted by lamps held in the hands of twelve life-size statues of the deities of Olympus; its secret recesses are full of valuables which have long been advertised as missing: and its walls are covered with pictures of a most degrading type.'

He replaced the three trumpets in the chest, closed the lid, and sat down.

'What have you to say, Epaphroditus? Death is your due.'

In such words did Tiberius suggest to the Governor of Syria the hospitality which the latter had at an earlier stage in the interview said that he had expected, and which Caesar had assured him would come in due time. Confronted with the final outcome of his worst forebodings, the man abandoned hope. But he still tried to save himself by impudence. 'The artist must not be trammelled,' he said with a faltering lisp.

Tiberius slowly descended from his throne: slowly, with neck stiff and upright, he came across the room to the place where Epaphroditus was standing. But the frown which habitually clouded the Emperor's brow had passed from it, and a faint smile played for a moment on his lips, seldom seen to relax their habitual severity. He laid his great hand in a friendly way on Epaphroditus' shoulder. The Governor of Syria conceived hope at this kindly gesture: his heart beat lighter with hope as he heard these words spoken in tones of good fellowship.

'Your plea is so singular a one, Epaphroditus, that it shall avail you. You shall taste, good Governor of Syria, of the mercy of Tiberius. We too love the arts. And we invite you to view these poor wall paintings.'

Epaphroditus felt an arm of iron linked in one of his own, and found himself being led in a friendly stroll to the further end of the room. Hope, which had died in the man's heart, came to life once more. The friendly tones of Tiberius again made themselves heard.

'We invite you, good Epaphroditus, to especially study this picture of "Hercules leaning on his Club". The hero is resting after his labours, you will observe, as you, after these tedious formalities, will be glad to rest after yours. Step nearer to it, good Governor of Syria. The fine finish calls for close inspection. You are short-sighted. Step nearer to it.'

Epaphroditus, now fully restored to hope, complied with a smile to his great Art Patron's request. He peered into the picture, bending forward and pretending to admire its beauties.

Suddenly he received a push from behind, given with such violence that he was hurled, with hands outstretched to save himself, full against the painting. The masterpiece so admired by Tiberius gave way before that sudden shock. It opened outwards. Epaphroditus felt himself falling headlong.

A last agonized look showed him the sea four hundred feet below.

RICHARD GARNETT

The Cupbearer

∞

The minister Photinius had fallen, to the joy of Constantinople. He had taken sanctuary in the immense monastery adjoining the Golden Gate in the twelfth region of the city, founded for a thousand monks by the patrician Studius, in the year 463. There he occupied himself with the concoction of poisons, the resource of fallen statesmen. When a defeated minister of our own day is indisposed to accept his discomfiture, he applies himself to poison the public mind, inciting the lower orders against the higher, and blowing up every smouldering ember of sedition he can discover, trusting that the conflagration thus kindled, though it consume the edifice of the State, will not fail to roast his own egg. Photinius' conceptions of mischief were less refined; he perfected his toxicological knowledge in the medical laboratory of the monastery, and sought eagerly for an opportunity of employing it; whether in an experiment upon the Emperor, or on his own successor, or on some other personage, circumstances must determine.

The sanctity of Studius' convent, and the strength of its monastic garrison, rendered it a safe refuge for disgraced courtiers, and in this thirtieth year of the Emperor Basil the Second (reckoning from his nominal accession) it harboured a legion of ex-prime ministers, patriarchs, archbishops, chief secretaries, hypati, anthypati, silentiarii, protospatharii, and even spatharocandidati. And this small army was nothing to the host that, maimed or blinded or tonsured or all three, dragged out their lives in monasteries or in dungeons or on rocky islets; and these again were few in comparison with the spirits of the traitors or the betrayed who wailed nightly amid the planes and cypresses of the Aretae, or stalked through the palatial apartments of verdantique and porphyry. But of those comparatively at liberty but whose liberty was circumscribed by the hallowed precincts of Studius, every soul was plotting. And never, perhaps, in the corrupt Byzantine Court, where true friendship had been unknown since Theodora quarrelled with Antonina, had so near an approach to it existed as in this asylum of villains. A sort of freemasonry came to prevail in the sanctuary: every one longed to know how his neighbour's plot throve, and grudged not to buy the knowledge by disclosing a little corner of his own.

Thus rendered communicative, their colloquies would travel back into the past, and as the veterans of intrigue fought their battles over again, the most experienced would learn things that made them open their eyes with amazement. 'Ah!' they would hear, 'that is just where you were mistaken. You had bought Eromenus, but so had I, and old Nicephorus had outbid us both.' 'You deemed the dancer Anthusa a sure card, and knew not of her secret infirmity, of which I had been apprised by her waiting woman.' 'Did you really know nothing of that sliding panel? And were you ignorant that whatever one says in the blue chamber is heard in the green?' 'Yes, I thought so too, and I spent a mint of money before finding out that the dog whose slaver that brazen impostor Panurgiades pretended to sell me was no more mad than he was.' After such rehearsals of future dialogues by the banks of Styx, the fallen statesmen were observed to appear exceedingly dejected, but the stimulus had become necessary to their existence. None gossipped so freely or disclosed so much as Photinius and his predecessor Eustathius, whom he had himself displaced—probably because Eustathius, believing in nothing in heaven or earth but gold, and labouring under an absolute privation of that metal, was regarded even by himself as an extinct volcano.

'Well,' observed he one day, when discoursing with Photinius in an unusually confidential mood, 'I am free to say that for my own part I don't think overmuch of poison. It has its advantages, to be sure, but to my mind the disadvantages are even more conspicuous.'

'For example?' enquired Photinius, who had the best reason for confiding in the efficacy of a drug administered with dexterity and discretion.

'Two people must be in the secret at least, if not three,' replied Eustathius, 'and cooks, as a rule, are a class of persons entirely unfit to be employed in affairs of State.'

'The Court physician,' suggested Photinius.

'Is only available,' answered Eustathius, 'in case his Majesty should send for him, which is most improbable. If he ever did, poison, praised be the Lord! would be totally unnecessary and entirely superfluous.'

'My dear friend,' said Photinius, venturing at this favourable moment on a question he had been dying to ask ever since he had been an inmate of the convent, 'would you mind telling me in confidence, did *you* ever administer any potion of a deleterious nature to his Sacred Majesty?'

'Never!' protested Eustathius, with fervour. 'I tried once, to be sure, but it was no use.'

'What was the impediment?'

'The perverse opposition of the cupbearer. It is idle attempting anything of the kind as long as she is about the Emperor.'

'*She!*' exclaimed Photinius.

'Don't you know *that*?' responded Eustathius, with an air and manner that plainly said, 'You don't know much.'

Humbled and ashamed, Photinius nevertheless wisely stooped to avow his nescience, and flattering his rival on his superior penetration, led him to divulge the State secret that the handsome cupbearer Helladius was but the disguise of the lovely Helladia, the object of Basil's tenderest affection, and whose romantic attachment to his person had already frustrated more conspiracies than the aged plotter could reckon up.

This intelligence made Photinius for a season exceedingly thoughtful. He had not deemed Basil of an amorous complexion. At length he sent for his daughter, the beautiful and virtuous Euprepia, who from time to time visited him in the monastery.

'Daughter,' he said, 'it appears to me that the time has now arrived when thou mayest with propriety present a petition to the Emperor on behalf of thy unfortunate father. Here is the document. It is, I flatter myself, composed with no ordinary address; nevertheless I will not conceal from thee that I place my hopes rather on thy beauty of person than on my beauty of style. Shake down thy hair and dishevel it, so!—that is excellent. Remember to tear thy robe some little in the poignancy of thy woe, and to lose a sandal. Tears and sobs of course thou hast always at command, but let not the frenzy of thy grief render thee wholly inarticulate. Here is a slight memorandum of what is most fitting for thee to say: thy old nurse's instructions will do the rest. Light a candle for St Sergius, and watch for a favourable opportunity.'

Euprepia was upright, candid, and loyal; but the best of women has something of the actress in her nature; and her histrionic talent was stimulated by her filial affection. Basil was for a moment fairly carried away by the consummate tact of her performance and the genuine feeling of her appeal; but he was himself again by the time he had finished perusing his late minister's long-winded and mendacious memorial.

'What manner of woman was thy mother?' he enquired kindly.

Euprepia was eloquent in praise of her deceased parent's perfections of mind and person.

'Then I can believe thee Photinius' daughter, which I might otherwise have doubted,' returned Basil. 'As concerns him, I can only say, if he feels himself innocent, let him come out of sanctuary, and stand his trial. But I will give thee a place at Court.'

This was about all that Photinius hoped to obtain, and he joyfully consented to his daughter's entering the Imperial Court, exulting at having got in the thin end of the wedge. She was attached to the person of the Emperor's sister-in-law, the 'Slayer of the Bulgarians' himself being a most determined bachelor.

Time wore on. Euprepia's opportunities of visiting her father were less frequent than formerly. At last she came, looking thoroughly miserable, distracted, and forlorn.

'What ails thee, child?' he enquired anxiously.

'Oh, father, in what a frightful position do I find myself!'

'Speak,' he said, 'and rely on my counsel.'

'When I entered the Court,' she proceeded, 'I found at first but one human creature I could love or trust, and he—let me so call him—seemed to make up for the deficiencies of all the rest. It was the cupbearer Helladius.'

'I hope he is still thy friend,' interrupted Photinius. 'The good graces of an Imperial cupbearer are always important, and I would have bought those of Helladius with a myriad of bezants.'

'They were not to be thus obtained, father,' said she. 'The purest disinterestedness, the noblest integrity, the most unselfish devotion, were the distinction of my friend. And such beauty! I cannot, I must not conceal that my heart was soon entirely his. But—most strange it seemed to me then—it was long impossible for me to tell whether Helladius loved me or loved me not. The most perfect sympathy existed between us: we seemed one heart and one soul: and yet, and yet, Helladius never gave the slightest indication of the sentiments which a young man might be supposed to entertain for a young girl. Vainly did I try every innocent wile that a modest maiden may permit herself: he was ever the friend, never the lover. At length, after long pining between despairing fondness and wounded pride, I myself turned away, and listened to one who left me in no doubt of the sincerity of his passion.'

'Who?'

'The Emperor! And, to shorten the story of my shame, I became his mistress.'

'The saints be praised!' shouted Photinius. 'O my incomparable daughter!'

'Father!' cried Euprepia, blushing and indignant. 'But let me hurry on with my wretched tale. In proportion as the Emperor's affection became more marked, Helladius, hitherto so buoyant and serene, became a visible prey to despondency. Some scornful beauty, I deemed, was inflicting on him the tortures he had previously inflicted upon me, and cured of my unhappy attachment, and entirely devoted to my Imperial lover, I did all in my power to encourage him. He received my comfort with gratitude, nor did it, as I had feared might happen, seem to excite the least lover-like feeling towards me on his own part.

' "Euprepia," he said only two days ago, "never in this Court have I met one like thee. Thou art the soul of honour and generosity. I can safely trust thee with a secret which my bursting heart can no longer retain, but which I dread to breathe even to myself. Know first I am not what I seem, I am a woman!' And opening his vest—'

'We know all about that already,' interrupted Photinius. 'Get on!'

'If thou knowest this already, father,' said the astonished Euprepia, 'thou wilt spare me the pain of entering further into Helladia's affection for Basil. Suffice that it was impassioned beyond description, and vied with whatever history or romance records. In her male costume she had accompanied the conqueror of the Bulgarians in his campaigns, she had fought in his battles; a gigantic foe, in act to strike him from behind, had fallen by her arrow; she had warded the poison-cup from his lips, and the assassin's dagger from his heart; she had rejected enormous wealth offered as a bribe for treachery, and lived only for the Emperor. "And now," she cried, "his love for me is cold, and he deserts me for another. Who she is I cannot find, else on her it were, not on him, that my vengeance should alight. Oh, Euprepia, I would tear her eyes from her head, were they beautiful as thine! But vengeance I must have. Basil must die. On the third day he expires by my hand, poisoned by the cup which I alone am trusted to offer him at the Imperial banquet where thou wilt be present. Thou shalt see his agonies and my triumph; and rejoice that thy friend has known how to avenge herself."

'Thou seest now, father, in how frightful a difficulty I am placed. All my entreaties and remonstrances have been in vain: at my threats Helladia merely laughs. I love Basil with my whole heart. Shall I look on and see him murdered? Shall I, having first unwittingly done my friend the most grievous injury, proceed further to betray her, and doom her to a cruel death? I might anticipate her fell purpose by slaying her, but for that I have neither strength nor courage. Many a time have I felt on the point of revealing everything to her, and offering myself as her victim, but for this also I lack fortitude. I might convey a warning to Basil, but Helladia's vengeance is unsleeping and nothing but her death or mine will screen him. Oh, father, father! What am I to do?'

'Nothing romantic or sentimental, I trust, dear child,' replied Photinius.

'Torture me not, father. I came to thee for counsel.'

'And counsel shalt thou have, but it must be the issue of mature deliberation. Thou mayest observe,' continued he with the air of a good man contending with adversity, 'how weak and miserable is man's estate even in the day of good fortune, how hard it is for purblind mortals to discern the right path, especially when two alluring routes are simultaneously presented for their decision! The most obvious and natural course, the one I should have adopted without hesitation half an hour ago, would be simply to let Helladia alone. Should she succeed—and

Heaven forbid else!—the knot is loosed in the simplest manner. Basil dies——'

'Father!'

'I am a favourite with his sister-in-law,' continued Photinius, entirely unconscious of his daughter's horror and agitation, 'who will govern in the name of her weak husband, and is moreover thy mistress. She recalls me to Court, and all is peace and joy. But then, Helladia may fail. In that case, when she has been executed——'

'Father, father!'

'We are exactly where we were, save for the hold thou hast established over the Emperor, which is of course invaluable. I cannot but feel that Heaven is good when I reflect how easily thou mightest have thrown thyself away upon a courtier. Now there is a much bolder game to play, which, relying on the protection of Providence, I feel half disposed to attempt. Thou mightiest betray Helladia.'

'Deliver my friend to the tormentors!'

'Then,' pursued Photinius, without hearing her, 'thy claim on the Emperor's gratitude is boundless, and if he has any sense of what is seemly—and he is what they call chivalrous—he will make thee his lawful consort. I father-in-law of an Emperor! My brain reels to think of it. I must be cool. I must not suffer myself to be dazzled or hurried away. Let me consider. Thus acting, thou puttest all to the hazard of the die. For if Helladia should deny everything, as of course she would, and the Emperor should foolishly scruple to put her to the rack, she might probably persuade him of her innocence, and where wouldst thou be then? It might almost be better to be beforehand, and poison Helladia herself, but I fear there is no time now. Thou hast no evidence but her threats, I suppose? Thou hast not caught her tampering with poisons? There can of course be nothing in writing. I dare say I could find something, if I had but time. Canst thou counterfeit her signature?'

But long ere this Euprepia, dissolved in tears, her bosom torn by convulsive sobs, had become as inattentive to her parent's discourse as he had been to her interjections. Photinius at last remarked her distress: he was by no means a bad father.

'Poor child,' he said, 'thy nerves are unstrung, and no wonder. It is a terrible risk to run. Even if thou saidest nothing, and Helladia under the torture accused thee of having been privy to her design, it might have a bad effect on the Emperor's mind. If he put thee to the torture too—but no! that's impossible. I feel faint and giddy, dear child, and unable to decide a point of such importance. Come to me at daybreak tomorrow.'

But Euprepia did not reappear, and Photinius spent the day in an agony of expectation, fearing that she had compromised herself by some imprudence. He gazed on the setting sun with uncontrollable impatience, knowing that it would shine on the Imperial banquet, where so much was to happen. Basil was, in fact, at that very moment seating himself among a brilliant assemblage. By his side stood a choir of musicians, among them Euprepia. Soon the cup was called for, and Helladia, in her masculine dress, stepped forward, darting a glance of sinister triumph at her friend. Silently, almost imperceptibly to the bulk of the company, Euprepia glided forward, and hissed rather than whispered in Helladia's ear, ere she could retire from the Emperor's side:

'Didst thou not say that if thou couldst discover her who had wronged thee, thou wouldst wreak thy vengeance on her, and molest Basil no further?'

'I did, and I meant it.'

'See that thou keepest thy word. I am she!' And snatching the cup from the table, she quaffed it to the last drop, and instantly expired in convulsions.

We pass over the dismay of the banqueters, the arrest and confession of Helladia, the general amazement at the revelation of her sex, the frantic grief of the Emperor.

Basil's sorrow was sincere and durable. On an early occasion he thus addressed his courtiers:

'I cannot determine which of these two women loved me best: she who gave her life for me, or she who would have taken mine. The first made the greater sacrifice; the second did most violence to her feelings. What say ye?'

The courtiers hesitated, feeling themselves incompetent judges in problems of this nature. At length the youngest exclaimed:

'O Emperor, how can we tell thee, unless we know what thou thinkest thyself?'

'What!' exclaimed Basil, 'an honest man in the Court of Byzantium! Let his mouth be filled with gold immediately!'

This operation having been performed, and the precious metal distributed in fees among the proper officers, Basil thus addressed the object of his favour:

'Manuel, thy name shall henceforth be Chrysostomus, in memory of what has just taken place. In further token of my approbation of thy honesty, I will confer upon thee the hand of the only other respectable person about the Court, namely, of Helladia. Take her, my son, and raise up a race of heroes! She shall be amply dowered out of what remains of the property of Photinius.'

'Gennadius,' whispered a cynical courtier to his neighbour, 'I hope thou admirest the magnanimity of our sovereign, who deems he is performing a most generous action in presenting Manuel with his cast-off mistress, who has tried to poison him, and with whom he has been at his wits' end what to do, and in dowering her at the expense of another.'

The snarl was just; but it is just also to acknowledge that Basil, as a prince born in the purple, had not the least idea that he was laying himself open to any such criticism. He actually did feel the manly glow of self-approbation which accompanies the performance of a good action: an emotion which no one else present, except Chrysostomus, was so much as able to conceive. It is further to be remarked that the old courtier who sneered at Chrysostomus was devoured by envy of his good fortune, and would have given his right eye to have been in his place.

'Chrysostomus,' pursued Basil, 'we must now think of the hapless Photinius. That unfortunate father is doubtless in an agony of grief which renders the forfeiture of the remains of his possessions indifferent to him. Thou, his successor therein, mayest be regarded as in some sort his son-in-law. Go, therefore, and comfort him, and report to me upon his condition.'

Chrysostomus accordingly proceeded to the monastery, where he was informed that Photinius had retired with his spiritual adviser, and could on no account be disturbed.

'It is on my head to see the Emperor's orders obeyed,' returned Chrysostomus, and forced the door. The bereaved parent was busily engaged in sticking pins into a wax effigy of Basil, under the direction of Panurgiades, already honourably mentioned in this history.

'Wretched old man!' exclaimed Chrysostomus, 'is this thy grief for thy daughter?'

'My grief is great,' answered Photinius, 'but my time is small. If I turn not every moment to account, I shall never be prime minister again. But all is over now. Thou wilt denounce me, of course. I will give thee a counsel. Say that thou didst arrive just as we were about to place the effigy of Basil before a slow fire, and melt it into a caldron of bubbling poison.'

'I shall report what I have seen,' replied Chrysostomus, 'neither more nor less. But I think I can assure thee that none will suffer for this mummery except Panurgiades, and that he will at most be whipped.'

'Chrysostomus,' said Basil, on receiving the report, 'lust of power, a fever in youth, is a leprosy in age. The hoary statesman out of place would sell his daughter, his country, his soul, to regain it: yea, he would part with his skin and his senses, were it possible to hold office without them. I commiserate Photinius, whose faculties are clearly on the decline; the day has been when he would not have wasted his time sticking pins into a waxen figure. I will give him some shadow of authority to amuse his old days and keep him out of mischief. The Abbot of Catangion is just dead. Photinius shall succeed him.'

So Photinius received the tonsure and the dignity, and made a very tolerable Abbot. It is even recorded to his honour that he bestowed a handsome funeral on his old enemy Eustathius.

Helladia made Chrysostomus an excellent wife, a little over-prudish, some thought. When, nearly two centuries afterwards, the Courts of Love came to be established in Provence, the question at issue between her and Euprepia was referred to those tribunals, which, finding the decision difficult, adjourned it for seven hundred years. That period having now expired, it is submitted to the British public.

EDGAR WALLACE

A Fragment of Medieval Life

ᛊᛟᛊ

Translated from the Norman-French and done into English by Edgar Wallace, who, not being absolutely certain how people talked and what figures of speech were employed in the eleventh century, puts himself on the safe side by employing the language of today

My name is Guy de Pomeroy Mompellier, and I was born in France, where my people had a couple of castles somewhere near Tours. My father, having murdered my mother's brother, got into trouble with the authorities, and at an early age I was brought to England—King William and my father being close friends. They had, in their youth, been engaged in a private murder which William thought out and my father executed.

We settled down in a castle in Kent, presented to father by the King as a slight token of his regard, and except for a little unpleasantness with a Baron, who lived on the other side of our hill, my boyhood was uneventful.

When I was fourteen years of age, my brother Henry, who was four years my senior, taking advantage of father's absence in town, seized the castle, proclaimed himself Earl of Wye, and roused all the countryside to take up arms against father.

I was only a nipper at the time, and mother shut me up in the east wing, and we barricaded the door; then Henry said that if we came out like sensible people he would not do much to us.

For four days my brother was Earl of Wye, but father, hearing of these happenings, hurried back with an army he'd picked up on the road, and storming the castle, caught Master Henry, who looked pretty sick, I can tell you, when they put a rope round his neck and chucked him off the battlements.

Father was very decent with the people that Henry had got over: he put a few in the dungeon under the larder, where they could smell the fat capons being prepared for Sunday dinner, and he hanged a few, but the majority he just kicked out. Henry was always a bit of a rotter (*belitre*), and spent his life bunching up trouble for himself.

Father made me his heir, and gave me a few kind words of advice. He also had a bit of the rope that hanged Henry hung up in my bedroom as a sort of a hint.

But he need not have done that, because I was not that sort of chap at all; and, anyway, I'd take jolly good care to put father out of the way before I started any hanky-panky (*projet*).

I was ever of a studious character, and under good Father Bernard de Bois, I perfected myself in the arts of peace, such as painting, music, and the illuminating of missals. I also took lessons from Giles de Crux in chemistry, on the use of samples, and the noxious and poisonous properties of plants.

About this time my father died suddenly after partaking of a dish of green salad, which, fortunately for me, I had refused.

My poor mother became demented, and made so many wild statements which coupled my study of chemistry with the preparation of the green salad, that I thought it best for all parties to put her away in a Nunnery, where she died soon after.

Having come to my own I looked about for a bride.

There was an awfully nice girl, the daughter of the Baron, whose estate was, as I have said, on the other side of the hill.

I saw her once when a few fellows and myself were out hawking, and I can tell you, she was a stunner.

It was a most unhappy circumstance that our families were not on good terms, but now father was dead I thought that it wouldn't be a bad sort of idea if I made friendly overtures.

I sent a messenger with a white flag to bear my cartel and my compliments to the Baron. William Blackfox, Earl of the Marsh, was the ridiculous title he gave himself, though he hadn't a bob in the world.

My messenger got away with his life.

I sent another—he didn't get away with anything. Now, I'm the mildest man in the world, and it takes a lot to get me rattled, but I must confess that this sort of thing gave me the hump.

I determined to go in person, but a shower of arrows and a smell of burning pitch made me turn home sadly and thoughtfully.

Of course I could have burnt him out, but I did not want to frizzle the girl too, and I was in despair, when chance threw me in the way of the old Baron. I was out hunting, and was quite alone, having outpaced my retinue, when I came suddenly upon him.

He knew me as soon as he spotted me.

He was alone too. As a matter of fact he hadn't got much of a force, and he kept them to hold the castle.

He whipped out his sword at the sight of me.

'Hullo! Green salad!' he said nastily.

'Look here, Baron,' I said. I was no match for him single-handed, and had to temporize. 'What's the good of throwing that sort of thing in a fellow's face. I want to be friends.'

'I daresay you do,' he sneered, snicking his bright sword left and right. 'I suppose you'd like to stand me a dish of green salad!'

'Let bygones be bygones,' I said severely. 'Let the dead bury their dead.'

'You're not going to bury me,' he said; 'not with green salad.'

And with that I could see him making preparations to out me.

'Half a moment,' I said. 'Before I die, I should like to make one matter square: it appears that there was some dispute about our boundary——'

'Dispute!' he roared; 'there's no disputing! Your filching, thieving, murdering father——'

'May he rest in peace,' I said piously.

'Before you poisoned him,' he went on, 'stole half my demesne!'

'That's true,' I said; 'and I would like to make restitution and give you proof.'

He dropped his sword-point whilst I fumbled in my doublet, and as I produced a paper he spurred his horse forward.

I handed him the paper with my left hand, and with my right I gave him twelve inches of good Genoese steel between the fourth and fifth rib.

He tried to get at me, but it was too late. In a few moments he was stretched on the grass breathing his last.

'I hope this will be a lesson to you,' I said. But he only swore at me, using expressions which no true knight would lower himself to repeat.

The way was now clear; with the old man out of the way, the castle and the girl were mine.

I got my chaps together, put the body of the Baron on a bier, and we started off home. When we were near the castle of the Baron, I gave my force the tip to look as if they were sorry, and marching slowly, singing a low, sad song, we made our way to the castle.

There was a great flutter of excitement, and the Baroness came to the outward wall.

I took off my helmet and bowed to her.

'What is wrong?' she asked.

'I'm afraid your dear husband has met with an accident,' I said sadly. 'I found him lying by the side of the road——'

I told her The Story, and down came the drawbridge, and we crossed it—me and my men.

It was one of those jerry-built castles that you sometimes meet in the suburbs, and I must confess I was rather disappointed with it.

It did not take the Baroness long to discover that she had been hoodwinked, and the row that woman made was simply unwomanly. I had all the difficulty in the world to persuade her to retire to a dungeon and think matters over.

Then I went in search of the girl.

I found her surrounded by her ladies. She was very pale, but perfectly calm.

'Hullo,' she said; 'it's you, is it?'

'It's me,' I admitted. 'I love you; will you be my bride?'

'What about father?' she said.

I told her of her father's sad end, and she brightened up wonderfully.

'Poor old chap,' she said, with that sympathetic tenderness which is so beautiful in women. 'He was an awful tartar, but I'm sorry for him. I shan't be sorry, though, to get away from here. Its awfully draughty.'

We were married the next week, and two oxen were roasted whole for the feeding of our wedding guests. They were the Baron's oxen.

ALDOUS HUXLEY

The Death of Lully

∞

The sea lay in a breathing calm, and the galley, bosomed in its transparent water, stirred rhythmically to the slow pulse of its sleeping life. Down below there, fathoms away through the crystal-clear Mediterranean, the shadow of the ship lazily swung, moving, a long dark patch, very slowly back and forth across the white sand of the sea-bottom—very slowly, a scarcely perceptible advance and recession of the green darkness. Fishes sometimes passed, now hanging poised with idly tremulous fins, now darting onwards, effortless and incredibly swift; and always, as it seemed, utterly aimless, whether they rested or whether they moved; as the life of angels their life seemed mysterious and unknowable.

All was silence on board the ship. In their fetid cage below decks the rowers slept where they sat, chained, on their narrow benches. On deck the sailors lay sleeping or sat in little groups playing at dice. The fore-part of the deck was reserved, it seemed, for passengers of distinction. Two figures, a man and a woman, were reclining there on couches, their faces and half-bared limbs flushed in the coloured shadow that was thrown by the great red awning stretched above them.

It was a nobleman, the sailors had heard, and his mistress that they had on board. They had taken their passage at Scanderoon, and were homeward bound for Spain. Proud as sin these Spaniards were; the man treated them like slaves or dogs. As for the woman, she was well enough, but they could find as good a face and pair of breasts in their native Genoa. If any one so much as looked at her from half the ship's length away it sent her possessor into a rage. He had struck one man for smiling at her. Damned Catalonian, as jealous as a stag; they wished him the stag's horns as well as its temper.

It was intensely hot even under the awning. The man woke from his uneasy sleep and reached out to where on a little table beside him stood a deep silver cup of mixed wine and water. He drank a gulp of it; it was as warm as blood and hardly cooled his throat. He turned over and, leaning on his elbow, looked at his companion. She on her back, quietly breathing through parted lips, still asleep. He leaned across and pinched her on the breast, so that she woke up with a sudden start and cry of pain.

'Why did you wake me?' she asked.

He laughed and shrugged his shoulders. He had, indeed, had no reason for doing so, except that he did not like it that she should be comfortably asleep, while he was awake and unpleasantly conscious of the heat.

'It is hotter than ever,' he said, with a kind of gloomy satisfaction at the thought that she would now have to suffer the same discomforts as himself. 'The wine scorches instead of cooling; the sun seems no lower down the sky.'

The woman pouted. 'You pinched me cruelly,' she said. 'And I still do not know why you wanted to wake me.'

He smiled again, this time with a good-humoured lasciviousness. 'I wanted to kiss you,' he said. He passed his hand over her body possessively, as a man might caress a dog.

Suddenly the quiet of the afternoon was shattered. A great clamour rose up, ragged and uneven, on the air. Shrill yells pierced the dull rumbling growl of bass voices, pierced the sound of beaten drums and hammered metal.

'What are they doing in the town?' asked the woman anxiously of her lover.

'God knows,' he answered. 'Perhaps the heathen hounds are making some trouble with our men.'

He got up and walked to the rail of the ship. A quarter of a mile away, across the smooth water of the bay, stood the little African town at which they had stopped to call. The sunlight showed everything with a hard and merciless definition. Sky, palms, white houses, domes, and towers seemed as though made from some hard enamelled metal. A ridge of low red hills rolled away to right and left. The sunshine gave to everything in the scene the same clarity of detail, so that to the eye of the onlooker there was no impression of distance. The whole thing seemed to be painted in flat upon a single plane.

The young man returned to his couch under the awning and lay down. It was hotter than ever, or seemed so, at least, since he had made the exertion of getting up. He thought of high cool pastures in the hills, with the pleasant sound of streams, far down and out of sight in their deep channels. He thought of winds that were fresh and scented—winds that were not mere breaths of dust and fire. He thought of the shade of cypresses, a narrow opaque strip of darkness; and he thought too of the green coolness, more diffused and fluid and transparent, of chestnut groves. And he thought of the people he remembered sitting under the trees—young people, gay and brightly dressed, whose life was all gaiety and deliciousness. There were the songs that they sang—he recalled the voices and the dancing of the strings. And there were perfumes and, when one drew closer, the faint intoxicating fragrance of a woman's body. He thought of the stories they told; one in particular came to his mind, a capital tale of a sorcerer who offered to change a peasant's wife into a mare, and how he gulled the husband and enjoyed the woman before his eyes, and the delightful excuses he made when she failed to change her shape. He smiled to himself at the thought of it, and stretching out a hand touched his mistress. Her bosom was soft to his fingers and damp with sweat; he had an unpleasant notion that she was melting in the heat.

'Why do you touch me?' she asked.

He made no reply, but turned away from her. He wondered how it would come to pass that people would rise again in the body. It seemed curious, considering the manifest activities of worms. And suppose one rose in the body that one possessed in age. He shuddered, picturing to himself what this woman would be like when she was sixty, seventy. She would be beyond words repulsive. Old men too were horrible. They stank, and their eyes were rheumy and rosiny, like the eyes of deer. He decided that he would kill himself before he grew old. He was eight-and-twenty now. He would give himself twelve years more. Then he would end it. His thoughts dimmed and faded away into sleep.

The woman looked at him as he slept. He was a good man, she thought, though sometimes cruel. He was different from all the other men she had known. Once, when she was sixteen and a beginner in the business of love, she had thought that all men were always drunk when they made love. They were all dirty and like beasts; she had felt herself superior to them. But this man was a nobleman. She could not understand him; his thoughts were always obscure. She felt herself infinitely inferior to him. She was afraid of him and his occasional cruelty; but still he was a good man, and he might do what he liked with her.

From far off came the sound of oars, a rhythmical splash and creak. Somebody shouted, and from startlingly close at hand one of the sailors hallooed back.

The young man woke up with a start.

'What is it?' he asked, turning with an angry look to the girl, as though he held her to be responsible for this breaking in upon his slumbers.

'The boat, I think,' she said. 'It must be coming back from the shore.'

The boat's crew came up over the side, and all the stagnant life of the ship flowed excitedly round them. They were the centre of a vortex towards which all were drawn. Even the young Catalonian, for all his hatred of these stinking Genoese shipmen, was sucked into the eddy. Everybody was talking at once, and in the general hubbub of question and answer there was nothing coherent to be made out. Piercingly distinct above all the noise came the voice of the little cabin-boy, who had been to shore with the boat's crew. He was running round to every one in turn repeating: 'I hit one of them. You know. I hit one. With a stone on the forehead. Didn't he bleed, ooh! didn't he just!' And he would dance with uncontrollable excitement.

The captain held up his hand and shouted for silence. 'One at a time, there,' he ordered, and when order had a little been restored, added grumblingly, 'Like a pack of dogs on a bone. You talk, boatswain.'

'I hit one of them,' said the boy. Somebody cuffed him over the head, and he relapsed into silence.

When the boatswain's story had rambled through labyrinths of digression, over countless obstacles of interruptions and emendations, to its conclusion, the Spaniard went back to join his companion under the awning. He had assumed again his habitual indifference.

'Nearly butchered,' he said languidly, in response to her eager questions. 'They'—he jerked a hand in the direction of the town—'they were pelting an old fellow who had come there preaching the Faith. Left him dead on the beach. Our men had to run for it.'

She could get no more out of him; he turned over and pretended to go to sleep.

Towards evening they received a visit from the captain. He was a large, handsome man, with gold ear-rings glinting from among a bush of black hair.

'Divine Providence,' he remarked sententiously, after the usual courtesies had passed, 'has called upon us to perform a very notable work.'

'Indeed?' said the young man.

'No less a work,' continued the captain, 'than to save from the clutches of the infidels and heathen the precious remains of a holy martyr.'

The captain let fall his pompous manner. It was evident that he had carefully prepared these pious sentences, they rolled so roundly off his tongue. But he was eager now to get on with his story, and it was in a homelier style that he went on: 'If you knew these seas as well as I—and it's near twenty years now that I've been sailing them—you'd have some knowledge of this same holy man that—God rot their souls for it!—these cursed Arabs have done to death here. I've heard of him more than once in my time, and not always well spoken of; for, to tell the honest truth, he does more harm with his preachments to good Christian traders than ever he did good to black-hearted heathen dogs. Leave the bees alone, I say, and if you can get a little honey out of them quietly, so much the better; but he goes about among the beehives with a pole, stirring up trouble for himself and others too. Leave them alone to their damnation, is what I say, and get what you can from them this side of hell. But, still, he has died a holy martyr's death. God rest his soul! A martyr is a wonderful thing, you know, and it's not for the likes of us to understand what they mean by it all.

'They do say, too, that he could make gold. And, to my mind, it would have been a thing more pleasing to God and man if he had stopped at home minting money for poor folks and dealing it round, so that there'd be no need to work any more and break oneself for a morsel of bread. Yes, he was great at gold-making and at the books too. They tell me he was called the Illuminated Doctor. But I know him still as plain Lully. I used to hear of him from my father, plain Lully, and no better once than he should have been.

'My father was a shipwright in Minorca in those days—how long since? Fifty, sixty years perhaps. He knew him then; he has often told me the tale. And a raffish young dog he was. Drinking, drabbing, and dicing he outdid them all, and between the bouts wrote poems, they say, which was more than the rest could do. But he gave it all up on the sudden. Gave away his lands, quitted his former companions, and turned hermit up in the hills, living alone like a fox in his burrow, high up above the vines. And all because of a woman and his own qualmish stomach.'

The shipmaster paused and helped himself to a little wine. 'And what did this woman do?' the girl asked curiously.

'Ah, it's not what she did but what she didn't do,' the captain answered, with a leer and wink. 'She kept him at his distance—all but once, all but once; and that was what put him on the road to being a martyr. But there, I'm outrunning myself. I must go more soberly.

'There was a lady of some consequence in the island—one of the Castellos, I think she was; her first name has quite slipped my memory— Anastasia, or something of the kind. Lully conceives a passion for her, and sighs and importunes her through I know not how many months and years. But her virtue stands steady as the judgement seat. Well, in the end, what happens was this. The story leaked out after it was all over, and he was turned hermit in the mountains. What happened, I say, was this. She tells him at last that he may come and see her, fixing some solitary twilight place and time, her own room at nightfall. You can guess how he washes and curls and scents himself, shaves his chin, chews anises, musks over whatever of the goat may cling about the body. Off he goes, dreaming swoons and ecstasies, foretasting inconceivable sweets. Arrived, he finds the lady a little melancholy—her settled humour, but a man might expect a smile at such a time. Still, nothing abashed, he falls at her feet and pours out his piteous case, telling her he has sighed through seven years, not closed an eye for above a hundred nights, is forepined to a shadow, and, in a word, will perish unless she show some mercy. She, still melancholy—her settled humour, mark you—makes answer that she is ready to yield, and that her body is entirely his. With that, she lets herself be done with as he pleases, but always sorrowfully. 'You are all mine,' says he—'all mine'—and unlaces her gorgeret to prove the same. But he was wrong. Another lover was already in her bosom, and his kisses had been passionate—oh, burning passionate, for he had kissed away half her left breast. From the nipple down it had all been gnawed away by a cancer.

'Bah, a man may see as bad as that any day in the street or at church-doors where beggars most congregate. I grant you that it is a nasty sight, worm-eaten flesh, but still—not enough, you will agree, to make yourself a hermit over. But there, I told you he had a queasiness of the stomach. But doubtless it was all in God's plan to make a holy martyr of him. But for that same queasiness of his, he would still be living there, a super-annuated rake; or else have died in very foul odour, instead of passing, all embalmed with sanctity, to Paradise Gate.

'I know not what happened to him between his hermithood and his quest for martyrdom. I saw him first a dozen years ago, down Tunis way. They were always clapping him into prison or pulling out his beard for preaching. This time, it seems, they have made a holy martyr of him, done the business thoroughly with no bungling. Well, may he pray for our souls at the throne of God. I go in secretly tonight to steal his body. It lies on the shore there beyond the jetty. It will be a notable work, I tell you, to bring back so precious a corpse to Christendom. A most notable work. . . .'

The captain rubbed his hands.

It was after midnight, but there was still a bustle of activity on board the galley. At any moment they were expecting the arrival of the boat with the corpse of the martyr. A couch, neatly draped in black, with at its head and foot candles burning two by two, had been set out on the poop for the reception of the body. The captain called the young Spaniard and his mistress to come and see the bier.

'That's a good bit of work for you,' he said, with justifiable pride. 'I defy anyone to make a more decent resting-place for a martyr than that is. It could hardly have been done better on shore, with every appliance at hand. But we sailors, you know, can make anything out of nothing. A truckle-bed, a strip of tarred canvas, and four tallow dips from the cabin lanterns—there you are, a bier for a king.'

He hurried away, and a little later the young man and the girl could hear him giving orders and cursing somewhere down below. The candles burned almost without a tremor in the windless air, and the reflections of the stars were long, thin tracks of fire along the utterly calm water.

'Were there but perfumed flowers and the sound of a lute,' said the young Spaniard, 'the night would tremble into passion of its own accord. Love should come unsought on such a night as this, among these black waters and the stars that sleep so peacefully on their bosom.'

He put his arm round the girl and bent his head to kiss her. But she averted her face. He could feel a shudder run her through the body.

'Not tonight,' she whispered. 'I think of the poor dead man. I would rather pray.'

'No, no,' he cried. 'Forget him. Remember only that we are alive, and that we have but little time and none to waste.'

He drew her into the shadow under the bulwark, and, sitting down on a coil of rope, crushed her body to his own and began kissing her with fury. She lay, at first, limp in his arms, but gradually she kindled to his passion.

A plash of oars announced the approach of the boat. The captain hallooed into the darkness: 'Did you find him?'

'Yes, we have him here,' came back the answer.

'Good. Bring him alongside and we'll hoist him up. We have the bier in readiness. He shall lie in state tonight.'

'But he's not dead,' shouted back the voice from the night.

'Not dead?' repeated the captain, thunderstruck. 'But what about the bier, then?'

A thin, feeble voice came back. 'Your work will not be wasted, my friend. It will be but a short time before I need your bier.'

The captain, a little abashed, answered in a gentler tone, 'We thought, holy father, that the heathens had done their worst and that Almighty God had already given you the martyr's crown.'

By this time the boat had emerged from the darkness. In the stern sheets an old man was lying, his white hair and beard stained with blood, his Dominican's robe torn and fouled with dust. At the sight of him, the captain pulled off his cap and dropped upon his knees.

'Give us your blessing, holy father,' he begged.

The old man raised his hand and wished him peace.

They lifted him on board and, at his own desire, laid him upon the bier which had been prepared for his dead body. 'It would be a waste of trouble,' he said, 'to put me anywhere else, seeing I shall in any case be lying there so soon.'

So there he lay, very still under the four candles. One might have taken him for dead already, but that his eyes, when he opened them, shone so brightly.

He dismissed from the poop every one except the young Spaniard. 'We are countrymen,' he said, 'and of noble blood, both of us. I would rather have you near me than anyone else.'

The sailors knelt for a blessing and disappeared; soon they could be heard weighing the anchor; it was safest to be off before day. Like mourners at either side of the lighted bier crouched the Spaniard and his mistress. The body of the old man, who was not yet dead, lay quiet under the candles. The martyr was silent for some time, but at last he opened his eyes and looked at the young man and the woman.

'I too,' he said, 'was in love, once. In this year falls the jubilee of my last earthly passion; fifty years have run since last I longed after the flesh—fifty years since God opened my eyes to the hideousness of the corruption that man has brought upon himself.

'You are young, and your bodies are clean and straight, with no blotch or ulcer or leprous taint to mar their much-desired beauty; but because of your outward pride, your souls, it may be, fester inwardly the more.

'And yet God made all perfect; it is but accident and the evil of will that causes defaults. All metals should be gold, were it not that their elements willed evilly in their desire to combine. And so with men: the burning sulphur of passion, the salt of wisdom, the nimble mercurial soul should come together to make a golden being, incorruptible and rustless. But the elements mingle jarringly, not in a pure harmony of love, and gold is rare, while lead and iron and poisonous brass that leaves a taste as of remorse behind it are everywhere common.

'God opened my eyes to it before my youth had too utterly wasted itself to rottenness. It was half a hundred years ago, but I see her still, my Ambrosia, with her white, sad face and her naked body and that monstrous ill eating away at her breast.

'I have lived since then trying to amend the evil, trying to restore, as far as my poor powers would go, some measure of original perfection to the corrupted world. I have striven to give to all metals their true nature, to make true gold from the false, the unreal, the accidental metals, lead and copper and tin and iron. And I have essayed that more difficult alchemy, the transformation of men. I die now in my effort to purge away that most foul dross of misbelief from the souls of these heathen men. Have I achieved anything? I know not.'

The galley was moving now, its head turned seaward. The candles shivered in the wind of its speed, casting uncertain, changing shadows upon his face. There was a long silence on the poop. The oars creaked and splashed. Sometimes a shout would come up from below, orders given by the overseer of the slaves, a curse, the sound of a blow. The old man spoke again, more weakly now, as though to himself.

'I have had eighty years of it,' he said—'eighty years in the midst of this corroding sea of hatred and strife. A man has need to keep pure and unalloyed his core of gold, that little centre of perfection with which all, even in this declination of time, are born. All other metal, though it be as tough as steel, as shining-hard as brass, will melt before the devouring bitterness of life. Hatred, lust, anger—the vile passions will corrode your will of iron, the warlike pomp of your front of brass. It needs the golden perfection of pure love and pure knowledge to withstand them.

'God has willed that I should be the stone—weak, indeed, in virtue— that has touched and transformed at least a little of baser metal into the gold that is above corruption. But it is hard work—thankless work. Man has made a hell of his world, and has set up gods of pain to rule it. Goatish gods, that revel and feast on the agony of it all, poring over the tortured world, like those hateful lovers, whose lust burns darkly into cruelty.

'Fever goads us through life in a delirium of madness. Thirsting for the swamps of evil whence the fever came, thirsting for the mirages of his own delirium, man rushes headlong he knows not whither. And all the time a devouring cancer gnaws at his entrails. It will kill him in the end, when even the ghastly inspiration of fever will not be enough to whip him on. He will lie there, cumbering the earth, a heap of rottenness and pain, until at last the cleansing fire comes to sweep the horror away.

'Fever and cancer; acids that burn and corrode. . . . I have had eighty years of it. Thank God, it is the end.'

It was already dawn; the candles were hardly visible now in the light, faded to nothing, like souls in prosperity. In a little while the old man was asleep.

The captain tiptoed up on to the poop and drew the young Spaniard aside for a confidential talk.

'Do you think he will die today?' he asked.

The young man nodded.

'God rest his soul,' said the captain piously. 'But do you think it would be best to take his body to Minorca or to Genoa? At Minorca they would give much to have their own patron martyr. At the same time it would add to the glory of Genoa to possess so holy a relic, though he is in no way connected with the place. It's there is my difficulty. Suppose, you see, that my people of Genoa did not want the body, he being from Minorca and not one of them. I should look a fool then, bringing it in in state. Oh, it's hard, it's hard. There's so much to think about. I am not sure but what I hadn't better put in at Minorca first. What do you think?'

The Spaniard shrugged his shoulders. 'I have no advice to offer.'

'Lord,' said the captain as he bustled away, 'life is a tangled knot to unravel.'

FRED URQUHART

Pretty Prickly English Rose

∽∞∾

I was but eleven when I was appointit maid in waiting to the young Queen Joan. I should have been still at home learning my letters in our castle in Angus, yon dreich cauld place on a crag jutting into the North Sea, but my grandfather, the auld earl, said it was time I went to court to learn deportment. He was aye called the auld earl to differentiate him from my father, who was an earl too, having inherited the title from his mother. My grandsire, the Earl of Auchencairn, had married the Countess of Bervie, uniting two great houses.

Phemie, my nurse who had been my mother's maid, protested it was not fitting that I go to court. 'The puir bairn should bide here among the folk she kens, not among strangers where she'll be laughed at.'

'Blethers, wifie,' the auld earl said. 'She maun face up to her disability sometime, and the sooner the better.'

'She's nae fit yet to ging into the great wide world, and don't you try to force her, Geordie Hepburn,' said Phemie, who was a Hepburn too, though born on the wrong side of the blanket; although it was seldom admitted, she was, in truth, my grandsire's half-sister. 'Ye'll live to regret it.'

'Regret? What is regret, woman? 'Tis a word that means nothing to me. Or any Hepburn.'

'To any male Hepburn mayhap,' Phemie said, 'Ye can speak for yersel', ye stiff-neckit auld blaggart. But I'm speakin' for this bairn, who's nae auld enough yet to speak for hersel'. I tell ye, if she gings to the coort she'll be laughed at.'

'She'll live long after she's laughed at,' he said, smoothing my hair. 'Will ye no', Grizel, my wee doo?'

And so indeed it has come to pass. I'm an auld woman now, aulder far than Phemie was then, and I have learnit to thole folks' spite and malice and their cruel laughter. I have lived long after a wheen of them that mockit me are dead and gone. Yet it was hard when first I came among strangers to keep a proud face and smile as Phemie and my grandsire had tellt me.

It was because she was never among the mockers that I loved and worshipped Joan Beaufort. I saw her first at Falkland Palace when my grandsire brought me into her presence. My father, Lord Bervie, was a hostage in England at the time: like several other great Scottish lords he had been sent to the court of Henry VI to languish there while our King James tried by divers new taxations to pay the money for his own ransom. So it fell upon my grandsire to introduce me into the court of King James and Queen Joan.

I mind fine how there was a silence as awesome as the tolling of the death bell as the auld earl led me down the length of the great hall to where she was sitting. We walkit slowly. This was not because of my grandsire's great age but because I could not move at a quicker pace. Seldom have I been more aware than I was yon day of my dragging leg, my long arms reaching to ablow my knees, my right shoulder higher than the other one, and my twisted body. The expanse of polished floor seemed endless as I hirpled along, crab-like, one hand in my grandsire's, the other holding the ebony stick that is aye my truest assistant.

The silence was broken by some titters from the half-dozen braw ladies surrounding Queen Joan. I could feel the blush burning my cheek-bones, but I keepit my head as high as I could manage because of my hunch. My grandsire pressit my hand so hard that it hurt, and I bared my teeth in a smile to keep from yelping.

The Queen turned and glowered at her ladies, but before they became quiet a voice I recognized as my cousin Kate's said clearly, 'I declare to Our Lady, 'tis wee humpy Hep!'

The Queen rose and came to greet us. 'Welcome, Lady Grizel,' she said, bending to kiss me on both cheeks. 'I heard tell from my Lord Auchencairn that you were bonnie, but I scarce expectit you to be as bonnie as this. You have a most beautiful face, my love.'

'Your Majesty is very beautiful yourself,' I said. And I blushit at having been so bold, and I hid my face against my shoulder.

Truly she was the brawest lady. I need not describe her, except to affirm that she had the longest, the most beautiful golden hair and a milk-and-roseblush complexion, for King James has proclaimed rightly to the world in the long poem he scribed after he first saw her from his prison window in Windsor Castle that she was his 'milk white dove' and 'the fairest and the freshest young flower that ever I saw.'

'I am glad, Lord Auchencairn, that you have placed this little maid in my care,' Queen Joan said, giving her hand to my grandsire to kiss. 'I shall nurture her right well and see no harm comes to sully her innocence.'

My grandsire bowed fell deeply and said, 'I can never thank Your Majesty enough.'

There was a skirl from amidst the assembled ladies, and Lady Catherine

Douglas rushit forth in a swirl of green silk skirts, swept Queen Joan a bit curtsey, saying, 'By your leave, madam,' and then flung her arms about me and cried, 'Sweet coz! So ye have come to grace the Queen's court?'

My mother and Kate Douglas's were sisters. I never likit Kate, and she never likit me. She is dead now, and it ill befits me to decry her memory. She was hailed as a heroine at one time, and although I have doubts about her heroics, thinking them false, I will not stoop to hinder her from being given some merit in the history of our native land.

In that year 1424, when I came to court, Kate was nineteen years auld: a big sonsy quaen with an out-thrust bust, copper-coloured hair and sandy eyelashes. Her eyes were as green as her gown. She kissit me and said to the Queen, 'She's my puir wee coz, madam, and by your leave I will take her under my wing.'

'Ay, she's your wee coz, Kate,' my grandsire said. 'I trust you will remember that and aye treat her with decorum. I trust too that ye will never forget the honour of the houses of Hepburn and Douglas.'

'There will be scant need for you to take Lady Grizel under your wing, Lady Catherine,' the Queen said. 'She is already under mine, and she shall remain there.' And she took my hand and said, 'Come, my love, bid farewell to your grandfather and then I will show you the apartments that have been assigned for you and your attendants.'

Can it be wondered that I loved Joan Beaufort then? She was aye the kindest and most considerate of mistresses, and never once did she give me a harsh word. Others were not so kind. For longer years than I care to remember, I was ridiculed and never called anything but 'Wee humpy Hep'. It is only in recent times that folk have startit to call me 'the auld crookit Countess'. In those early days, though, apart from the servants, and they were nae often sae polite, the only folk that ever called me 'Lady Grizel' were the Queen and Master Oswald Graham.

Oswald was a big fine-looking childe of nineteen or twenty when I first set eyes on him and fell in love. My heart has been turned by divers braw bucks in my long lifetime—not that it has ever done me any good, for who could look at me without scorn or repulsion or, less often, pity when nothing preventit me from wearing my heart on my sleeve? The fact that I was of great wealth and the owner of vast estates, and a Countess forby, made not a blade of difference to them. But Oswald was a man apart, and he aye treatit me with gentleness and courtesy, and there was never any glint of pity or patronage in his manner towards me. I fell in love with him that first day, and although he has been dead now for these many years I love him still, and I shall never forget his dreadful death. Ofttimes I awaken throughout the night and a great wave of horror engulfs me, so that I cannot sleep again, and I lie for hours in stomach-churning misery.

Oswald was a gentleman-in-waiting to King James. His Majesty himself I scarce ever saw in my first weeks at court. When I did he was aye fell polite and treatit me as if I looked like everyone else. He was that occupied, though, with affairs of state that he had little time for any of his courtiers, above all a misshapit bairn. He was a handsome man, King James, and in that first year when he came home from long captivity in England, after marrying the Lady Joan Beaufort, a cousin of King Henry V, he gladdened the eyes of all beholders. But he did not gladden their hearts when he imposed a tax of twelve pennies on ilka pound of the wealth of his subjects. This tax was levied to help pay his ransom money to the rapacious English. No tax had ever been levied in the regency of the King's uncle, the auld Duke of Albany, so even the most loyal courtiers winced and whinged about paying, for loyalty is stricken hard when it comes to the purse and the pocket.

Fewer ladies cast lovesick eyes at the King's broad shoulders and there was fainter applause when he showed his strength at the games he was fond of playing. And there were many murmurs of discontent when he arrestit the auld Earl of Lennox and Sir Robert Graham of Kincardine and castit them intil prison. What King James had against Sir Robert, a fine upright gentleman and a cousin of Master Oswald Graham, nobody knew but His Majesty himself, though many folk made wild conjectures. But it was evident to all that what he had agin Lord Lennox, who was sib to me on my mother's side, was his relationship to the House of Albany. Lord Lennox's daughter was married to Duke Murdoch of Albany, the King's cousin, and it was against the Albany door that the King laid his capture by the English when, a bit lad of twelve, he was on his way to France to receive further education. The King blamit his uncle, the auld Duke, who ruled Scotland for most of the seventeen years James was a prisoner, for not paying a ransom to free him instead of garring him pay this ransom on his lone after he wed Lady Joan, which was what he truly believed although it was the whole Scottish nation that bore the brunt of it. So there was small doubt that his spite against the Albany family causit him to take it out on Lord Lennox, who was a harmless enough auld childe but owned great estates that King James coveted.

I was fond of Lord Lennox, whom I had been brought up to call Uncle Duncan, and I was fell upset at his incarceration. I was in attendance on Queen Joan when she pleadit with the King for him and Sir Robert, and well I remember how the King lookit at her with a hard eye and said, 'I would be grateful, my dearest wife, if ye would not question my authority. Do not fash your sweet head with affairs of state. Now, loveliest of all the English roses, if I summon Master Oswald to attend me at the tennis court, will you and your ladies pleasure us by watching our game?'

But Master Oswald sent back the servant to crave His Majesty's pardon. He was ill abed and unable to play. A day or two afterwards Oswald's illness causit him to leave the palace and ging to his estate near Dunfermline to recuperate. He did not come back to court for a goodly while, by which time Sir Robert Graham had been given his freedom. My puir auld Uncle Duncan was not given his, however, and a wheen months later—six or seven if I recollect properly—the King's hatred for the House of Albany was given full rein and he arrestit Duke Murdoch and his sons, Walter and Alexander. The Albanys and Lord Lennox were tried before an assize of noblemen and, sib though they were to the King himself as well as to most of those that sat in judgement upon them, they were proclaimed guilty of divers false charges. When they were led out to have their heads choppit off on the Heading Hill at Stirling there was much weeping among the spectators and there was no doubt that the King was the most unpopular man in his ain kingdom yon day.

I weepit sore for my Uncle Duncan, especially when King James seized his estates and the Lennox earldom and keepit them for himself. Still anon, folk forget things quickly: far ower quickly it seems to me. Before long the King was popular again, and ladies like my glaikit cousin, Kate Douglas, never stoppit talking about his good looks and his great strength. Even then Kate could not help but show she was in love with His Majesty. It was not seemly to show sic an unbridled passion, howsomever, and so Kate was forced to dissemble and act as though she were in love with divers other young gallants of the court.

If I never knew the blessings of marriage, neither did the Lady Catherine Douglas, who was sic a braw hussy. Kate had an eye for the lads, and the lads had an eye for Kate. Yet though they daffit often enough with her none was ever willing to offer his hand in wedlock. Kate was ower bold even for the boldest of them. She was a domineering wench, and her clippit tongue made the callants steer clear of entering into any long alliance with her. Besides, she had little tocher. However much some of her sweethearts may have lusted after her, none was willing for wedlock when all she had to offer was a small estate in Perthshire and a wee pickle siller.

Yet, though I was resigned to the fate of being an auld maid, me that had great wealth, Kate was not. Like myself, she turned her gaze and the full force of her passion upon Master Oswald Graham when he returned to court after the long mysterious malady that folk whispered about ahint their hands. None ever whispered about it to me, and it was not until long after that I learnit that Master Oswald's illness had been a diplomatic one, him being fell fond of both King James and his cousin Sir Robert and not wanting to take sides with either in whatever was the cause of the strife between them.

Master Oswald found himself in just as sore a pickle when both Lady Kate and I set our caps at him. Puir Oswald, he was hard put to it whiles to keep his dignity and not take to flight like a hare fornent the hunter. Not that he was worriet about myself. After all, I was little more than a bairn, and disabled forby, so all he needit to do was treat me like a favourite sister. And this he did by teaching me the French tongue, playing chess with me for hours on end, reading aloud when I was ower tired to hold the volume myself, and singing ballads in his sweet tenor voice. But when it came to dealing with Kate Douglas, he was like a flounder in deeper, stronger and darker waters. Kate was not content to be treatit like a favourite sister. She hankered not only for Oswald's bonnie countenance and big, braw body; she had her eyes on his wealth and great estates, and both of her grasping hands were ready to stretch out and hold them. Howsomever, Oswald was no fool, and after polite dalliance with Kate, which he could not well prevent since they were in sic close contact at court, he was forced to show that if he was for wedlock, Lady Catherine Douglas was not the lady he would share it with.

Kate was sic a brazen limmer she could not comprehend that Master Oswald was not hers for the plucking, and he had to spell it out in large ciphers. Kate took the rebuff with an ill grace. She had had rebuffs before, and she was to have rebuffs aplenty in the years to come; but this ane from Master Oswald she took extreme sore to heart. From gey near swooning with ecstasy whenever she was in his presence, she startit to behave like a cock at a grossit, doing her utmost to damage him with tongue, teeth, and claws.

I shall never forget one instance of Kate's unrelenting hatred of Oswald Graham. This is partly because it showed me for the first time that Joan Beaufort was not the gentle, sweet creature she had aye appeared hitherto. The Queen had occasion to borrow one of Master Oswald's grooms to tend her favourite horse for a week while its regular attendant was sick. The groom tended the beast well but, alas, one hot morning when he led it into the palace yard for the Queen to mount, the horse misbehavit badly, prancing and rearing. The groom, a fair-headed young childe who was also suffering from the heat, lost his temper and struck the beast across the hurdies with his whip. Little did he ken that the Queen was already standing inside the palace door, in the shade, waiting for her favourite.

'How dare ye presume to correct our palfrey, serf!' Queen Joan cried. 'For this effrontery you will receive correction from your own whip.'

And she called the seneschal and ordered the groom to be given twelve stripes. She then mountit her horse and her ladies mounted theirs. I was, of course, unable to ride, so I was helpit intil my litter. As I settled myself

I heard Lady Catherine Douglas say, 'The knave has neither beggit Your Majesty's pardon nor pleadit for clemency.'

In truth the groom had not uttered a word, though his face flamed with shame at his nakitness being exposed fornent so many fine ladies when he was held by two of the palace guards, his hose were tooken down and he was bendit over.

The Queen said, 'Neither he has. Good seneschal, give the varlet as many stripes as you wish until he begs for mercy.'

And she urgit on her palfrey. As she did so, Lady Kate said, with a sly glance through the curtains of my litter, ''Tis a pity the knave's master is not in his place. It would serve him well for allowing sic a scurvy creature to attend til Your Majesty's valuable palfrey.'

'Have no fear, Kate,' the Queen said. 'Master Graham will feel the lash of my tongue ere this day is out.'

'Guidsakes, madam, I never did think ye were sae prickly,' Kate said with a skirl.

'I can be far pricklier than this gin I have the provocation,' the Queen said.

When she came back from her ride the Queen sent for Master Oswald. Howsomever, he had forestalled her. King James himself appeared with Oswald when he came that even to her summons, the King, who was the shorter, with his arm around Oswald's waist.

'What ails ye, my lovely rose?' said King James in a silky voice. 'This is not a bonnie tale I hear from our Oswald. Was the heat ower strong to withhold my lady's ire this forenoon? Oswald tells me his serf is in a fell bad way and will need doctoring for a good wheen weeks.'

The Queen said nothing. She turnit the side of her face to him and said to Kate, 'Lady Catherine, will ye be good enough to read us some stanzas from His Majesty's long poem? Yon poem he wrote to me when he was a poor prisoner in the tower at Windsor?'

And so we all had perforce to listen to Kate mouthing the lines we were ower often deeved with. I watched the Queen's face when Kate recited:

> And therewith cast I down mine eyes again
> Where as I saw, walking under the Tower,
> Full secretly new coming here to Pleyne,
> The fairest or the freshest young flower
> That ever I saw, me thought, before that hour,
> For which, sudden abate, anon astart
> The blood of all my body to my heart.

Without looking in the King's direction, the Queen interruptit, 'Mayhap the King's grace does not recollect that Tower? 'Twas so many years ago, and times and feelings change.'

The King laughed heartily and said, 'I recollect yon Tower full well, my lovely English rose. My feelings havena changit with the years, even though we are in a caulder climate now. This was ower het a day for Scotland's clime, and 'tis a great pity that your ladyship's humour was nettled by sic a trivial occurrence. Will ye forget it, my love, and dance a measure?'

He clappit his hands and the musicians struck up a gay, lilting tune. The King walkit smilingly towards Kate, saying, 'If the Queen's Majesty will give her hand to Master Oswald, I will give mine to Lady Catherine. Come, my loves!'

As I watchit Queen Joan treading the steps of the dance with Oswald I realized for the first time that, under all her sweetness, she was as ruthless as the King; and her ruthlessness was stronger and more subtle than his. Besides, she was more devious and this fooled folk the more easily into thinking she was as beautiful inside as she was out. It was the first time I ever criticized my mistress in my mind, but it was not to be the last.

And so in this way the years went by. The court moved from Falkland to Stirling, from Dunfermline to Edinburgh, from Jedburgh to Linlithgow. We bided for a few months in ilka palace or castle, and then movit on to another when the effluvium from the privies became ower noxious for our noble nostrils. In between whiles I would ging home to Angus for a wheen weeks, sometimes months, to supervise the running of my estates. My father died in his English captivity when I was thirteen and I became Countess of Bervie. My grandfather died of a broken heart seven months afterwards, and so I became Countess of Auchencairn as well. Often after that I was aware of King James looking at me with a speculative eye, and fine I knew he was considering whether it might not be worth his while for something to happen to me, an accident mayhap, so that, since I had no close kin and he was my guardian, he could appropriate my estates and the two earldoms as he had already appropriated the earldoms of Lennox, Mar, Fife, and Strathearn, all of which had been forfeitit to the Crown because of divers mishaps.

But I am thankful to say that whatever evil may have passit through His Majesty's mind it never came to fruition, and I hirpled on happily, devoted to the Queen and her bairns, and aye in love with Master Oswald and grateful for every small crumb of comfort and joy he gavit me.

During all this time the Queen gave birth ilka year, except in those in which she had miscarriages. By the time I was three and twenty, in the year 1436, she and King James had six daughters, but only the one son, little Jamie, who was the delight of his mother's and his father's eyes and mine.

And during all these years, too, Kate Douglas nurtured her viperish spite against Oswald Graham, a spite that came to sic a woeful pass in the hinterend. She and I hatit each other more and more until, after it was all over, I would gladly have stranglit her—except that strangling was ower gentle a death for sic a spawn of the Devil.

I was sixteen when Phemie died of an ailment of the breast that she had striven to hide from everybody until the last few weeks when the puir soul was in agony and could not prevent herself from crying out at odd moments when she thought nobody was hearkening til her. Another by-blow of the Hepburns took her place as my maid and companion and faithful guardian. Elspeth was three or four years aulder than me, and we never doubtit that my father and hers were one and the same; but Elspeth, thanks to the blessings of the Lord, was not burdened by any of my infirmities, and she had a nippier tongue than mine intil the bargain. She was aye as able as Phemie had been to protect me from folks' backbitings until she died three years syne.

And as Elspeth was a never-failing prop, against which I leanit with safety, so was I, for a long time, a prop for my favourite, next to Prince Jamie, of all the royal bairns, the Lady Joan Stuart. Joan is the third daughter of King James and Queen Joan and, like myself, she is sore afflictit; she is deaf and dumb, puir bairn, even though of unsurpassable beauty like her mother. Since she marriet James Douglas, Lord Dalkeith, a second or third cousin of my own and Kate's, in 1458, I have seen little of her; for after Dalkeith was made Earl of Morton on their wedding day, she has chosen to settle in his castle at Dunbar and cloister herself far from court life. But when she was a bairn she and I were aye together, sic close companions that my dear coz Kate usit often to skirl to her cronies, ''Tis a question of the blind leading the lame. Or is it the dumb leading the half-daft? 'Tis certain anyway that the two dafties seem to crave no other company than their own. Doubtless 'twould be a guid thing if they would hie intil a nunnery thegither—'twould be a great blessing to their families.'

Yet neither Princess Joan nor I had any notion to comply with Lady Kate's wish. As she grew older Joan did not lippen so much on my help and company, for a deaf and dumb bairn of her own age was brought to court to be her maid-in-waiting. This was only as it should be. I was fell glad for Joan's sake, and it enablit me to turn and fasten all my affection on her brother Jamie, who was also afflictit with the strawberry birthmark that, later, made folk call him James of the fiery face.

Like me, Oswald Graham was devoted to the royal bairns, and he dandled them in his arms when they were wee, wipit their noses and their bottoms, helpit to teach them to walk and read and talk, and held them on their ponies and taught them games, later on, when their father was ower busy with affairs of state to attend til his parental duties. Like me, too, Oswald had a favourite. His was the auldest lassie, Margaret. He and she were fell chief. Often on wet and snowy winter days, when the King and Queen were out on wolf hunts and other ploys, and he and I were left ahint to play with the bairns I have heard the Princess Margaret say, 'I full intend to marry ye, Oswald, when I grow up. I'd fain be your wee wifie.'

This was not to be, howsomever. Whilst still a bairn Margaret was affianced to the Dauphin of France. Although she was tellt at the time that she would be going to Paris one day to be a Queen, Margaret as often as not forgot it and keepit on saying she would marry Oswald. Then in 1436, the time came for her to ging to France to be wed. She was but twelve years auld. She gret full sore when she was told and, I think she kent then, puir lambie, she would never see Scotland nor any of us again. Queen Joan was wishful for me to accompany the Princess to be her chief lady-in-waiting, and I was agreeable, for I would have likit fine to see Paris. But the King objectit because he did not consider me auld enough nor experienced enough to be Margaret's guardian. King James and Queen Joan argued about this, but the matter was settled with finality when the royal physician decreed that I would never have the strength to survive the long sea voyage, far less the stamina and cunning necessary to combat the rigours and intrigues of the French court. Ofttimes I have thought King James garred the physician give this verdict because he was feared I might have the wrong kind of influence over the Princess or that I might die and leave the lassie alone and defenceless in yon foreign clime —though this, doubtless, would have givit him a grand opportunity to seize my estates and titles for himself. It is a pity His Majesty thought so poorly of me, for disablit though I am with my hunch and my lameness, I am yet of fell strong constitution, witness the fact that now, as I tell this tale, I am in my eighty-eighth year.

Master Oswald Graham was also wishful to accompany the Princess to Paris, but here again King James puttit down his foot. He tellt Oswald his proper place was here in Scotland to attend on himself, and he refusit to give him permission to ging with Margaret. When Oswald objectit, saying he was Margaret's auldest friend of mature years and near like being her foster-father, and that he felt it only right he should guide her through the mazes of the French Valois family and all their complicated alliances, King James said he would hear no more of it. His Majesty waxed so wroth that he said if Oswald persistit he would bring down upon him the law he had made forbidding certain folk, mainly clerics addicted to the Pope in Rome, to leave Scotland and take their siller to a foreign clime.

Master Oswald said he was no cleric and that if he wantit to bide for a while in France he saw no reason why he should be stoppit. Because of this the King and Oswald castit out, and Oswald retired in a huff to his estates in Galloway, swearing he would never show face at court again. Puir man, 'tis a great pity he never keepit that vow.

Howsomever, after a couple of months he and the King's Majesty became reconcilit, and Oswald came back to court in time to see young Margaret set sail from Leith for her new life as the Dauphin's bride. Puir Margaret, like Oswald she was born under an evil star, for she was illservit by yon awful-like man who was to become King Louis XI of France and has aye been known, rightly, as the Spider King. His treatment of the puir lassie was devilish, and it was a blessing in a way that she did not live long enough to become Queen of France.

And so, in this fashion, we came to the last winter of the King's reign. By then, after all their catterbattering about Margaret's marriage to the Dauphin, King James and Queen Joan were not cohabiting. And for a while now Kate Douglas had achievit her ambition and was become the King's mistress. Queen Joan Beaufort consolit herself with Sir James Stewart, who was aye called 'The Black Knight of Lorn'. And I will say in her favour that she was more discreet about her dealings with Sir James than the King was about his with Lady Kate.

What the King's feelings were about Sir James I know not. On the surface he aye greetit him in a friendly manner, and they oft played tennis thegither. Indeed, it was Sir James now that the King played with more often than he played with Master Oswald. As for the Queen and Kate, they were like sisters, living in each other's laps and seldom apart, except when Kate was in the King's bed.

I, alas, had nobody to console myself with. Although Oswald had come back from Galloway, his manner and way of life seemed to have changit. He was not so gay and blithe as was his wont in past years. He broodit greatly, and he was prone to sudden fits of black rage that ill became him. Mark ye, he seldom showed this ill-humour fornent me; we still played chess, and he would sing and read to me as of yore. Yet I could not but notice that every wee whilie he would appear to forget what he was doing and stare into vacancy like a bullock in a pen waiting for the slaughterman's mallet.

In December 1436 the court movit to Perth where the King was to preside over a council receiving a papal legate. Perth Castle was in a sorry state, so we took up residence in the Black Friars' monastery outside the town. The monastery was a bit crampit for sic a large assembly, even though the friars cooried six and seven in a cell to give the courtiers more space. I shared a room with Kate Douglas and Elspeth and Kate's maid, Kirsten. It was a fell squeeze, but the weather was cauld and, the firing in the monastery being austere, none of us mindit much, except Kate who was aware of the rest of us watching ilka time she slippit out to ging to the King's chamber.

Whilst waiting for the great council to start, the King amusit himself by playing tennis with Oswald and the Black Knight of Lorn and his private chamberlain, Sir Robert Stuart, the grandson of the auld Earl of Atholl, who was of the blood royal and some folk said should have been king instead of King James. The King had gotten ower stout for his age—he was but thirty-eight—and he gave himself over to the tennis in an attempt to lessen his corpulence. He and his three attendants appeared to get on full well while playing their games, but they played with sic vigour, especially the King's grace, who did not seem aware of his own great strength, that they keepit sending the tennis-balls into an underground vault where they rolled down a drain intil the stagnant waters of the moat. So, incensed at losing sae many balls, for he was a most parsimonious childe, the King had the drain vault closit up. This was an act that was to cost him unco dear in the long run.

Another manner the King found of amusing himself was to visit a Celtic seeress to have his fortune told. But he was not so amusit when she tellt him he would be kilt before long and that his body would have sixteen wounds. He tried to laugh off her warning, gave her a paltry pickle siller for her pains, and in a gay court that Christmastide His Majesty was the gayest person.

In the New Year I fell ill. I shivered with cauld and was hot with fever, turn and turn about, and so perforce I had to take to my bed. Day after day, week after week, I lay there wishing I could have this illness in the wider space of my own chamber in my dreich castle in Angus. But I could not quit the court, for by then snow had fallen heavily and the roads were blockit. It would have been ill advisit for me to travel in sic a feeble state, even though I would have been well happit up in a litter. Sometimes Oswald came to see me, and he would sit and glower in atween reading or singing to me. He would give me some news of the court's ongoings: how many were away on a wolf hunt, what had been said at the last meeting of the great council, news of the royal bairns who had been forbiddit to come and see me in case they caught an infection, and sic like bits of homely gossip. But most of my news was gleanit from Elspeth, who keepit me well nourished with titbits.

And so it was she who came bursting into my room about midnight on 21 February and tellt me that King James had been murdered.

'Oh, my leddy!' she cried, for she never took advantage of being born on the wrong side of the blanket. 'Sic terrible ongoings there has been this nicht. The puir King has been stabbit to death. 'Tis a wonder ye didna hear all the fearsome clamour. 'Twas enough to waken the dead— though nothing will waken the King now, puir unfortunate gentleman. His Majesty was preparin' to retire, and the Queen and Lady Kate were helpin' to remove his harness—for he'd been joustin' this even with Master Oswald on horseback in the tilt-yard ootby—and me and Lady Kate's maidie, Kirsten, were helpin' and wonderin', I micht as well say, which o' the ladies the King was for beddin' down with, when there was an unco clamour outside. Sic clashin' o' swords and armour clankin' and blood-curdlin' shoutin'. Lady Kate rushit to the door to bar it, but lo and behold there was no bar til it. Some callant had removit the bar aforehand. Lady Kate tried to sneck the door by puttin' her arm where the bar should be, but she wasna long in snatchin' it awa' and skirlin' like a scaddit cat when the first sword came through the opening and gave her a bit cut.

'Meantime the King had tore up the boards leading down intil the vault and squeezit through the hole. The Queen and Kirsten and me put back the boards and stood on them, and Kirsten and me startit to unplait Her Majesty's hair as though it was the maist natural thing to do: us gettin' her ready for her bed, like. Sir Robert Graham and Sir Robert Stuart and some other callants came burstin' intil the room. "Whaur's the King?" demandit Sir Robert Graham. "His hour has come." "His Majesty has retired to bed," says Queen Joan. "And how dare ye burst like this, sir, intil a lady's chamber? I shall have you arrestit at once. Call the guard, Lady Catherine." But Lady Kate was that fashed with the cut on her airm she took no notice. She kept skirlin': "Oh, my puir airm! Oh, my bonnie airm! I'll be scarred for life." '

Elspeth paused to draw breath. I was sitting up in bed, my hands clutchit to my throat. 'What happened next?' I managed to croak.

'The twa Sir Roberts and the other callants withdrew,' Elspeth said. 'They rushit away to the King's bedchamber, and we could hear them yowlin' like werewolves. Oh, 'twas unco uncanny, my leddy, and I hope I'll never live to experience the like again. The Queen says to me, "Elspeth, go and rouse the palace guard. And you, Kirsten, ging doon intil the vault and see that His Majesty has managed to escape through the drain intil the moat." But before we could move a foot's turn the King came back up through the hole in the floor. He hadna been able to get out intil the moat, he had forgotten he had blockit up the drain on account o' his lost tennis balls.'

Elspeth put her hands to her forehead and swayit from side to side, 'Oh, my leddy!' she keened. 'I hardly daur tell ye what happened next. Sir Robert Graham and the other callants came rushin' back intil the room, and when they saw the King they fell upon him and startit to hack intil him with their swords. At that moment, too, Master Oswald Graham came burstin' into the chamber, wavin' his sword and shoutin', "Stop! Stop!" And he thrust himself fornent the King and tried to parry the sword thrusts. The Queen, too, got atween the King and the murderers, and she got a sair cut on the shoulder. But 'twas of no avail. The King was struck doon, and he is lyin' there now as dead as an auld stag, with sixteen wounds in his body.'

'Sweet Christ in Heaven,' I whispered, and I crossed myself. 'What will happen next?'

'I havena finished, my leddy,' Elspeth said. 'Ye have not heard the half o' it. And this ye will never believe.'

'They have not killt the Queen, too?' I cried.

'No, she is well enough except for the cut on her shoother, and it will heal in time.'

'Well, what?' I shouted. 'Tell me, woman!'

Elspeth bowed her head and claspit her hands thegither. 'Ye must be brave, my lambie,' she said. ''Tis dreadful news I am about to tell ye.'

'What?' I sank back against the pillows, sweat pouring down my face.

''Tis Master Oswald,' Elspeth whispered. 'The puir gentleman. He has been arrestit and cast intil a dungeon. Lady Kate has accusit him of being in tow with the King's murderers. In troth, she vows he is the one who arrangit the whole affair, and that he appearit at the last minute to pretend he wasna in league with them. Lady Kate vows 'twas he who put the planks across the moat to help the murderers get intil the monastery, and 'twas he who took all the bars offen the doors. She doesna seem to think it might have been Sir Robert Stuart, who, God kens, has mair cause for wantin' the King out of the way.'

'Kate said that?' I cried. 'How does she . . .?'

'Ye may weel ask, my lambie,' Elspeth said. 'Kate Douglas—and even though she's your ain cousin—is an ill-gettit trollop who'll stop at nothin' once she has got her knife in. And she has got her knife intil Master Oswald right up til the verra hilt. And forby, she has the Queen's ear. Why she should have it is beyond my puir feeble comprehension, for I would ha'e thocht Queen Joan would have loathit the sicht o' her for flauntin' herself sae blatant-like as the King's light o' love. But women are kittle-cattle, as you and I ken full well. Anyways, Kate's lyin' tongue has won the day, and puir Master Oswald is chainit up in ain o' the cells.'

Oswald protestit his innocence time and time again, but Joan Beaufort was adamant. I realized then that she had aye been jealous of poor Oswald. And Kate Douglas was there at her side, whispering foul lies into her lug.

Within a week or two Sir Robert Graham and Sir Robert Stuart and the five others who had helpit to stab the King's grace were roundit up and lodged in the dungeons of Edinburgh Castle. For Queen Joan and her court quittit Perth immediately after the murder, and preparations were made to crown wee Jamie, who was just six years auld, as King James II in Holyrood Abbey.

I was still ill, and I was sore of heart because of Oswald. Kate Douglas kept coming to my chamber and telling me about his guilt. Time after time I tellt the lying limmer to begone from my sight, but she keepit on pricking at me, skirling like a loonie-wifie when I pressed my hands ower my ears to shut out her taunts.

A month after the murder the auld Earl of Atholl, who was near seventy, was arrestit because folk said he aspired to be king and was at the head of the conspiracy to put himself on the throne. He was draggit in chains at a horse's tail to Edinburgh, and a couple of days after that he and Sir Robert Stuart, Sir Robert Graham and Master Oswald Graham and the five others were put on trial.

By this time I was out of bed and hirpling about, but my heart was heavy and I would fain have been a thousand miles from Edinburgh town. I askit Queen Joan if I could have her gracious permission to retire to Angus so that I might recover from my long illness. But before Her Majesty could say anything, the bold Kate Douglas cried, 'And is Wee Humpy not to be in attendance on Your Grace on the great day, madam?'

'I fear Lady Grizel is too delicate and too gentle-natured to watch the proceedings,' the Queen said.

'Havers, madam!' said Kate. 'Wee Grizel is a Hepburn, and the Hepburns are well kent for their toughness. Am I not sib to them? Fine I ken how hard yon family can be, and if I can watch the trial so can she.'

'Lady Grizel has had the flux,' the Queen said. 'We deem it hardly fit that she should witness . . .'

Kate cried, 'Oswald Graham spurnit puir wee Grizel, my lady. 'Tis only right she should see the blaggart suffer his just punishment.'

'In that case,' the Queen said, 'you will attend upon us tomorrow morn, Lady Grizel, when the culprits are brought to justice.'

I pleadit and pleadit, but next morning at eleven o'clock, the bitterest morning of my life, I was in the Queen's train when we went into the great hall of the castle. I stood behind Her Majesty's chair. Kate Douglas, who had her arm full covered with bandages, was allowed to sit on a stool beside the Queen. There was no need for her arm to be bandaged, for her wound had healit up, Kirsten had tellt Elspeth. Yet Kate was full determined to make the most of it, to appear the heroine that earnit her the nickname 'Kate Bar the Door' till the end of her days.

I was sae weak with sorrow, I had to lean heavy on my stick with one hand and cling to Elspeth's arm with the other. I near swooned when the prisoners were brought up from the dungeons. My poor Oswald lookit like a ghost, with a growth of beard on his white, hollow-cheekit face. I shut my eyes and hopit he would not see me coorying there ahint the Queen's chair.

Sir William Crichton, the Keeper of Edinburgh Castle, a perfidious rogue who had aye lackeyed to King James, then read out the charges against the nine gentlemen. He endit his harangue with, 'Ye shall suffer the utmost rigours of execution, and may God have mercy on your souls —for not a body in Scotland has.'

Sir William bowed to the Queen. She inclined her head graciously to him, and she said, 'Let the trial begin.'

While a lawyer-mannie, brought there to give an air of verisimilitude to the proceedings, was pleading for the prisoners, I heard Kate Douglas say, 'You remember, madam, the end of King Edward the Second of England? Remember what his wife causit to be done til him when he displeasit her with yon callant Hugh Despenser? It seems a goodly fate for Master Oswald Graham.'

'Am I likely to forget it, Lady Catherine?' Joan Beaufort said. 'King Edward was my great-great-grandsire.'

I did not think of it then, for I was that overwhelmit by pain and misery, but oft I have thought about it since. King Edward may have been Queen Joan's great-great-grandsire, but equally his wife, Queen Isabella, her that folk cried 'The She Wolf of France', was Queen Joan's great-great-grandmother. Is there small wonder then that from that day I hatit Joan Beaufort with a hatred that coroded my entire being?

Suddenly there were screams, and it was me that was screaming. I screamed and screamed. When I came to, I was in my own bed, with Elspeth attending to me. The tears streamed down her face as she tellt me what had happened to the prisoners, as tellt to her by some castle guard who had been in the torture chamber after the trial.

And that eve Kate Douglas came and repeatit the tale, telling me that a red-hot crown had been placed on Lord Atholl's head. I buried my head ablow the bed-clothes and refusit to listen. Elspeth said afterwards that the common folk who, God kens, are not over blessit with the finer feelings, were outragit at the cruelties perpetrated on the murderers, and many said that when Sir Robert Graham cried in his death agony that they had killed James Stuart because he was a cruel king, he might well have addit that James had been wed to an even crueller queen.

A wheen weeks later when I was riding with the Queen in a litter through the gate of Edinburgh Castle I saw Lord Atholl's skull, with the iron crown embeddit intil it, stuck on a spike above the portcullis. I spewed all over Joan Beaufort's rose-pink satin gown.

Ye may wonder why I continued in the service of that foul woman for so many years afterwards. She had been sae nice to me when I was a bairn I could scarce credit yon lovely rose-bloom face could hide sic a charnel house of horror. I never forgave her for what she causit to be done to Oswald, that kindly gentle soul, and I hatit her ever after. Yet the truth is that I was full fond of her poor afflictit bairn, wee Jamie with the fiery birthmark on his face, the new King, and I felt it was my bounden duty to bide beside him and try to instil some goodness intil him, hoping he would not inherit the cruel streaks of his mother and her ancestress, Isabella of France. And I regret to say that I gloatit with the fullest pleasure when, a year or two afterwards, Joan Beaufort was glaikit enough to destroy her own position as the Queen Dowager of Scotland by taking her paramour, Sir James Stewart, the Black Knight of Lorn, as her second husband. I had not the least pickle of pity for her downfall.

ARTHUR GRAY

The Sacrist of Saint Radegund

ဢ

On a certain day in mid-June in the year 1431 the tolling of the bell in
St Radegund's church tower announced to the neighbours of the Priory
that a nun was to be buried that day.

In an interval between church services the nuns wander in the garden,
which is also the graveyard of St Radegund's, and lies sequestered next
the chancel walls. Today they are drawn thither by a new-made empty
grave; for a funeral is a mildly exciting incident in conventual routine.
But three sisters sit in the cloister on the stone bench next the chapter
door. Also a small novice is curled up on the paved floor with her back
against the bench. The day is warm, and the church wall casts a grateful
shadow where they sit. And, because labour and silence are enjoined in
the cloister, they rest, and two of them gossip, and Agnes Senclowe, the
novice, listens and lays to heart.

The two who gossip are Joan Sudbury, succentrix, and Elizabeth Daveys,
who is older than Joan, and holds no office in the monastery. With them
sits, and half dozes, Emma Denton, who is very old and very infirm. She
does not gossip, for she has hardly spoken a word of sense these forty
years past. She is a heavy affliction to the cloister society. She lives
mainly in the infirmary, and does not attend church. She knows when it
is the hour for a meal, and she knows very little else. If she speaks an
intelligible word, it is about something that happened forty years ago. She
remembers the great pestilence in 1390.

What ailed poor sister Emma to bring her to this sad pass? When she
was young she was something of a religious enthusiast, and because
enthusiasm was rare in the cloister, she was promoted by her sisters to
high station. When they made her Sacrist she had her one and dearest
wish. To have the charge of the beautiful church, of the books, vestments,
and jewels of the sanctuary, to live in the holy place, with holy thoughts
for companions, and in the unfailing round of holy duties—was not that
a happy lot? Dignified too the office was; for in the little cloister world the
Prioress herself was scarcely a greater lady than the Sacrist. The Sacrist
did not sleep with the other nuns in the dormitory; her constant duties
did not allow her ordinarily to take her meals in the refectory. Like the

Prioress, she had her own servant to attend her, her own house to dwell in. Her habitation was built against the northern chancel wall, and consisted of two chambers. From the upper room, through a hole pierced in the wall, she watched the never-dying light that hung before the High Altar.

But it was not good to be Sacrist for long. The unvarying routine of duty produces torpidity; holy thoughts uncommunicated end in cessation of thought; the solitude was deadly. The office was not coveted by the sisterhood, and was seldom held for more than a year or two together. Wherefore they rejoiced when Emma Denton held it for nine years. For nine years she trimmed the sacred lamp. During nine years her own light dwindled out, and at last the world became dark to sister Emma.

The crazy belfry rocked with the swaying of the bell, which, being cracked, was doubly dolorous. The sound of it roused old sister Emma to a dim consciousness of what was passing, and she spoke to nobody in particular.

'The bell,' she said, 'the bell again! Last week it tolled, and we buried two. Now there are two more in the dead-house.'

'The saints protect us!' said sister Joan; 'she is at her old talk of the pestilence year.'

'It was Assumption Day,' continued the old nun, 'when we buried them. We had no Mass that day. Today it is the cellaress and sister Margery Cailly—God pardon her for a sinful woman. No; Margery is sick, not dead; and I forget, I forget.'

'Margery Cailly,' cried Joan Sudbury, 'what quoth she of Margery Cailly, that goes to her grave today? Margery Cailly, that has been our most religious Sacrist ever since yonder poor thing fell beside her wits.'

'Religious you may call her,' said Elizabeth Daveys, 'but God knows, and sister Emma knows, that of her which we know not. Thirty years have I lived in St Radegund's, and I remember not the time when any but Margery was our Sacrist, and well I know that the sacristy has been her prison all those days. But I have heard sister Emma say in her dull way that Margery once knew the convent prison too.'

'Well, twelve years I have spent here, and never had speech with the Sacrist. Once I was alone in the church when it was dark, and the day-light only lingered aloft in the roof, and of a sudden I lighted on her in the chancel, busied in her office. Her pale face in her black hood showed like a spirit's, and I thought it was the blessed Radegund that had come down from her window, and I was horribly afraid.'

'I think that from the sacristy window her eye followed me about the garden as I walked there,' said Elizabeth. 'It follows me still, and it makes my flesh creep. What good woman would shun her sisters so? Heaven rest her soul, for be sure she has much to answer for. If she has confessed herself, it is not to our confessor or the Prioress, for I think she has hardly spoken these many years to any but Alice Portress that waits on her.'

'Yes, Alice was with her at the end. It was Alice that dug the grave; Alice rings the knell; Alice laid her out in her Sacrist's chamber, and she has placed two white roses on the dead woman's breast.'

'Roses?' said Elizabeth Daveys; 'roses are not for dead nuns. Whence got she roses?'

'That I can tell you,' said the novice, glad to take her part in the conversation, 'for Alice told me herself. She got them from the churchyard of St Peter's on the hill.'

The office for the dead was said, the empty grave was filled, and Alice the Portress was closeted with the Prioress.

'To you, lady Prioress—not to the Nuns in Chapter—I confess the sin of my youth; not to them, nor yet to you while sister Margery lived. She is gone, and why should I remain? Forty years we shared the secret. She is past censure or forgiveness. On me let the blame rest. I ask no pardon, but only to be dismissed from the house of St Radegund, that I have so unworthily served.

'There is none but myself and poor sister Emma that remembers St Radegund's before the pestilence year. I was but a child then, and my mother was Portress before me. My mother often brought me to the lodge, and I used to play with the novices, or sit at the gate when my mother was away. Margery had but lately come to St Radegund's—seventeen, perhaps, or eighteen years of age she was. Hers was a proud family—the Caillys of Trumpington, and they were rich, and good to St Radegund's. They are gone and forgotten now, but often have I heard old Thomas Key tell of them, for he was a Trumpington man, and he knew the De Freviles of Shelford too. There are De Freviles at Shelford yet, but I think that none there remembers young Nicholas De Frevile that was Sir Robert's son.

'I had a child's thought—that Margery was the most beautiful creature in the wide world—most beautiful and best. And because she was young and fair and gracious in speech even our hard sisters loved her, and thought it pity of the world when her fair tresses were shorn and she took the ugly veil. For Margery was not religious. God pardon me for my sinful words, but I think she was meant for better things than religion and a cloister. And though she was good and kind to all, Margery did not take to our sisters. There was some trouble—I know not what, for she never told—and for some family reason she was sent to St Radegund's, and ill she liked it. So she went about her work in cloister and church, grieving; and there was talk of her among the sisters. Some thought, some said, that they knew, but Margery said nothing.

'It is all forgotten now, for the pestilence wiped out the memory of those days. Scarcely twelve months had gone since she took the veil when Margery Cailly disappeared from the Priory. You may think what babble of tongues there was in our parlour—how they who were wisest had always known how it would be, and the rest rebuked them for not telling them beforehand. And so for another twelvemonth she was lost to us, and some sisters, who were kind, hoped that she would come back, and some who were kinder, hoped she would not.

'Then, one day in the year before the pestilence, comes an apparitor with our lost Margery, and a letter to the Prioress from the Lord Bishop of Ely. The letter is to say that the Archbishop of Canterbury, in his visitation of Lincoln diocese, has found Margery there, living a secular life; and because secular life is sin to those who have entered the religious order, he commits her to his brother of Ely, in order that the lost sheep may be restored to the fold where she was professed. And his Lordship of Ely—Heaven help him for a blundering bachelor!—directs that she shall be committed to the convent prison-house until she repents of her wickedness, and when she is loosed from it, shall make public confession in Chapter, and implore the pardon of the sisters for her enormities.

'Our Prioress was kinder to Margery than the Bishop meant—who could not be kind to her? Her prison life was no longer than would satisfy the Bishop's enquiries, and as for the confession in the Chapter-house— it never happened. There were some, though they liked not confession for themselves, who thought an opportunity was missed, and blamed the Prioress; for cloister talk is dull if we know not one another's failings. Still, the sisters were kind to Margery, and very kind when they wanted to get the secret from her. But she said never a word about it, unless it were to the Prioress. Beautiful she was as ever, but grief and humiliation were on her, heavy as death, and because she confided in none, she lost the friendship of the sisters. To me, who was but a child, she would talk, but scarcely to another, and her talk with me was never about herself.

'One other there was with whom sometimes she had speech, and that was old Thomas Key, maltster and trusty servant in general matters of the Priory. Him she had known in happier days when he was a tenant of the Caillys at Trumpington. Her family was too proud and too pious to remember the disgraced nun, and they never visited her; but from Thomas she learnt something of home and the outside world.

'Then came the dreadful year when the pestilence raged in Cambridge town. The nuns had been used to get leave from the Prioress to go out into the town, but there was no gadding now. The gate was closely barred, and none were admitted from outside except Thomas Key. We carried the Host in procession about the Nuns' Croft and—laud be to the saints!—it protected our precincts from the contagion. And while the sinful world without died like the beasts that perish, we sat secure, but frightened, in our cloister, and blessed our glorious saint for extending the protection of her prayers over the pious few who did her service in St Radegund's.

'You have heard how the parish clergy died that year. One, two, sometimes three died in one parish, and the Bishop found it hard to provide successors. Boys that had barely taken the tonsure a week before were sent in haste to anoint the sick and bury the dead in places where the plague had left an unshepherded flock. Sir John Dekyn, priest of St Peter's church on the hill, was one that died, and his successor did not live a fortnight after him. Then we heard from Thomas Key that a mere youth had taken the place, one Sir Nicholas of the Shelford De Frevile family, who had but lately been ordered priest at Ely. And we were told that he worked with a feverous zeal among the poor, the sick, and the dying of his parish.

'Now when this news was brought to sister Margery by Thomas Key, it was to her as a summons from death to life. Her eye brightened and her cheek glowed when she heard of the heroic goodness of this young priest. While the sisters shuddered and shrank at each morning's fatal news, she was consumed with a passionate desire to know what was passing in the plague-stricken town, and she plied my mother and Thomas Key with incessant questionings. "Who was sick of the townsfolk? Were any of the clergy visited? How went it with the poor in St Peter's, where the pestilence was hottest?" For some weeks she heard that the light burned still at night in St Peter's parsonage, and that the priest was unscathed, incessant in his ministrations and blessed by his parishioners. And it seemed as though the sickness was abating.

'Then, late one afternoon in early August, there came a call for Margery. Thomas Key brought it, and whether it was his own tidings, or a message from some other, I cannot say; Margery never told me. But this I know, that she took me apart in the cloister and spoke to me, and she was terribly moved and her voice was choked. "Little Alice," she said, "as you love me, get me the gate-key after Lauds tonight. It is life or death to me to go out into the town. Only do it, and say nothing—no, not to your mother." Young as I was, I knew how the nuns were used to humour my mother into letting them pass the gate; but that was in daytime. At night, in our besieged state, with the death-bells tolling all around, it seemed a terrible thing to venture. But I asked no questions. Say it was the recklessness of a girl—say it was the love that I bore to Margery. I stole the key and gave it to her after sundown.

'What happened afterwards I will tell you as it was told to me by

Thomas Key, who waited for her outside the gate. They passed along the dark, deserted streets. The plague-fires burnt low in the middle of the roadway, but there were none to tend them, and no living thing they saw but the starving dogs, herded at barred doors. They crossed the bridge and mounted to St Peter's church. The priest's manse—you know it—is a low house next the church. A white rose, still in flower, clambered on its walls, and, half hidden by its sprays, a taper gleamed through the open window; but there was no sound of life within. They pushed open the door and entered.

'Stretched on his pallet, forsaken and untended, lay the young priest of St Peter's, the pangs of death upon him. Margery threw herself on her knees by his bedside, and Thomas watched and waited. For a time there was silence, for Margery had no voice to pray. Only at times the dying man grumbled and wandered in his talk; but little he said that Thomas understood.

'Then after a long time, he stirred himself uneasily and uttered one word, "Margery". And she—alas the day!—put out her arm and laid it on his shoulder. In an instant the dying man half raised himself on his bed and turned his eyes on her, and there was recognition in them. And one arm he threw about her neck, and felt blindly for the fair locks that had been shorn long since, and he said heavily and painfully, "Margery, *belle amie*, let us go to the pool above the mill, where the great pike lie, and sun and shadow lie on the deep water." So Thomas knew that they were boy and girl again by the old mill at Trumpington.

'That was all, and the end came soon. They two laid him decently beneath his white sheet, and Margery plucked two white roses from the spray that straggled across his window, and laid them on the dead man's breast. So they left him, with the candle still burning out into the dark.

'There was a horrible dread in St Radegund's when, four days later, sister Margery sickened of the pestilence; and it was worse when we learnt soon after that Thomas Key was visited—then that he was dead. That was the beginning of our sorrows. You have heard, Lady Prioress, how three sisters died before August was out, how most of the others deserted the house, and some never returned to it. Our prayers were unheard, and to us who remained it seemed as if the saints slept, or God were dead.

'So it happened that when the plague abated, and the first meeting was held in St Radegund's Chapter-house, about St Luke's day in the autumn, there were only three to attend it—the Prioress, the Sacrist (Emma Denton), and Margery Cailly. For—wonderful it seems—Margery, who least needed to live, was the one spared of those who were taken with the pestilence. Presently some old sisters returned, and new ones took the place of the departed. But the sword of the pestilence cut off the memory of the old days, and the sins and sufferings, the virtues and the victories of the former sisterhood were a forgotten dream when the cloister filled again. So when Emma Denton passed into her lethargy, and Margery Cailly earnestly petitioned to fill her place in the Sacristy, there was not a sister to question her character and devoutness.

'Not yesterday, but forty years ago, Margery Cailly passed out of life; for you know that, save to me, she has spoken few words since. And though I have waited on her for most of those years she never breathed to me the name of Nicholas De Frevile, never hinted at the story of her unhappy girlhood. But once in the springtime, just after she entered her Sacrist prison-house, she entreated me to plant a white rose-bush on the grave of the young priest of St Peter's. I did so, and have renewed it since, and one day, by your grace, I shall plant a spray of the same roses where she lies apart from him. I have confessed my wrong in stealing the key and bringing death into the cloister. If you can forgive me, so; if not, all I ask is that you let your sinful servant depart in peace.'

There is a curious aperture in the outer northern wall of the chancel of the nuns' church which is now Jesus College Chapel. If it is examined its purpose is evident. It was the lychnoscope, through which the Sacrist watched by night the light before the High Altar. It is the sole abiding memorial of Margery Cailly, Sacrist of St Radegund.

ROBERT LOUIS STEVENSON

A Lodging for the Night

এ৪১

It was late in November 1456. The snow fell over Paris with rigorous, relentless persistence; sometimes the wind made a sally and scattered it in flying vortices; sometimes there was a lull, and flake after flake descended out of black night air, silent, circuitous, interminable. To poor people, looking up under moist eyebrows, it seemed a wonder where it all came from. Master Francis Villon had propounded an alternative that afternoon, at a tavern window: was it only Pagan Jupiter plucking geese upon Olympus? or were the holy angels moulting? He was only a poor Master of Arts, he went on; and as the question somewhat touched upon divinity, he durst not venture to conclude. A silly old priest from Montargis, who was among the company, treated the young rascal to a bottle of wine in honour of the jest and the grimaces with which it was accompanied, and swore on his own white beard that he had been just such another irreverent dog when he was Villon's age.

The air was raw and pointed, but not far below freezing; and the flakes were large, damp, and adhesive. The whole city was sheeted up. An army might have marched from end to end and not a footfall given the alarm. If there were any belated birds in heaven, they saw the island like a large white patch, and the bridges like slim white spars, on the black ground of the river. High up overhead the snow settled among the tracery of the cathedral towers. Many a niche was drifted full; many a statue wore a long white bonnet on its grotesque or sainted head. The gargoyles had been transformed into great false noses, drooping towards the point. The crockets were like upright pillows swollen on one side. In the intervals of the wind, there was a dull sound of dripping about the precincts of the church.

The cemetery of St John had taken its own share of the snow. All the graves were decently covered; tall white housetops stood around in grave array; worthy burghers were long ago in bed, benightcapped like their domiciles; there was no light in all the neighbourhood but a little peep from a lamp that hung swinging in the church choir, and tossed the shadows to and fro in time to its oscillations. The clock was hard on ten when the patrol went by with halberds and a lantern, beating their hands; and they saw nothing suspicious about the cemetery of St John.

Yet there was a small house, backed up against the cemetery wall, which was still awake, and awake to evil purpose, in that snoring district. There was not much to betray it from without; only a stream of warm vapour from the chimney-top, a patch where the snow melted on the roof, and a few half-obliterated footprints at the door. But within, behind the shuttered windows, Master Francis Villon the poet, and some of the thievish crew with whom he consorted, were keeping the night alive and passing round the bottle.

A great pile of living embers diffused a strong and ruddy glow from the arched chimney. Before this straddled Dom Nicolas, the Picardy monk, with his skirts picked up and his fat legs bared to the comfortable warmth. His dilated shadow cut the room in half; and the firelight only escaped on either side of his broad person, and in a little pool between his outspread feet. His face had the beery, bruised appearance of the continual drinker's; it was covered with a network of congested veins, purple in ordinary circumstances, but now pale violet, for even with his back to the fire the cold pinched him on the other side. His cowl had half fallen back, and made a strange excrescence on either side of his bull neck. So he straddled, grumbling, and cut the room in half with shadow of his portly frame.

On the right, Villon and Guy Tabary were huddled together over a scrap of parchment; Villon making a ballade which he was to call the 'Ballade of Roast Fish', and Tabary spluttering admiration at his shoulder. The poet was a rag of a man, dark, little, and lean, with hollow cheeks and thin black locks. He carried his four-and-twenty years with feverish animation. Greed had made folds about his eyes, evil smiles had puckered his mouth. The wolf and pig struggled together in his face. It was an eloquent, sharp, ugly, earthly countenance. His hands were small and prehensile, with fingers knotted like a cord; and they were continually flickering in front of him in violent and expressive pantomime. As for Tabary, a broad, complacent, admiring imbecility breathed from his squash nose and slobbering lips: he had become a thief, just as he might have become the most decent of burgesses, by the imperious chance that rules the lives of human geese and human donkeys.

At the monk's other hand, Montigny and Thevenin Pensete played a game of chance. About the first there clung some flavour of good birth and training, as about a fallen angel; something long, lithe, and courtly in the person; something aquiline and darkling in the face. Thevenin, poor soul, was in great feather: he had done a good stroke of knavery that afternoon in the Faubourg St Jacques, and all night he had been gaining from Montigny. A flat smile illuminated his face; his bald head shone rosily in a garland of red curls; his little protuberant stomach shook with silent chucklings as he swept in his gains.

'Doubles or quits?' said Thevenin.

Montigny nodded grimly.

'*Some may prefer to dine in state,*' wrote Villon, '*On bread and cheese on silver plate.* Or—or—help me out, Guido!'

Tabary giggled.

'*Or parsley on a golden dish,*' scribbled the poet.

The wind was freshening without; it drove the snow before it, and sometimes raised its voice in a victorious whoop, and made sepulchral grumblings in the chimney. The cold was growing sharper as the night went on. Villon, protruding his lips, imitated the gust with something between a whistle and a groan. It was an eerie, uncomfortable talent of the poet's much detested by the Picardy monk.

'Can't you hear it rattle in the gibbet?' said Villon. 'They are all dancing the devil's jig on nothing, up there. You may dance, my gallants, you'll be none the warmer! Whew! what a gust! Down went somebody just now! A medlar the fewer on the three-legged medlar-tree!—I say, Dom Nicolas, it'll be cold tonight on the St Denis Road?' he asked.

Dom Nicolas winked both his big eyes, and seemed to choke upon his Adam's apple. Montfaucon, the great grisly Paris gibbet, stood hard by the St Denis Road, and the pleasantry touched him on the raw. As for Tabary, he laughed immoderately over the medlars; he had never heard anything more light-hearted; and he held his sides and crowed. Villon fetched him a fillip on the nose, which turned his mirth into an attack of coughing.

'Oh, stop that row,' said Villon, 'and think of rhymes to "fish".'

'Doubles or quits,' said Montigny doggedly.

'With all my heart,' quoth Thevenin.

'Is there any more in that bottle?' asked the monk.

'Open another,' said Villon. 'How do you ever hope to fill that big hogshead, your body, with little things like bottles? And how do you expect to get to heaven? How many angels, do you fancy, can be spared to carry up a single monk from Picardy? Or do you think yourself another Elias—and they'll send the coach for you?'

'*Hominibus impossibile,*' replied the monk, as he filled his glass.

Tabary was in ecstasies.

Villon filliped his nose again.

'Laugh at my jokes, if you like,' he said.

'It was very good,' objected Tabary.

Villon made a face at him. 'Think of rhymes to "fish",' he said. 'What have you to do with Latin? You'll wish you knew none of it at the great assizes, when the devil calls for Guido Tabary, clericus—the devil with the hump-back and red-hot finger-nails. Talking of the devil,' he added in a whisper, 'look at Montigny!'

All three peered covertly at the gamester. He did not seem to be enjoying his luck. His mouth was a little to a side; one nostril nearly shut, and the other much inflated. The black dog was on his back, as people say, in terrifying nursery metaphor; and he breathed hard under the gruesome burden.

'He looks as if he could knife him,' whispered Tabary, with round eyes.

The monk shuddered, and turned his face and spread his open hands to the red embers. It was the cold that thus affected Dom Nicolas, and not any excess of moral sensibility.

'Come now,' said Villon—'about this ballade. How does it run so far?' And, beating time with his hand, he read it aloud to Tabary.

They were interrupted at the fourth rhyme by a brief and fatal movement among the gamesters. The round was completed, and Thevenin was just opening his mouth to claim another victory, when Montigny leaped up, swift as an adder, and stabbed him to the heart. The blow took effect before he had time to utter a cry, before he had time to move. A tremor or two convulsed his frame; his hands opened and shut, his heels rattled on the floor; then his head rolled backward over one shoulder with the eyes wide open; and Thevenin Pensete's spirit had returned to Him who made it.

Every one sprang to his feet; but the business was over in two twos. The four living fellows looked at each other in rather a ghastly fashion; the dead man contemplating a corner of the roof with a singular and ugly leer.

'My God!' said Tabary; and he began to pray in Latin.

Villon broke out into hysterical laughter. He came a step forward and ducked a ridiculous bow at Thevenin, and laughed still louder. Then he sat down suddenly, all of a heap, upon a stool, and continued laughing bitterly as though he would shake himself to pieces.

Montigny recovered his composure first.

'Let's see what he has about him,' he remarked; and he picked the dead man's pockets with a practised hand and divided the money into four equal portions on the table. 'There's for you,' he said.

The monk received his share with a deep sigh, and a single stealthy glance at the dead Thevenin, who was beginning to sink into himself and topple sideways off the chair.

'We're all in for it,' cried Villon, swallowing his mirth. 'It's a hanging

job for every man jack of us that's here—not to speak of those who aren't.'
He made a shocking gesture in the air with his raised right hand, and put
out his tongue and threw his head on one side, so as to counterfeit the
appearance of one who has been hanged. Then he pocketed his share of
the spoil, and executed a shuffle with his feet as if to restore the circulation.

Tabary was the last to help himself; he made a dash at the money, and
retired to the other end of the apartment.

Montigny stuck Thevenin upright in the chair, and drew out the
dagger, which was followed by a jet of blood.

'You fellows had better be moving,' he said, as he wiped the blade on
his victim's doublet.

'I think we had,' returned Villon, with a gulp. 'Damn his fat head!' he
broke out. 'It sticks in my throat like phlegm. What right has a man to
have red hair when he is dead?' And he fell all of a heap again upon the
stool, and fairly covered his face with his hands.

Montigny and Dom Nicolas laughed aloud, even Tabary feebly chiming
in.

'Cry baby,' said the monk.

'I always said he was a woman,' added Montigny, with a sneer. 'Sit
up, can't you?' he went on, giving another shake to the murdered body.
'Tread out that fire, Nick!'

But Nick was better employed; he was quietly taking Villon's purse, as
the poet sat, limp and trembling, on the stool where he had been making
a ballade not three minutes before. Montigny and Tabary dumbly demanded a share of the booty, which the monk silently promised as he
passed the little bag into the bosom of his gown. In many ways an artistic
nature unfits a man for practical existence.

No sooner had the theft been accomplished than Villon shook himself,
jumped to his feet, and began helping to scatter and extinguish the
embers. Meanwhile Montigny opened the door and cautiously peered
into the street. The coast was clear; there was no meddlesome patrol in
sight. Still it was judged wiser to slip out severally; and as Villon was
himself in a hurry to escape from the neighbourhood of the dead Thevenin,
and the rest were in a still greater hurry to get rid of him before he should
discover the loss of his money, he was the first by general consent to issue
forth into the street.

The wind had triumphed and swept all the clouds from heaven. Only
a few vapours, as thin as moonlight, fleeting rapidly across the stars. It
was bitter cold; and by a common optical effect, things seemed almost
more definite than in the broadest daylight. The sleeping city was absolutely
still: a company of white hoods, a field full of little Alps, below the
twinkling stars. Villon cursed his fortune. Would it were still snowing!

Now, wherever he went, he left an indelible trail behind him on the glittering streets; wherever he went he was still tethered to the house by the cemetery of St John; wherever he went he must weave, with his own plodding feet, the rope that bound him to the crime and would bind him to the gallows. The leer of the dead man came back to him with a new significance. He snapped his fingers as if to pluck up his own spirits, and, choosing a street at random, stepped boldly forward in the snow.

Two things preoccupied him as he went: the aspect of the gallows at Montfaucon in this bright windy phase of the night's existence, for one; and for another, the look of the dead man with his bald head and garland of red curls. Both struck cold upon his heart, and he kept quickening his pace as if he could escape from unpleasant thoughts by mere fleetness of foot. Sometimes he looked back over his shoulder with a sudden nervous jerk; but he was the only moving thing in the white streets, except when the wind swooped round a corner and threw up the snow, which was beginning to freeze, in spouts of glittering dust.

Suddenly he saw, a long way before him, a black clump and a couple of lanterns. The clump was in motion, and the lanterns swung as though carried by men walking. It was a patrol. And though it was merely cross-ing his line of march, he judged it wiser to get out of eyeshot as speedily as he could. He was not in the humour to be challenged, and he was conscious of making a very conspicuous mark upon the snow. Just on his left hand there stood a great hotel, with some turrets and a large porch before the door; it was half-ruinous, he remembered, and had long stood empty; and so he made three steps of it and jumped in to the shelter of the porch. It was pretty dark inside, after the glimmer of the snowy streets, and he was groping forward with outspread hands, when he stumbled over some substance which offered an indescribable mixture of resistances, hard and soft, firm and loose. His heart gave a leap, and he sprang two steps back and stared dreadfully at the obstacle. Then he gave a little laugh of relief. It was only a woman, and she dead. He knelt beside her to make sure upon this latter point. She was freezing cold, and rigid like a stick. A little ragged finery fluttered in the wind about her hair, and her cheeks had been heavily rouged that same afternoon. Her pockets were quite empty; but in her stocking, underneath the garter, Villon found two of the small coins that went by the name of whites. It was little enough; but it was always something; and the poet was moved with a deep sense of pathos that she should have died before she had spent her money. That seemed to him a dark and pitiable mystery; and he looked from the coins in his hand to the dead woman, and back again to the coins, shaking his head over the riddle of man's life. Henry V of England, dying at Vincennes just after he had conquered France, and this poor jade

cut off by a cold draught in a great man's doorway, before she had time
to spend her couple of whites—it seemed a cruel way to carry on the
world. Two whites would have taken such a little while to squander; and
yet it would have been one more good taste in the mouth, one more
smack of the lips, before the devil got the soul, and the body was left to
birds and vermin. He would like to use all his tallow before the light was
blown out and the lantern broken.

While these thoughts were passing through his mind, he was feeling,
half mechanically, for his purse. Suddenly his heart stopped beating;
a feeling of cold scales passed up the back of his legs, and a cold blow
seemed to fall upon his scalp. He stood petrified for a moment; then he
felt again with one feverish movement; and then his loss burst upon him,
and he was covered at once with perspiration. To spendthrifts money is
so living and actual—it is such a thin veil between them and their pleasures!
There is only one limit to their fortune—that of time; and a spendthrift
with only a few crowns is the Emperor of Rome until they are spent. For
such a person to lose his money is to suffer the most shocking reverse,
and fall from heaven to hell, from all to nothing, in a breath. And all the
more if he has put his head in the halter for it; if he may be hanged
tomorrow for that same purse, so dearly earned, so foolishly departed!
Villon stood and cursed; he threw the two whites into the street; he shook
his fist at heaven; he stamped, and was not horrified to find himself tramp-
ling the poor corpse. Then he began rapidly to retrace his steps towards
the house beside the cemetery. He had forgotten all fear of the patrol,
which was long gone by at any rate, and had no idea but that of his lost
purse. It was in vain that he looked right and left upon the snow: nothing
was to be seen. He had not dropped it in the streets. Had it fallen in the
house? He would have liked dearly to go in and see; but the idea of the
grisly occupant unmanned him. And he saw besides, as he drew near, that
their efforts to put out the fire had been unsuccessful; on the contrary, it
had broken into a blaze, and a changeful light played in the chinks of door
and window, and revived his terror for the authorities and Paris gibbet.

He returned to the hotel with the porch, and groped about upon the
snow for the money he had thrown away in his childish passion. But he
could only find one white; the other had probably struck sideways and
sunk deeply in. With a single white in his pocket, all his projects for a
rousing night in some wild tavern vanished utterly away. And it was not
only pleasure that fled laughing from his grasp; positive discomfort, positive
pain, attacked him as he stood ruefully before the porch. His perspiration
had dried upon him; and though the wind had now fallen, a binding frost
was setting in stronger with every hour, and he felt benumbed and sick
at heart. What was to be done? Late as was the hour, improbable as was

success, he would try the house of his adopted father, the chaplain of St Benoît.

He ran there all the way, and knocked timidly. There was no answer. He knocked again and again, taking heart with every stroke; and at last steps were heard approaching from within. A barred wicket fell open in the iron-studded door, and emitted a gush of yellow light.

'Hold up your face to the wicket,' said the chaplain from within.

'It's only me,' whimpered Villon.

'Oh, it's only you, is it?' returned the chaplain; and he cursed him with foul unpriestly oaths for disturbing him at such an hour, and bade him be off to hell, where he came from.

'My hands are blue to the wrist,' pleaded Villon; 'my feet are dead and full of twinges; my nose aches with the sharp air; the cold lies at my heart. I may be dead before morning. Only this once, father, and before God I will never ask again!'

'You should have come earlier,' said the ecclesiastic coolly. 'Young men require a lesson now and then.' He shut the wicket and retired deliberately into the interior of the house.

Villon was beside himself; he beat upon the door with his hands and feet, and shouted hoarsely after the chaplain.

'Wormy old fox!' he cried. 'If I had my hand under your twist, I would send you flying headlong into the bottomless pit.'

A door shut in the interior, faintly audible to the poet down long passages. He passed his hand over his mouth with an oath. And then the humour of the situation struck him, and he laughed and looked lightly up to heaven, where the stars seemed to be winking over his discomfiture.

What was to be done? It looked very like a night in the frosty streets. The idea of the dead woman popped into his imagination, and gave him a hearty fright; what had happened to her in the early night might very well happen to him before the morning. And he so young! and with such immense possibilities of disorderly amusement before him! He felt quite pathetic over the notion of his own fate, as if it had been some one else's, and made a little imaginative vignette of the scene in the morning when they should find his body.

He passed all his chances under review, turning the white between his thumb and forefinger. Unfortunately he was on bad terms with some old friends who would once have taken pity on him in such a plight. He had lampooned them in verses, he had beaten and cheated them; and yet now, when he was in so close a pinch, he thought there was at least one who might perhaps relent. It was a chance. It was worth trying at least, and he would go and see.

On the way, two little accidents happened to him which coloured his musings in a very different manner. For, first, he fell in with the track of a patrol, and walked in it for some hundred yards, although it lay out of his direction. And this spirited him up; at least he had confused his trail; for he was still possessed with the idea of people tracking him all about Paris over the snow, and collaring him next morning before he was awake. The other matter affected him very differently. He passed a street corner, where, not so long before, a woman and her child had been devoured by wolves. This was just the kind of weather, he reflected, when wolves might take it into their heads to enter Paris again; and a lone man in these deserted streets would run the chance of something worse than a mere scare. He stopped and looked upon the place with an unpleasant interest—it was a centre where several lanes intersected each other; and he looked down them all one after another, and held his breath to listen, lest he should detect some galloping black things on the snow or hear the sound of howling between him and the river. He remembered his mother telling him the story and pointing out the spot, while he was yet a child. His mother! If he only knew where she lived, he might make sure at least of shelter. He determined he would enquire upon the morrow; nay, he would go and see her too, poor old girl! So thinking, he arrived at his destination—his last hope for the night.

The house was quite dark, like its neighbours; and yet after a few taps, he heard a movement overhead, a door opening, and a cautious voice asking who was there. The poet named himself in a loud whisper, and waited not without some trepidation, the result. Nor had he to wait long. A window was suddenly opened, and a pailful of slops splashed down upon the doorstep. Villon had not been unprepared for something of the sort, and had put himself as much in shelter as the nature of the porch admitted; but for all that, he was deplorably drenched below the waist. His hose began to freeze almost at once. Death from cold and exposure stared him in the face; he remembered he was of phthisical tendency, and began coughing tentatively. But the gravity of the danger steadied his nerves. He stopped a few hundred yards from the door where he had been so rudely used, and reflected with his finger to his nose. He could only see one way of getting a lodging, and that was to take it. He had noticed a house not far away, which looked as if it might be easily broken into, and thither he betook himself promptly, entertaining himself on the way with the idea of a room still hot, with a table still loaded with the remains of supper, where he might pass the rest of the black hours, and whence he should issue, on the morrow, with an armful of valuable plate. He even considered on what viands and what wines he should prefer; and as he was calling the roll of his favourite dainties, roast fish presented itself to his mind with an odd mixture of amusement and horror.

'I shall never finish that ballade,' he thought to himself; and then, with another shudder at the recollection, 'Oh, damn his fat head!' he repeated fervently, and spat upon the snow.

The house in question looked dark at first sight; but as Villon made a preliminary inspection in search of the handiest point of attack, a little twinkle of light caught his eye from behind a curtained window.

'The devil!' he thought. 'People awake! Some student or some saint, confound the crew! Can't they get drunk and lie in bed snoring like their neighbours? What's the good of curfew, and poor devils of bell-ringers jumping at a rope's end in bell-towers? What's the use of day, if people sit up all night? The gripes to them!' He grinned as he saw where his logic was leading him. 'Every man to his business, after all,' added he, 'and if they're awake, by the Lord, I may come by a supper honestly for this once, and cheat the devil.'

He went boldly to the door and knocked with an assured hand. On both previous occasions, he had knocked timidly and with some dread of attracting notice; but now when he had just discarded the thought of a burglarious entry, knocking at a door seemed a mighty simple and innocent proceeding. The sound of his blows echoed through the house with thin, phantasmal reverberations, as though it were quite empty; but these had scarcely died away before a measured tread drew near, a couple of bolts were withdrawn, and one wing was opened broadly, as though no guile or fear of guile were known to those within. A tall figure of a man, muscular and spare, but a little bent, confronted Villon. The head was massive in bulk, but finely sculptured; the nose blunt at the bottom, but refining upward to where it joined a pair of strong and honest eyebrows; the mouth and eyes surrounded with delicate markings, and the whole face based upon a thick white beard, boldly and squarely trimmed. Seen as it was by the light of a flickering hand-lamp, it looked perhaps nobler than it had a right to do; but it was a fine face, honourable rather than intelligent, strong, simple, and righteous.

'You knock late, sir,' said the old man, in resonant, courteous tones.

Villon cringed, and brought up many servile words of apology; at a crisis of this sort, the beggar was uppermost in him, and the man of genius hid his head with confusion.

'You are cold,' repeated the old man, 'and hungry? Well, step in.' And he ordered him into the house with a noble enough gesture.

'Some great seigneur,' thought Villon, as his host, setting down the lamp on the flagged pavement of the entry, shot the bolts once more into their places.

'You will pardon me if I go in front,' he said, when this was done; and he preceded the poet upstairs into a large apartment, warmed with a pan of charcoal and lit by a great lamp hanging from the roof. It was very bare of furniture: only some gold plate on a sideboard; some folios; and a stand of armour between the windows. Some smart tapestry hung upon the walls, representing the crucifixion of our Lord in one piece, and in another a scene of shepherds and shepherdesses by a running stream. Over the chimney was a shield of arms.

'Will you seat yourself,' said the old man, 'and forgive me if I leave you? I am alone in my house tonight, and if you are to eat I must forage for you myself.'

No sooner was his host gone than Villon leaped from the chair on which he had just seated himself, and began examining the room, with the stealth and passion of a cat. He weighed the gold flagons in his hand, opened all the folios, and investigated the arms upon the shield, and the stuff with which the seats were lined. He raised the window curtains, and saw that the windows were set with rich stained glass in figures, so far as he could see, of martial import. Then he stood in the middle of the room, drew a long breath, and retaining it with puffed cheeks, looked round and round him, turning on his heels, as if to impress every feature of the apartment on his memory.

'Seven pieces of plate,' he said. 'If there had been ten, I would have risked it. A fine house, and a fine old master, so help me all the saints!'

And just then, hearing the old man's tread returning along the corridor, he stole back to his chair, and began humbly toasting his wet legs before the charcoal pan.

His entertainer had a plate of meat in one hand and a jug of wine in the other. He set down the plate upon the table, motioning Villon to draw in his chair, and, going to the sideboard, brought back two goblets, which he filled.

'I drink to your better fortune,' he said, gravely touching Villon's cup with his own.

'To our better acquaintance,' said the poet, growing bold. A mere man of the people would have been awed by the courtesy of the old seigneur, but Villon was hardened in that matter; he had made mirth for great lords before now, and found them as black rascals as himself. And so he devoted himself to the viands with a ravenous gusto, while the old man, leaning backward, watched him with steady, curious eyes.

'You have blood on your shoulder, my man,' he said.

Montigny must have laid his wet right hand upon him as he left the house. He cursed Montigny in his heart.

'It was none of my shedding,' he stammered.

'I had not supposed so,' returned his host quietly. 'A brawl?'

'Well, something of that sort,' Villon admitted, with a quaver.

'Perhaps a fellow murdered?'

'Oh no, not murdered,' said the poet, more and more confused. 'It was all fair play—murdered by accident. I had no hand in it, God strike me dead!' he added fervently.

'One rogue the fewer, I dare say,' observed the master of the house.

'You may dare to say that,' agreed Villon, infinitely relieved. 'As big a rogue as there is between here and Jerusalem. He turned up his toes like a lamb. But it was a nasty thing to look at. I dare say you've seen dead men in your time, my lord?' he added, glancing at the armour.

'Many,' said the old man. 'I have followed the wars, as you imagine.'

Villon laid down his knife and fork, which he had just taken up again. 'Were any of them bald?' he asked.

'Oh yes, and with hair as white as mine.'

'I don't think I should mind the white so much,' said Villon. 'His was red.' And he had a return of his shuddering and tendency to laughter, which he drowned with a great draught of wine. 'I'm a little put out when I think of it,' he went on. 'I knew him—damn him! And then the cold gives a man fancies—or the fancies give a man cold, I don't know which.'

'Have you any money?' asked the old man.

'I have one white,' returned the poet, laughing. 'I got it out of a dead jade's stocking in a porch. She was as dead as Cæsar, poor wench, and as cold as a church, with bits of ribbon sticking in her hair. This is a hard world in winter for wolves and wenches and poor rogues like me.'

'I,' said the old man, 'am Enguerrand de la Feuillée, seigneur de Brisetout, bailly du Patatrac. Who and what may you be?'

Villon rose and made a suitable reverence. 'I am called Francis Villon,' he said, 'a poor Master of Arts of this university. I know some Latin, and a deal of vice. I can make chansons, ballades, lais, virelais, and roundels, and I am very fond of wine. I was born in a garret, and I shall not improbably die upon the gallows. I may add, my lord, that from this night forward I am your lordship's very obsequious servant to command.'

'No servant of mine,' said the knight; 'my guest for this evening, and no more.'

'A very grateful guest,' said Villon politely; and he drank in dumb show to his entertainer.

'You are shrewd,' began the old man, tapping his forehead, 'very shrewd; you have learning; you are a clerk; and yet you take a small piece of money off a dead woman in the street. Is it not a kind of theft?'

'It is a kind of theft much practised in the wars, my lord.'

'The wars are the field of honour,' returned the old man proudly. 'There a man plays his life upon the cast; he fights in the name of his lord the king, his Lord God, and all their lordships the holy saints and angels.'

'Put it,' said Villon, 'that I were really a thief, should I not play my life also, and against heavier odds?'

'For gain, but not for honour.'

'Gain?' repeated Villon, with a shrug. 'Gain! The poor fellow wants supper, and takes it. So does the soldier in a campaign. Why, what are all these requisitions we hear so much about? If they are not gain to those who take them, they are loss enough to the others. The men-at-arms drink by a good fire, while the burgher bites his nails to buy them wine and wood. I have seen a good many ploughmen swinging on trees about the country, ay, I have seen thirty on one elm, and a very poor figure they made; and when I asked someone how all these came to be hanged, I was told it was because they could not scrape together enough crowns to satisfy the men-at-arms.'

'These things are a necessity of war, which the low-born must endure with constancy. It is true that some captains drive over hard; there are spirits in every rank not easily moved by pity; and indeed many follow arms who are no better than brigands.'

'You see,' said the poet, 'you cannot separate the soldier from the brigand; and what is a thief but an isolated brigand with circumspect manners? I steal a couple of mutton chops, without so much as disturbing people's sleep; the farmer grumbles a bit, but sups none the less wholesomely on what remains. You come up blowing gloriously on a trumpet, take away the whole sheep, and beat the farmer pitifully into the bargain. I have no trumpet; I am only Tom, Dick, or Harry; I am a rogue and a dog, and hanging's too good for me—with all my heart; but just you ask the farmer which of us he prefers, just find out which of us he lies awake to curse on cold nights.'

'Look at us two,' said his lordship. 'I am old, strong, and honoured. If I were turned from my house tomorrow, hundreds would be proud to shelter me. Poor people would go out and pass the night in the streets with their children, if I merely hinted that I wished to be alone. And I find you up, wandering homeless, and picking farthings off dead women by the wayside! I fear no man and nothing; I have seen you tremble and lose countenance at a word. I wait God's summons contentedly in my own house, or, if it please the king to call me out again, upon the field of battle. You look for the gallows; a rough, swift death, without hope or honour. Is there no difference between these two?'

'As far as to the moon,' Villon acquiesced. 'But if I had been born lord of Brisetout, and you had been the poor scholar Francis, would the difference have been any the less? Should not I have been warming my knees at this charcoal pan, and would not you have been groping for farthings in the snow? Should not I have been the soldier, and you the thief?'

'A thief!' cried the old man. 'I a thief! If you understood your words, you would repent them.'

Villon turned out his hands with a gesture of inimitable impudence. 'If your lordship had done me the honour to follow my argument!' he said.

'I do you too much honour in submitting to your presence,' said the knight. 'Learn to curb your tongue when you speak with old and honourable men, or some one hastier than I may reprove you in a sharper fashion.' And he rose and paced the lower end of the apartment, struggling with anger and antipathy. Villon surreptitiously refilled his cup, and settled himself more comfortably in the chair, crossing his knees and leaning his head upon one hand and the elbow against the back of the chair. He was now replete and warm; and he was in nowise frightened for his host, having gauged him as justly as was possible between two such different characters. The night was far spent, and in a very comfortable fashion after all; and he felt morally certain of a safe departure on the morrow.

'Tell me one thing,' said the old man, pausing in his walk. 'Are you really a thief?'

'I claim the sacred rights of hospitality,' returned the poet. 'My lord, I am.'

'You are very young,' the knight continued.

'I should never have been so old,' replied Villon, showing his fingers, 'if I had not helped myself with these ten talents. They have been my nursing mothers and my nursing fathers.'

'You may still repent and change.'

'I repent daily,' said the poet. 'There are few people more given to repentance than poor Francis. As for change, let somebody change my circumstances. A man must continue to eat, if it were only that he may continue to repent.'

'The change must begin in the heart,' returned the old man solemnly.

'My dear lord,' answered Villon, 'do you really fancy that I steal for pleasure? I hate stealing, like any other piece of work or of danger. My teeth chatter when I see a gallows. But I must eat, I must drink, I must mix in society of some sort. What the devil! Man is not a solitary animal— *Cui Deus fœminam tradit*. Make me king's pantler—make me abbot of St Denis; make me bailly of the Patatrac; and then I shall be changed indeed. But as long as you leave me the poor scholar Francis Villon, without a farthing, why, of course, I remain the same.'

'The grace of God is all-powerful.'

'I should be a heretic to question it,' said Francis. 'It has made you lord of Brisetout and bailly of the Patatrac; it has given me nothing but the quick wits under my hat and these ten toes upon my hands. May I help myself to wine? I thank you respectfully. By God's grace, you have a very superior vintage.'

The lord of Brisetout walked to and fro with his hands behind his back. Perhaps he was not yet quite settled in his mind about the parallel between thieves and soldiers; perhaps Villon had interested him by some cross-thread of sympathy; perhaps his wits were simply muddled by so much unfamiliar reasoning; but whatever the cause, he somehow yearned to convert the young man to a better way of thinking, and could not make up his mind to drive him forth again into the street.

'There is something more than I can understand in this,' he said at length. 'Your mouth is full of subtleties, and the devil has led you very far astray; but the devil is only a very weak spirit before God's truth, and all his subtleties vanish at a word of true honour, like darkness at morning. Listen to me once more. I learned long ago that a gentleman should live chivalrously and lovingly to God, and the king, and his lady; and though I have seen many strange things done, I have still striven to command my ways upon that rule. It is not only written in all noble histories, but in every man's heart, if he will take care to read. You speak of food and wine, and I know very well that hunger is a difficult trial to endure; but you do not speak of other wants; you say nothing of honour, of faith to God and other men, of courtesy, of love without reproach. It may be that I am not very wise—and yet I think I am—but you seem to me like one who has lost his way and made a great error in life. You are attending to the little wants, and you have totally forgotten the great and only real ones, like a man who should be doctoring a toothache on the Judgement Day. For such things as honour and love and faith are not only nobler than food and drink, but indeed I think that we desire them more, and suffer more sharply for their absence. I speak to you as I think you will most easily understand me. Are you not, while careful to fill your belly, disregarding another appetite in your heart, which spoils the pleasure of your life and keeps you continually wretched?'

Villon was sensibly nettled under all this sermonizing. 'You think I have no sense of honour!' he cried. 'I'm poor enough, God knows! It's hard to see rich people with their gloves, and you blowing in your hands. An empty belly is a bitter thing, although you speak so lightly of it. If you had had as many as I, perhaps you would change your tune. Anyway I'm a thief—make the most of that—but I'm not a devil from hell, God strike me dead. I would have you to know I've an honour of my own, as good as yours, though I don't prate about it all day long, as if it was a God's miracle to have any. It seems quite natural to me; I keep it in its box till it's wanted. Why now, look you here, how long have I been in this room with you? Did you not tell me you were alone in the house? Look at your gold plate! You're strong, if you like, but you're old and unarmed, and I have my knife. What did I want but a jerk of the elbow and here would have been you with the cold steel in your bowels, and there would have been me, linking in the streets, with an armful of gold cups! Did you suppose I hadn't wit enough to see that? And I scorned the action. There are your damned goblets, as safe as in a church; there are you, with your heart ticking as good as new; and here am I, ready to go out again as poor as I came in, with my one white that you threw in my teeth! And you think I have no sense of honour—God strike me dead!'

The old man stretched out his right arm. 'I will tell you what you are,' he said. 'You are a rogue, my man, an impudent and a black-hearted rogue and vagabond. I have passed an hour with you. Oh! believe me, I feel myself disgraced! And you have eaten and drunk at my table. But now I am sick at your presence; the day has come, and the night-bird should be off to his roost. Will you go before, or after?'

'Which you please,' returned the poet, rising. 'I believe you to be strictly honourable.' He thoughtfully emptied his cup. 'I wish I could add you were intelligent,' he went on, knocking on his head with his knuckles. 'Age, age! the brains stiff and rheumatic.'

The old man preceded him from a point of self-respect; Villon followed, whistling, with his thumbs in his girdle.

'God pity you,' said the lord of Brisetout at the door.

'Goodbye, papa,' returned Villon, with a yawn. 'Many thanks for the cold mutton.'

The door closed behind him. The dawn was breaking over the white roofs. A chill, uncomfortable morning ushered in the day. Villon stood and heartily stretched himself in the middle of the road.

'A very dull old gentleman,' he thought. 'I wonder what his goblets may be worth.'

MARJORIE BOWEN

Twilight

అం

Three women stood before a marble-margined pool in the grounds of the
Ducal palace at Ferrara; behind them three cypresses waved against a
purple sky from which the sun was beginning to fade; at the base of these
trees grew laurel, ilex, and rose bushes. Round the pool was a sweep of
smooth green across which the light wind lifted and chased the red, white
and pink rose leaves.

Beyond the pool the gardens descended, terrace on terrace of opulent
trees and flowers; behind the pool the square strength of the palace rose,
with winding steps leading to balustraded balconies. Further still, beyond
palace and garden, hung vineyard and cornfield in the last warm haze of
heat.

All was spacious, noble, silent; ambrosial scents rose from the heated
earth—the scent of pine, lily, rose, and grape.

The centre woman of the three who stood by the pool was the Spanish
Duchess, Lucrezia, daughter of the Borgia Pope. The other two held her
up under the arms, for her limbs were weak beneath her.

The pool was spread with the thick-veined leaves of water-lilies and
upright plants with succulent stalks broke the surface of the water. In
between the sky was reflected placidly, and the Duchess looked down at
the counterfeit of her face as clearly given as if in a hand-mirror.

It was no longer a young face; beauty was painted on it skilfully; false
red, false white, bleached hair cunningly dyed, faded eyes darkened on
brow and lash, lips glistening with red ointment, the lost loveliness of
throat and shoulders concealed under a lace of gold and pearls, made her
look like a portrait of a fair woman, painted crudely.

And, also like one composed for her picture, her face was expressionless save for a certain air of gentleness, which seemed as false as everything else about her—false and exquisite, inscrutable and alluring—alluring
still with a certain sickly and tainted charm, slightly revolting as were the
perfumes of her unguents when compared to the pure scents of trees and
flowers. Her women had painted faces, too, but they were plainly gowned,

one in violet, one in crimson, while the Duchess blazed in every device of splendour.

Her dress, of citron-coloured velvet, trailed about her in huge folds, her bodice and her enormous sleeves sparkled with tight-sewn jewels; her hair was twisted into plaits and curls and ringlets; in her ears were pearls so large that they touched her shoulders.

She trembled in her splendour and her knees bent; the two women stood silent, holding her up—they were little more than slaves.

She continued to gaze at the reflection of herself; in the water she was fair enough. Presently she moistened her painted lips with a quick movement of her tongue.

'Will you go in, Madonna?' asked one of the women.

The Duchess shook her head; the pearls tinkled among the dyed curls.

'Leave me here,' she said.

She drew herself from their support and sank heavily and wearily on the marble rim of the pool.

'Bring me my cloak.'

They fetched it from a seat among the laurels; it was white velvet, unwieldy with silver and crimson embroidery.

Lucrezia drew it round her shoulders with a little shudder.

'Leave me here,' she repeated.

They moved obediently across the soft grass and disappeared up the laurel-shaded steps that led to the terraces before the high-built palace.

The Duchess lifted her stiff fingers, that were rendered almost useless by the load of gems on them, to her breast.

Trails of pink vapour, mere wraiths of clouds began to float about the west; the long Italian twilight had fallen.

A young man parted the bushes and stepped on to the grass; he carried a lute slung by a red ribbon across his violet jacket; he moved delicately, as if reverent of the great beauty of the hour.

Lucrezia turned her head and watched him with weary eyes.

He came lightly nearer, not seeing her. A flock of homing doves passed over his head; he swung on his heel to look at them and the reluctantly departing sunshine was golden on his upturned face.

Lucrezia still watched him, intently, narrowly; he came nearer again, saw her, and paused in confusion, pulling off his black velvet cap.

'Come here,' she said in a chill, hoarse voice.

He obeyed with an exquisite swiftness and fell on one knee before her; his dropped hand touched the ground a pace beyond the furthest-flung edge of her gown.

'Who are you?' she asked.

'Ormfredo Orsini, one of the Duke's gentlemen, Madonna,' he answered.

He looked at her frankly surprised to see her alone in the garden at the turn of the day. He was used to see her surrounded by her poets, her courtiers, her women; she was the goddess of a cultured court and persistently worshipped.

'One of the Orsini,' she said. 'Get up from your knees.'

He thought she was thinking of her degraded lineage, of the bad, bad blood in her veins. As he rose he considered these things for the first time. She had lived decorously at Ferrara for twenty-one years, nearly the whole of his lifetime; but he had heard tales, though he had never dwelt on them.

'You look as if you were afraid of me——'

'Afraid of you—I, Madonna?'

'Sit down,' she said.

He seated himself on the marble rim and stared at her; his fresh face wore a puzzled expression.

'What do you want of me, Madonna?' he asked.

'Ahè!' she cried. 'How very young you are, Orsini!'

Her eyes flickered over him impatiently, greedily; the twilight was beginning to fall over her, a merciful veil; but he saw her for the first time as an old woman. Slightly he drew back, and his lute touched the marble rim as he moved, and the strings jangled.

'When I was your age,' she said, 'I had been betrothed to one man and married to another, and soon I was wedded to a third. I have forgotten all of them.'

'You have been so long our lady here,' he answered. 'You may well have forgotten the world, Madonna, beyond Ferrara.'

'You are a Roman?'

'Yes, Madonna.'

She put out her right hand and clasped his arm.

'Oh, for an hour of Rome!—in the old days!'

Her whole face, with its artificial beauty and undisguisable look of age, was close to his; he felt the sense of her as the sense of something evil.

She was no longer the honoured Duchess of Ferrara, but Lucrezia, the Borgia's lure, Cesare's sister, Alessandro's daughter, the heroine of a thousand orgies, the inspiration of a hundred crimes.

The force with which this feeling came over him made him shiver; he shrank beneath her hand.

'Have you heard things of me?' she asked in a piercing voice.

'There is no one in Italy who has not heard of you, Madonna.'

'That is no answer, Orsini. And I do not want your barren flatteries.'

'You are the Duke's wife,' he said, 'and I am the servant of the Duke.'

'Does that mean that you must lie to me?'

She leant even nearer to him; her whitened chin, circled by the stiff goldwork of her collar, touched his shoulder.

'Tell me I am beautiful,' she said. 'I must hear that once more—from young lips.'

'You are beautiful, Madonna.'

She moved back and her eyes flared.

'Did I not say I would not have your flatteries?'

'What, then, was your meaning?'

'Ten years ago you would not have asked; no man would have asked. I am old. Lucrezia old!—ah, Gods above!'

'You are beautiful,' he repeated. 'But how should I dare to touch you with my mouth?'

'You would have dared, if you had thought me desirable,' she answered hoarsely. 'You cannot guess how beautiful I was—before you were born, Orsini.'

He felt a sudden pity for her; the glamour of her fame clung round her and gilded her. Was not this a woman who had been the fairest in Italy seated beside him?

He raised her hand and kissed the palm, the only part that was not hidden with jewels.

'You are sorry for me,' she said.

Orsini started at her quick reading of his thoughts.

'I am the last of my family,' she added. 'And sick. Did you know that I was sick, Orsini?'

'Nay, Madonna.'

'For weeks I have been sick. And wearying for Rome.'

'Rome,' he ventured, 'is different now, Madonna.'

'Ahè!' she wailed. 'And I am different also.'

Her hand lay on his knee; he looked at it and wondered if the things he had heard of her were true. She had been the beloved child of her father, the old Pope, rotten with bitter wickedness; she had been the friend of her brother, the dreadful Cesare—her other brother, Francesco, and her second husband—was it not supposed that she knew how both had died?

But for twenty-one years she had lived in Ferrara, patroness of poet and painter, companion of such as the courteous gentle Venetian, Pietro Bembo.

And Alfonso d'Este, her husband, had found no fault with her; as far as the world could see, there had been no fault to find.

Ormfredo Orsini stared at the hand sparkling on his knee and wondered.

'Suppose that I was to make you my father confessor?' she said. The white mantle had fallen apart and the bosom of her gown glittered, even in the twilight.

'What sins have you to confess, Madonna?' he questioned.

She peered at him sideways.

'A Pope's daughter should not be afraid of the Judgement of God,' she answered. 'And I am not. I shall relate my sins at the bar of Heaven and say I have repented—Ahè—if I was young again!'

'Your Highness has enjoyed the world,' said Orsini.

'Yea, the sun,' she replied, 'but not the twilight.'

'The twilight?'

'It has been twilight now for many years,' she said, 'ever since I came to Ferrara.'

The moon was rising behind the cypress trees, a slip of glowing light. Lucrezia took her chin in her hand and stared before her; a soft breeze stirred the tall reeds in the pool behind her and gently ruffled the surface of the water.

The breath of the night-smelling flowers pierced the slumbrous air; the palace showed a faint shape, a marvellous tint; remote it looked and uncertain in outline.

Lucrezia was motionless; her garments were dim, yet glittering, her face a blur; she seemed the ruin of beauty and graciousness, a fair thing dropped suddenly into decay.

Orsini rose and stepped away from her; the perfume of her unguents offended him. He found something horrible in the memory of former allurement that clung to her; ghosts seemed to crowd round her and pluck at her, like fierce birds at carrion.

He caught the glitter of her eyes through the dusk; she was surely evil, bad to the inmost core of her heart; her stale beauty reeked of dead abomination. . . . Why had he never noticed it before?

The ready wit of his rank and blood failed him; he turned away towards the cypress trees.

The Duchess made no attempt to detain him; she did not move from her crouching, watchful attitude.

When he reached the belt of laurels he looked back and saw her dark shape still against the waters of the pool that were beginning to be touched with the argent glimmer of the rising moon. He hurried on, continually catching the strings of his lute against the boughs of the flowering shrubs; he tried to laugh at himself for being afraid of an old, sick woman; he tried to ridicule himself for believing that the admired Duchess, for so long a decorous great lady, could in truth be a creature of evil.

But the conviction flashed into his heart was too deep to be uprooted.

She had not spoken to him like a Duchess of Ferrara, but rather as the wanton Spaniard whose excesses had bewildered and sickened Rome.

A notable misgiving was upon him; he had heard great men praise her,

Ludovico Ariosto, Cardinal Ippolito's secretary and the noble Venetian Bembo; he had himself admired her remote and refined splendour. Yet, because of these few moments of close talk with her, because of a near gaze into her face, he felt that she was something horrible, the poisoned offshoot of a bad race.

He thought that there was death on her glistening painted lips, and that if he had kissed them he would have died, as so many of her lovers were reputed to have died.

He parted the cool leaves and blossoms and came on to the borders of a lake that lay placid under the darkling sky.

It was very lonely; bats twinkled past with a black flap of wings; the moon had burnt the heavens clear of stars; her pure light began to fill the dusk. Orsini moved softly, with no comfort in his heart.

The stillness was intense; he could hear his own footfall, the soft leather on the soft grass. He looked up and down the silence of the lake.

Then suddenly he glanced over his shoulder. Lucrezia Borgia was standing close behind him; when he turned her face looked straight into his.

He moaned with terror and stood rigid; awful it seemed to him that she should track him so stealthily and be so near to him in this silence and he never know of her presence.

'Eh, Madonna!' he said.

'Eh, Orsini,' she answered in a thin voice, and at the sound of it he stepped away, till his foot was almost in the lake.

His unwarrantable horror of her increased, as he found that the glowing twilight had confused him; for, whereas at first he had thought she was the same as when he had left her seated by the pool, royal in dress and bearing, he saw now that she was leaning on a stick, that her figure had fallen together, that her face was yellow as a church candle, and that her head was bound with plasters, from the under edge of which her eyes twinkled, small and lurid.

She wore a loose gown of scarlet brocade that hung open on her arms that showed lean and dry; the round bones at her wrist gleamed white under the tight skin, and she wore no rings.

'Madonna, you are ill,' muttered Ormfredo Orsini. He wondered how long he had been wandering in the garden.

'Very ill,' she said. 'But talk to me of Rome. You are the only Roman at the Court, Orsini.'

'Madonna, I know nothing of Rome,' he answered, 'save our palace there and sundry streets——'

She raised one hand from the stick and clutched his arm.

'Will you hear me confess?' she asked. 'All my beautiful sins that I cannot tell the priest? All we did in those days of youth before this dimness at Ferrara?'

'Confess to God,' he answered, trembling violently.

Lucrezia drew nearer.

'All the secrets Cesare taught me,' she whispered. 'Shall I make you heir to them?'

'Christ save me,' he said, 'from the Duke of Valentinois' secrets!'

'Who taught you to fear my family?' she questioned with a cunning accent. 'Will you hear how the Pope feasted with his Hebes and Ganymedes? Will you hear how we lived in the Vatican?'

Orsini tried to shake her arm off; anger rose to equal his fear.

'Weed without root or flower, fruitless uselessness!' he said hoarsely. 'Let me free of your spells!'

She loosed his arm and seemed to recede from him without movement; the plasters round her head showed ghastly white, and he saw all the wrinkles round her drooped lips and the bleached ugliness of her bare throat.

'Will you not hear of Rome?' she insisted in a wailing whisper. He fled from her, crashing through the bushes.

Swiftly and desperately he ran across the lawns and groves, up the winding steps to the terraces before the palace, beating the twilight with his outstretched hands as if it was an obstacle in his way.

Stumbling and breathless, he gained the painted corridors that were lit with a hasty blaze of wax light. Women were running to and fro, and he saw a priest carrying the Holy Eucharist cross a distant door.

One of these women he stopped.

'The Duchess——' he began, panting.

She laid her finger on her lip.

'They carried her in from the garden an hour ago; they bled and plastered her, but she died—before she could swallow the wafer—(hush! she was not thinking of holy things, Orsini!)—ten minutes ago——'

RAFAEL SABATINI

The Scapulary

ᘓᘒᘒ

The uneasiness that had been disturbing Gaspard de Putanges ever since
the King had visited the wounded Admiral de Coligny reached a climax
that night when he found his way barred by armed men at the Porte St
Denis, and a password was demanded of him.

'Password!' cried that Huguenot gentleman in amazement. 'Is a password necessary before a man can leave Paris? And why, if you please? Are
we suddenly at war?'

'Those are the orders,' the officer stiffly answered.

'Whose orders?' M. de Putanges was impatient.

'I owe you no account, sir. You will give me the word of the night, or
you may return home and wait until morning.'

Perforce he must turn his horse about, and, with his groom at his heels,
ride back by the way that he had come. The vexation which at any time
he must have felt at this unwarranted interference with his movements
was now swollen by misgivings.

He was one of the host of Huguenot gentlemen brought to Paris for the
nuptials of the King of Navarre with the sister of the King of France,
a marriage which the pacifists of both parties had hoped would heal the
feud between the Catholic and Protestant factions. But the Guisard attempt upon the life of Coligny, the great Huguenot leader, had now
rudely dashed this hope. M. de Putanges had been one of that flock of
Huguenot gentlemen who that day had thronged the wounded Admiral's
antechamber, when the epileptic king and the sleepy-eyed Queen Mother
came to pay their visit of sympathy. He had listened with misgivings to
the braggart threats of his co-religionists; with increased misgivings he
had observed the open hostility of their bearing towards Catherine de'
Medici what time she stood amongst them with Anjou, whilst the king
her son was closeted with his dear gossip, M. de Coligny. These hot-
headed fools, he felt, were fanning a fire that might presently blaze out to
consume them all. Already the Hotel de Guise was in a state of fortification,
filled with armed men ripe for any mischief, whilst others of the Guisard
faction were abroad exciting public feeling with fantastic stories of
the Huguenot peril, stories which gathered colour from the turbulent,

thrasonical bearing of these Huguenots, enraged by the attempt upon their leader's life.

And now, finding the gates of Paris barred for no apparent reason, it seemed to M. de Putanges that the danger he had been apprehending was close upon them, though in what form he could not yet discern.

It was therefore as well that he should be forced to postpone his journey into the country, however necessary, and that he should remain to watch over his wife.

Now this was a consideration in which M. de Putanges discovered a certain humour. He was a thoughtful gentleman with a lively sense of irony, and it amused him after a certain bitter fashion to observe his own mechanical obedience to his sense of duty towards the cold, arrogant, discontented lady who bore his name. You know, of course, of the beautiful Madame de Putanges and of the profound impression which her beauty and wit had made upon the Court of France upon this her first appearance there. It is even rumoured—and not at all difficult to believe—that amongst those who prostrated themselves in worship before her was the very bridegroom Henri of Navarre himself. But at least the lady was virtuous— her one saving grace in her husband's eyes—and of a mind that was not to be discomposed by the flattery of even a royal wooing. It is not without humour that the only quality M. de Putanges could find to commend in her was that same cold aloofness which so embittered him. It was a paradox upon which he had found occasion to comment to his friend and cousin Stanislas de la Vauvraye.

'The gods who cast her in a mould so fair have given her for heart a stone.'

That was the formula in which habitually he expressed it to Stanislas, conscious that it sounded like a line from a play.

'I curse her for the very quality that makes my honour safe; because this quality that in another woman would be a virtue, is almost a vice in her.'

And Stanislas, the gay trifler who turned all things to cynical jest, had merely laughed. 'Be content, Gaspard, with a blessing denied most husbands.'

Yet however little M. de Putanges might count himself blessed as a husband, he was fully conscious of a husband's duty, and his first thought now that he suspected trouble was for his wife.

He made his way to the *Veau qui Tète*, where his horses were stabled and his grooms were housed, and, having dismounted there, set out for the small house he temporarily occupied close by, in the Rue Bellerose. The summer night had closed down by then, but there was a fair moon that rendered the use of flambeaux unnecessary. As he stepped out into the street he came upon a man bent double under a load of pikes. It was an odd sight and M. de Putanges stood arrested by it, watching the fellow as he staggered down the narrow street until he was absorbed by the shadows of the night. Yet even as he vanished a second man similarly laden came stumbling past. M. de Putanges fell into step beside the fellow.

'Whither are you carrying that arsenal?' he demanded.

The perspiring hind looked up from under his sinister load, to answer this brisk authoritative questioner.

'It is for the entertainment at the Louvre, Monsieur.'

As a gentleman in the train of the King of Navarre, M. de Putanges was bidden to all court functions. Yet here was one of which he had not so much as heard, and well might he ask himself what entertainment was this that was being kept so secret and in which pikes were to be employed. It was difficult to suppose that their purpose could be festive.

With ever-mounting uneasiness M. de Putanges lengthened his stride for home. But at the corner, where the Rue Bellerose cuts across the more important Rue St Antoine, he ran into a group of men on the threshold of an imposing house that was all in darkness. At a glance he perceived that all were armed beyond the habit of peaceful citizens. Headpiece and corselet glinted lividly in the moonlight, and as he approached he caught from one on a note of sinister laughter the word '*Parpaillots*'—the nickname bestowed on members of the Huguenot party.

That was enough for M. de Putanges. Boldly—he was a man who never lacked for boldness—he mingled with the little throng. Others were joining it at every moment and already some were passing into the house, so that his coming was hardly observed; it would be assumed that he was one of themselves, bidden like them to this assembly. Perceiving this, and overhearing from one of those beside him the boast that by morning there would not be a whole heretic skin in Paris, M. de Putanges determined to push on and obtain more complete knowledge of what might be preparing. Heedless, then, of risks, he thrust forward to the threshold and attaching himself to a little knot of gentlemen in the act of entering, he went in with them, and up a broad, scantily-lighted staircase. At the stairhead the foremost of the company knocked upon a double door. One of its leaves was half opened and there ensued between someone within and each of those who sought admittance a preliminary exchange of confused murmurs. M. de Putanges pushed nearer, straining his ears, and at last from one whose mutter was louder than the others he caught the words: 'France and the Faith.' A moment later he was giving the same countersign.

He won through into a spacious gallery that was tolerably lighted by four great girandoles, and already thronged by men from every walk of life. All were armed and all were excited. From what he observed, from what was said to him even—for there was none here to recognize this gentleman of Béarn, or to suppose him other than one of themselves—he quickly came to understand that the thing preparing was no less than a massacre of the Huguenots in Paris.

Anon, when the room was filled almost to the extent of its capacity and the doors were closed, a lean, fiery-eyed preaching friar, in the black and white habit of St Dominic, mounted a table and delivered thence at length what might be called a 'sermon of the faith', a fierce denunciation of heresy.

'Hack it down, branch by branch, tear it up by the roots; extirpate from the land this pestilential growth, this upas tree that poisons the very air we breathe. About it, my children! Be stern and diligent and unsparing in this holy work!'

Those terrible final words of incitement were ringing in his ears when M. de Putanges quitted at last that chamber, with its physically and morally mephitic atmosphere, and was borne out upon a brawling, seething human torrent, into the clean air of early dawn. The human mass broke into packs which turned away in one direction and another to the cry—fierce and menacing as the baying of hounds upon a scent—of 'Parpaillots! Parpaillots! Kill! Kill!'

He won free of them at length, forearmed at least by knowledge of what to expect—by knowledge and something more. In that chamber, whilst the sermon had been preaching, someone had tied a strip of white calico to his left arm and set in his hat a cross made of two short pieces of white ribbon. These were the insignia of the slayers and in themselves would have afforded him immunity in the open streets, but that upon an impulse of unreasoning disgust he tore one and the other from him and flung them in the kennel. This as he plunged at last down the Rue Bellerose towards home.

In a measure as the sounds of his late companions receded from him, ahead of him grew an ominous rumble, coming from the quays and the neighbourhood of the river; and then, a crackling volley of musketry rang out abruptly from the direction of the Louvre.

M. de Putanges stood still and wondered a moment whether the faint flush in the sky was a herald of the early summer dawn or the reflection of fire. He became suddenly aware that the Rue Bellerose was astir with flitting shadows. He came upon a man setting a mark in chalk upon a door, and, peering, beheld another similarly engaged on the opposite side of the street. He understood the meaning of it, and doubting if he would be alive by morning, he considered almost dispassionately—for he was a dispassionate man despite his southern blood—that it but remained him to seek his wife and await his fate beside her. If they had never known how to conduct their joint life becomingly, at least he hoped they would know how to die becomingly together.

In this frame of mind he reached his own threshold to find one of the doorposts bearing the chalk sign that marked the inmates down for slaughter. How well informed were these cursed *papegots*, he thought; how well considered and well organized was their bloody work! With his sleeve he rubbed out the sign of doom. So much, at least, he could do in self-defence. And then he stood arrested, with pulses faintly quickening. Within the shadows of the deep porch something stirred, and at the same moment upon the deep-set inner door fell a triple knock three times repeated.

M. de Putanges stepped forward. Instantly the thing within the porch swung about and sprang to meet him, taking definite shape. It was a tall cowled figure in the white habit and black scapulary of a brother of St Dominic. M. de Putanges realized that he came no more than in time, if, indeed, he did not already come too late. He felt his heart tightening at the thought, tightening with icy dread for the cold termagant he loved. And then, to increase his fears, the friar pronounced his name and so proclaimed that he was not there by chance nor attracted by the mark of doom upon the doorpost, but with sure knowledge of the house's inmates.

'Monsieur de Putanges . . .' the man exclaimed, and got no further, for utterance and breath were abruptly choked by the iron fingers that locked themselves about his throat.

The fellow writhed and struggled, thrusting out a leg to trip his aggressor, clawing fiercely at the hands that were crushing out his life, and tearing them with his nails. But tear as he might, those hands would not relax their deadly grip. Soon his struggles weakened; soon they became mere twitchings. His body sagged together like an empty sack. He went down in a heap, dragging his assailant with him, and lay still at last.

M. de Putanges stood over the fallen friar, breathing hard; the beads of sweat upon his brow resulted partly from the exertion, partly from horror of a thing done in such a cold, relentless fashion. Mastering himself at last, he went forward towards the door. To his surprise he found it open, and remembered then that the friar had knocked upon it a moment ago. The latch was worked by a cord from the floor above and must have been so worked in answer to that knock.

M. de Putanges paused in the act of entering. It was not well—particularly considering that his house had been marked—to leave that thing in the porch where it might be discovered. He went back and, taking the limp body by the arms, he dragged it across the threshold into the hall. And then the instinct of self-preservation that had guided him so far urged yet another step. In that *papegot* livery it was possible that he might find safety for himself and his wife from the perils that were so obviously closing round them. The faintly reflected radiance of the moon afforded sufficient light for the simple task. In a moment he had unknotted the man's girdle and relieved him of habit and scapulary. Over his own clothes he donned the Dominican's loose white habit, and drew the cowl over his head. Then he closed the door, and set foot upon the narrow stairs, even as from the street outside a sudden shrieking of women was silenced by a fusillade. The massacre, he perceived, was in full swing already.

Overhead a light gleamed faintly; looking up from the darkness that encompassed him, he beheld in the feeble aureole of a candle, which she was holding, the face of Madame de Putanges. She was peering down, seeking with her glance to pierce the darkness of the staircase. He must speak at once, lest the sight of his monkish figure should alarm her. But even as he conceived the thought her silvery voice, strained now on an anxious note, forestalled him.

'How late you are, Stanislas!' she said.

In the gloom of the staircase, Gaspard de Putanges stood petrified, whilst through his mind that welcome meant for another echoed and re-echoed:

'How late you are, *Stanislas!*'

After a moment our gentleman's wits resumed their wonted function, sharpened now perhaps beyond their usual keenness. He scarcely needed to ask himself who could be this Stanislas for whom she mistook him. It could be none other than his dear friend and cousin, that gay trifler and libertine Stanislas de la Vauvraye. And she was expecting him at a time when she must suppose M. de Putanges already far from Paris. Was it possible, he asked himself, that he had thus by chance stumbled upon the explanation of her cold aloofness towards himself? Was Stanislas the Judas who simulated friendship in order that he might the more conveniently betray? In a flash a score of trifling incidents were suddenly remembered and connected to flood the mind of M. de Putanges with the light of revelation.

'Stanislas! Why don't you answer?' rang the impatient voice he knew so well.

Whence this absolute assurance of hers that he was the man she expected? He bethought him of that curious triple knock thrice repeated with which the friar had sought admittance, and in response to which the door-latch had been so promptly lifted from above. It was a signal, of course. But, then, the friar . . .

On a sudden suspicion amounting almost to certainty, M. de Putanges stepped back.

'Stanislas!' came again Madame de Putanges' call.

'A moment,' he answered to quiet her. 'I am coming.'

He ran his fingers swiftly over the face and head of the man he had choked. Here was no tonsure; no shaven cheeks. The hair of the head was full and crisp; moustachios bristled on the lip, and a little peaked beard sprouted from the chin: there was a jewel in the left ear which, like the rest, was as it should be with Stanislas de la Vauvraye. The garments were soft and silken—a slashed doublet and the rest. No doubt remained. All that remained was the mystery of how Stanislas should have come to be muffled in that monkish robe. That, however, could wait. It mattered little in comparison with all the rest. In his cursory examination of the body, M. de Putanges had ascertained that the man still breathed. He had not quite choked out his life. It was perhaps as well.

He locked the street-door and pocketed the key; then he went up the stairs with a step that was as firm and steady as his purpose.

When at last his cowled monkish figure came within the circle of light of the waiting woman's candle, she started back.

'Why, what is this? Who are you, sir?' Then remembering the covenanted knock to which she had opened, she partly explained the monkish travesty. 'Why do you come thus, Stanislas?'

'That,' was the quiet answer, 'we may ask presently of Stanislas himself.'

She staggered at the sound of her husband's voice, and a deathly pallor overspread her face. Her dark eyes opened wide in terror: her lips parted, but instead of speech they uttered a mere inarticulate sound of fear and horror, almost piteous to hear. Than she recovered, as he flung back his cowl and smiled upon her, grim, and white-faced as herself.

'You!' she gasped.

He observed that she was dressed for travelling, cloaked and hatted, and that a valise stood beside her at the stairhead.

'I thought you were gone to Poldarnes,' she said stupidly, a mere uncontrolled utterance of her mind.

'I am sorry, madame, to discompose you by so inopportune a return. Circumstances compelled it, as I will presently explain.'

Then at last the termagant in her recovered the sway momentarily extinguished by surprise.

'It needs no explanation, sir,' she answered with angry scorn, and angry unreason. 'You returned to spy upon me.'

'O fie, madame! To accuse me of that! But how unjust, how foolishly unjust! Had you but honoured me with your confidence, had you but informed me of your intent to elope with my dear friend and kinsman, I should have left you a clear field. I am too fond of you both to attempt against you so cruel a frustration of your designs.'

'You mean that you would have been glad to be rid of me!' was her fierce reproach.

M. de Putanges laughed his bitter amusement that even in such a situation she must make her own the grievance.

'Not glad, perhaps: but fortunate, madame,' said he. 'Once the hurt and humiliation of it were overcome, I might have seen in my dear kinsman Stanislas my best of friends.'

She fastened swiftly upon this admission. 'That is why I am leaving you.'

'But, of course. Do I not perceive it? I have, madame, a more sympathetic understanding than you have ever done me the justice to suppose.'

She answered by no more than an inarticulate expression of contempt.

'I shall hope to prove it to you tonight, madame,' he insisted. 'You may come to account it as fortunate for yourself as it is unfortunate for me that I returned. If you will be so good as to step into the salon, I will fetch my dear cousin Stanislas.'

This was to arouse her alarms afresh. 'You will fetch him!' she gasped. 'Where . . . where is he?'

'He is below.'

She was suddenly suspicious, glaring, fierce as a tigress. 'What have you done to him?'

'Oh, be reassured!—no permanent injury. He is a little . . . out of breath, shall we say? But that will pass.'

'Let me go to him.'

He barred her way. 'It is not necessary. He shall come to you. In no case can you depart just now. The streets are not safe. Listen!' Vague, hideous sounds of the foul business that was afoot penetrated the silence in which they stood, and filled her with wonder and some dread. He enlightened her. 'They are murdering the Huguenots in Paris tonight. That is what brought me back, supposing you would need protection, never dreaming that you would have the dear, brave Stanislas so near at hand.'

He took advantage of her amazement to add: 'If you will forgive the impertinence of the question, how long has M. de la Vauvraye been your lover, madame?'

She flushed to her eyes. 'Why must you insult me?'

'You are susceptible then to insult?'

Her anger, goaded by his cold mockery, raced on. 'Should I have stayed a single moment under your roof after taking a lover?' she demanded passionately. 'Do you think me capable of that?'

'I see,' said M. de Putanges, and sighed. 'You possess a casuist's mind, madame. You swallow camels yet strain at gnats.'

'Maybe. But I am honest, monsieur.'

'After your own fashion, madame; strictly after your own fashion.'

'After my own fashion, if you will. You have a great gift for mockery, monsieur.'

'Be thankful, madame. The man who does not know how and when to laugh, sometimes does very foolish and very painful things. And so, you were saying that you are honest . . .'

'I think I prove it. Having taken my resolve, I am leaving your house tonight. I am going away with M. de la Vauvraye.'

She was defiant. He bowed to her will.

'Why, so you shall, for me. It remains to take measures for your safety, considering what is happening in the street. Be so good as to wait in the salon, whilst I fetch this dear Stanislas.'

He took the candle from her limp hand and went down the stairs again. He found his kinsman sitting up, in bewilderment—a bewilderment which the sight of M. de Putanges transmuted into stark fear.

But M. de Putanges was of a reassuring urbanity. 'Be good enough to step up to the salon with me, my dear cousin. Henriette desires a word with you, whereafter you may carry out or not your fond intentions. That will depend. Meanwhile, be thankful that I have returned in time, as I gather from what madame tells me. If the harm were already beyond repair I should for my honour's sake be compelled to renew, and this time to complete, the strangling of you. That would be a painful matter for us both, for I have a horror of violence, as you know, my dear Stanislas. Fortunately the affair may still be amicably decided. The circumstances of the night are particularly propitious. Be so good, then, as to come up with me.'

M. de la Vauvraye, his mind benumbed by this extraordinary turn of what should have been so simple and delightful an adventure, followed his kinsman up the stairs with an obedience of such utter helplessness as to be almost entirely mechanical.

In the salon they found Madame awaiting them. She stood by the heavy, carved table of dark oak well in the light of the dozen tapers that burned in a gilded candlebranch. Her beautiful face was pale and haggard, yet neither so pale nor so haggard as that of the lover she now confronted in circumstances so vastly different from all that she had expected of this night.

She had declared to him but yesterday that it was to be the most fateful night of her life, and in that, indeed, she appeared to have been a true prophet.

Of the three, the only one at ease—outwardly at least—was M. de Putanges. Tall, erect, virilely handsome, a man in the very flower of his age, bronzed, intrepid and aquiline of countenance, he made the golden-headed trifler la Vauvraye seem so sickly and effeminate by contrast that it was difficult to discern how any woman could have come to prefer him.

'Be seated, pray.' Suavity itself, M. de Putanges waved each in turn to a chair. 'I shall not keep you long.'

La Vauvraye, under the woman's eyes, realizing that he was in danger of cutting a poor, unheroic figure, summoned impudence to his aid. 'You relieve my fears, monsieur.' With a shrug and a half-laugh he sank to an armchair. He was not as successful as he thought, for his crumpled ruff and dishevelled hair lent him an appearance which dignity but made the more ridiculous.

'Your fears?' quoth M. de Putanges.

'Of being wearied by futile recriminations. Weariness is the thing in life that I most dread.'

'Reassure yourself. You are in no danger of it. Indeed, the night should bring a surfeit of excitement even to such a glutton for high adventure.'

La Vauvraye stifled a yawn. 'If you will but explain,' he begged on a half plaintive note.

'Of course. But first a question: How came you by this Dominican frock in which I found you?'

'Oh, that!' M. de la Vauvraye was casual. 'You will know by now what is afoot tonight. I, too, discovered it as I was coming here. At the end of the Rue Bellerose I came upon a shaveling I had met once or twice at Court. Unfortunately for him, he knew me too, and calling me by name threatened me with a heretic's doom. It was rash. I had a dagger . . . I silenced that too garrulous friar, and left him in a doorway.'

'Possessing yourself of his habit?'

'Why, yes. I realized that on such a night his scapulary would possess the virtues that are claimed for scapularies, would be a panoply against all perils. You would seem to have been of the same mind yourself when, taking me unawares, you became possessed of it and donned it in your turn.'

'And then, dear cousin? Continue, pray.'

'What more is there to tell?'

'Why, that you counted upon the frock to shield not only yourself but this Huguenot lady with whom you had planned an elopement for tonight.'

'Oh, that, of course.' The nonchalance was almost overdone.

'It was my notion, too. Since we are both agreed as to its excellence, there is no reason not to act upon it.'

The affected languor passed from M. de la Vauvraye's eyes. They became alert and keen to the point almost of anxiety. M. de Putanges proceeded to explain himself.

'Madame de Putanges must be saved,' he said quietly. 'That, you will realize, is the paramount consideration. She must be carried out of reach of this bloodshed before it is too late. Escorted by a brother of St Dominic the thing is easy. We have but to determine which of us shall assume the frock and take her hence, which shall remain behind to die nobly for her sake. Since we both love her, the decision should be easy, for neither of us should hesitate to remain. Rather do I fear a generous emulation as to which shall go.' He dealt so delicately in irony that neither of his listeners could be certain that he was ironical.

With blanched cheeks and bulging eyes, M. de la Vauvraye stared at his cousin, waiting. The handsome woman seated in the high-backed chair looked on with parted lips, her breathing quickened, and waited, too.

M. de Putanges resumed: 'But for what is taking place in Paris tonight, my dear cousin, it might be necessary to choose some other way of resolving this painful situation. As it is, our common concern for Madame de Putanges points the way clearly. One of us goes. The other stays behind to die. Ideal and complete solution. It but remains to determine our respective parts.' He paused and the silence within that room was tense and heavy. Outside was uproar—the baying of the fanatical mob, the crash of shivered timbers and the screams of luckless victims.

M. de la Vauvraye moistened his parched lips with his tongue.

'And how . . . how is that to be determined?' he asked in a quavering, high-pitched voice that cracked on the last word.

'What way but one could gallantry conceive? It is for Madame de Putanges to choose.' He bowed deferentially, as he consigned into her hands that monstrous decision.

M. de la Vauvraye expressed in a gasp his immense relief, never doubting Madame's choice, never observing the stark horror that distorted her lovely face, nor how that horror deepened at the sound he made.

But M. de Putanges had something yet to add. He was being extremely subtle, where neither of his listeners suspected subtlety.

'Thus, madame, you shall have at last the choice that should have been yours at first. It is in your power tonight to repair the injustice that was done you three years ago when, unconsulted, you were driven into a wedlock of arrangement.' He sighed. 'Perhaps had I wooed you first and wed you afterwards all might have been well with us. I realize tonight the profound mistake of reversing that natural course of things, and I am almost glad of the opportunity to correct it, by giving you now, late though it be, the choice that is a woman's right. So pronounce, madame. Determine of your own free will which of us shall assume the frock and bear you hence, which shall remain to die.'

On that he ended, looking with solemn inscrutable glance deep into the eyes that stared at him out of his wife's white, horror-stricken face. He knew—unless he knew nothing of human nature—that the woman did not live who could take upon her conscience the responsibility of such a choice: and upon that knowledge he was boldly trading.

To help him to his deep end, came the bleating voice of his foolish rival, urgent with an obvious meaning that he dared not actually express in words. 'Henriette! Henriette!'

M. de Putanges observed her shrink and cringe before that appeal, as if conscious only of the meanness and cowardice that inspired it. Upon something of that kind, too, had he counted. He intended to compel her tonight to look, as she had never yet troubled to look, into his soul. And he intended no less to lay bare the little soul of the trifler for whom she would have left him. She should have the free choice he offered her; but first there should be full revelation to guide her in her choice.

'Well, madame?' He loosened the girdle of his monkish habit, as if to stress his readiness for renunciation at her bidding. 'Which of us shall wear the frock, the scapulary of salvation—more potently protecting tonight than any consecrated prophylactic?'

And again from M. de la Vauvraye—now half risen—came the appeal —'Henriette!'

Madame de Putanges looked with the glance of a hunted creature from one to the other of those men. Then a shudder ran through her; she twisted and untwisted the fingers of her interlocked hands, and faintly moaned.

'I cannot! I cannot!' she cried out at last. 'No—no! I cannot have the blood of either of you on my soul.'

M. de la Vauvraye flung himself back in his chair, his lip in his teeth, his hands clenched, whilst M. de Putanges smiled with wistful understanding.

'Yet consider, madame,' he urged her, 'that unless you choose we are likely all three of us to perish here together.'

'I can't! I can't! I will not choose. It is monstrous to demand it of me,' she cried.

'Demand?' echoed M. de Putanges. 'Oh, madame, I do not demand; I invite. But to spare your feelings, since you prove so tender in this matter, it shall be determined otherwise; and that I must deplore, because in no other way could your future happiness be assured. Still . . .' He broke off, and flung the Dominican frock and scapulary, of which he had now divested himself, upon the table; then he turned to a tall press that stood against the wall, and took from one of its drawers a dice-box. Rattling the cubes in their leathern container, he looked across at his kinsman.

'Come, cousin. You love a hazard. Here is one in which the stakes are something rare: love and life are here for one of us.'

M. de la Vauvraye, who had risen, shrank back in fear. If he loved hazards, he loved them not quite so hazardous.

'No, no,' he cried, thrusting out his hands in a violent gesture of denial. 'This is a horror.'

'What else remains?' asked M. de Putanges. 'Come, man. I'll lead the way.'

He threw recklessly as he spoke, thereby committing the other to the adventure. The three cubes scattered and rattled to a standstill on the polished oaken board. Madame's glittering eyes followed them and remained expressionless when the throw was revealed—two aces and a deuce.

M. de Putanges laughed softly, bitterly. 'That is my luck,' said he. 'You should have known that I was not to be feared, Stanislas. You'll scarce hesitate now to throw against me.'

Yet he was pleased enough in the heart of him. Fortune could not have served him better. Now that no risk remained M. de la Vauvraye would be as eager as before he had been hesitant, and so should make yet further self-revelation. And eager he was. It was his turn to laugh as with trembling fingers he gathered up the dice; a hectic flush tinged his cheek-bones as he threw—two fives and a four.

'Mine!' he cried, and snatched up the frock in exultation.

M. de Putanges bowed with quiet dignity.

'My congratulations, cousin; and to you, madame, since surely Fortune must have determined as you wish, yet dared not pronounce out of a generous thought for me. For that thought I thank you. What now remains for me will be the easier in the knowledge that it comes not from you, but from Fortune.' He became matter-of-fact and brisk. 'You had best make haste away. There is danger in delaying.'

'Indeed, indeed—he is right,' M. de la Vauvraye now urged her, as he shook out the folds of the frock and with trembling fingers knotted the girdle about his middle.

But Madame made no answer to either of them. With eyes of wonder she looked from one to the other, contrasting the noble calm in adversity of this husband she had never known, with the meanly selfish eagerness of the shallow courtier whom she had imagined that she loved. M. de Putanges had promised himself to afford her full revelation; and he had succeeded beyond his hopes.

M. de la Vauvraye took a step towards her. 'Come, Henriette,' he was beginning, when suddenly she laughed so oddly that he checked. She looked at him with shining eyes and oddly smiling lips.

'Do not wait for me, monsieur. All that it was yours to stake upon the throw was the scapulary that will shield your life. You have won that, and so you may win to safety. But I do not go with the scapulary. Goodnight and good fortune to you, M. de la Vauvraye.'

He stared at her, dumbfounded, stricken in his scanty wits and in his monstrous vanity.

'What?' he cried. 'You will remain?'

'With my husband, if you please, monsieur.'

Quiet and self-contained in the background beyond the table M. de Putanges looked on and wondered. M. de la Vauvraye rapped out an oath. Then anger bubbled to the surface of his shallow nature. His tone became vicious.

'Goodnight, madame.'

He turned on his heel; but as he reached the door her voice arrested him.

'M. de la Vauvraye!'

He turned again.

'A man in your place would have acted perhaps more generously. A man would have remembered all that is involved. A man would account that my final decision overrides the decision of the dice. He would have proffered the scapulary to the true winner. A man would have done that. But you, it seems, are something less. Since it has been tested, I am the more content to stay.'

He looked at her from the depths of the cowl which he had already pulled over his head, so that his face was no longer visible. Without answering her he went out, closing the door with expressive violence. They heard his steps go pattering down the stairs, and then the sound of them was drowned in the uproar from without.

They looked at each other across the table. M. de Putanges sighed as he spoke.

'Did I conceive him worthy of you, madame, I should not have suffered you to have had your way. Even now I doubt——'

'Gaspard,' she interrupted him, 'I am content. I have chosen, even though I have chosen too late.' She held out a hand to him, and he saw that she was weeping. He took the hand and very tenderly stooped to kiss it.

'My dear,' he said, 'it but remains for us to forgive each other.'

As he spoke the air was shaken by a sudden roar, deeper, fiercer, nearer than any that had gone before. She sank against him shuddering in sudden fear. 'What is that?'

'Wait.' Swiftly he quenched the lights. Then in the darkness he groped to the shutters, opened them and flung the window wide. Screened by the gloom above, they looked down into the seething, furious mob revealed by torchlight and the breaking day.

In the clutches of a knot of frenzied men, a Dominican friar was struggling wildly.

'Let us look at your face,' they were howling. 'Let us make sure that you are what you seem—that you are not the villain who murdered Father Gerbier and stole his frock.'

The cowl was suddenly torn back, and the livid, distorted face of Stanislas de la Vauvraye was revealed.

'We have found you at last, you murthering heretic!' cried a voice and a pike was swung above the Huguenot's luckless head. A roar arose:

'Kill! Kill! Death to the Parpaillot!'

M. de Putanges pulled his wife back and closed the window, and the shutters. She clung to him in the dark.

'My God!' she moaned. 'Had he been generous: had I been mad . . .' She said no more; she fell on her knees to pray.

Through the remainder of that night of St Bartholomew they waited hand in hand for death. It was, they said thereafter, their true nuptial night. But death did not seek them. M. de Putanges, you will remember, had rubbed the sign of doom from the doorpost with his sleeve, and the house was not molested. Three days later, when the reaction had set in, they quietly slipped out of Paris unchallenged, and made their way back to Béarn having found each other.

STANLEY J. WEYMAN

The King's Stratagem

∞

In the days when Henry the Fourth of France was as yet King of Navarre only, and in that little kingdom of hills and woods which occupies the south-western corner of the larger country, was with difficulty supporting the Huguenot cause against the French court and the Catholic League— in the days when every little moated town, from the Dordogne to the Pyrenees, was a bone of contention between the young king and the crafty queen-mother, Catherine de Medicis, a conference between these warring personages took place in the picturesque town of La Réole. And great was the fame of it.

La Réole still rises grey, time-worn, and half-ruined on a lofty cliff above the broad green waters of the Garonne, forty-odd miles from Bordeaux. It is a small place now, but in the days of which we are speaking it was important, strongly fortified, and guarded by a castle which looked down on some hundreds of red-tiled roofs, rising in terraces from the river. As the meeting-place of the two sovereigns it was for the time as gay as Paris itself. Catherine had brought with her a bevy of fair maids of honour, and trusted more perhaps in the effect of their charms than in her own diplomacy. But the peaceful appearance of the town was as delusive as the smooth bosom of the Gironde; for even while every other house in its streets rang with music and silvery laughter, each party was ready to fly to arms at a word if it saw that any advantage could be gained thereby.

On an evening shortly before the end of the conference two men were seated at play in a room, the deep-embrasured window of which looked down from a considerable height upon the river. The hour was late; below them the town lay silent. Outside, the moonlight fell bright and pure on sleeping fields, on vineyards, and dark far-spreading woods. Within the room a silver lamp suspended from the ceiling threw light upon the table, but left the further parts of the chamber in shadow. The walls were hung with faded tapestry, and on a low bedstead in one corner lay a handsome cloak, a sword, and one of the clumsy pistols of the period. Across a high-backed chair lay another cloak and sword, and on the window seat, beside a pair of saddle-bags, were strewn half a dozen trifles such as soldiers carried from camp to camp—a silver comfit-box, a jewelled dagger, a mask, a velvet cap.

The faces of the players, as they bent over the cards, were in shadow. One—a slight, dark man of middle height, with a weak chin—and a mouth that would have equally betrayed its weakness had it not been shaded by a dark moustache—seemed, from the occasional oaths which he let drop, to be losing heavily. Yet his opponent, a stouter and darker man, with a sword-cut across his left temple, and the swaggering air that has at all times marked the professional soldier, showed no signs of triumph or elation. On the contrary, though he kept silence, or spoke only a formal word or two, there was a gleam of anxiety and suppressed excitement in his eyes; and more than once he looked keenly at his companion, as if to judge of his feelings or to learn whether the time had come for some experiment which he meditated. But for this, an observer looking in through the window would have taken the two for that common conjunction—the hawk and the pigeon.

At last the younger player threw down his cards, with an exclamation. 'You have the luck of the evil one,' he said, bitterly. 'How much is that?'

'Two thousand crowns,' the other replied without emotion. 'You will play no more?'

'No! I wish to Heaven I had never played at all!' was the answer. As he spoke the loser rose, and moving to the window stood looking out. For a few moments the elder man remained in his seat, gazing furtively at him; at length he too rose, and, stepping softly to his companion, he touched him on the shoulder. 'Your pardon a moment, M. le Vicomte,' he said. 'Am I right in concluding that the loss of this sum will inconvenience you?'

'A thousand fiends!' the young gamester exclaimed, turning on him wrathfully. 'Is there any man whom the loss of two thousand crowns would not inconvenience? As for me——'

'For you,' the other continued smoothly, filling up the pause, 'shall I be wrong in supposing that it means something like ruin?'

'Well, sir, and if it does?' the young man retorted, and he drew himself up, his cheek a shade paler with passion. 'Depend upon it you shall be paid. Do not be afraid of that!'

'Gently, gently, my friend,' the winner answered, his patience in strong contrast to the other's violence. 'I had no intention of insulting you, believe me. Those who play with the Vicomte de Noirterre are not wont to doubt his honour. I spoke only in your own interest. It has occurred to me, Vicomte, that the matter may be arranged at less cost to yourself.'

'How?' was the curt question.

'May I speak freely?' The Vicomte shrugged his shoulders, and the other, taking silence for consent, proceeded: 'You, Vicomte, are governor of Lusigny for the King of Navarre; I, of Créance, for the King of France. Our towns lie but three leagues apart. Could I by any chance, say on one of these fine nights, make myself master of Lusigny, it would be worth more than two thousand crowns to me. Do you understand?'

'No,' the young man answered slowly, 'I do not.'

'Think over what I have said, then,' was the brief answer.

For a full minute there was silence in the room. The Vicomte gazed from the window with knitted brows and compressed lips, while his companion, seated near at hand, leant back in his chair, with an air of affected carefulness. Outside, the rattle of arms and hum of voices told that the watch were passing through the street. The church bell rang one o'clock. Suddenly the Vicomte burst into a forced laugh, and, turning, took up his cloak and sword. 'The trap was well laid, M. le Capitaine,' he said almost jovially; 'but I am still sober enough to take care of myself—and of Lusigny. I wish you goodnight. You shall have your money, do not fear.'

'Still, I am afraid it will cost you dearly,' the Captain answered, as he rose and moved towards the door to open it for his guest. And then, when his hand was already on the latch, he paused. 'My lord,' he said, 'what do you say to this, then? I will stake the two thousand crowns you have lost to me, and another thousand to boot—against your town. Oh, no one can hear us. If you win you go off a free man with my thousand. If you lose, you put me in possession—one of these fine nights. Now, that is an offer. What do you say to it? A single hand to decide.'

The younger man's face reddened. He turned; his eyes sought the table and the cards; he stood irresolute. The temptation came at an unfortunate moment; a moment when the excitement of play had given way to depression, and he saw nothing outside the door, on the latch of which his hand was laid, but the bleak reality of ruin. The temptation to return, the thought that by a single hand he might set himself right with the world was too much for him. Slowly—he came back to the table. 'Confound you!' he said passionately. 'I think you are the devil himself!'

'Don't talk child's talk!' the other answered coldly, drawing back as his victim advanced. 'If you do not like the offer you need not take it.'

But the young man was a born gambler, and his fingers had already closed on the cards. Picking them up idly he dropped them once, twice, thrice on the table, his eyes gleaming with the play-fever. 'If I win?' he said doubtfully. 'What then? Let us have it quite clearly.'

'You carry away a thousand crowns,' the Captain answered quietly. 'If you lose you contrive to leave one of the gates of Lusigny open for me before next full moon. That is all.'

'And what if I lose, and do not pay the forfeit?' the Vicomte asked, laughing weakly.

'I trust to your honour,' the Captain answered. And, strange as it may seem, he knew his man. The young noble of the day might betray his cause and his trust, but the debt of honour incurred at play was binding on him.

'Well,' said the Vicomte, with a deep breath, 'I agree. Who is to deal?'

'As you will,' the Captain replied, masking under an appearance of indifference an excitement which darkened his cheek, and caused the pulse in the old wound on his face to beat furiously.

'Then do you deal,' said the Vicomte.

'With your permission,' the Captain assented. And gathering the cards he dealt them with a practised hand, and pushed his opponent's six across to him.

The young man took up the hand and, as he sorted it, and looked from it to his companion's face, he repressed a groan with difficulty. The moonlight shining through the casement fell in silvery sheen on a few feet of the floor. With the light something of the silence and coolness of the night entered also, and appealed to him. For a few seconds he hesitated. He made even as if he would have replaced the hand on the table. But he had gone too far to retrace his steps with honour. It was too late, and with a muttered word, which his dry lips refused to articulate, he played the first card.

He took that trick and the next; they were secure.

'And now?' said the Captain who knew well where the pinch came. 'What next?'

The Vicomte compressed his lips. Two courses were open to him. By adopting one he could almost for certain win one more trick. By the other he might just possibly win two tricks. He was a gamester, he adopted the latter course. In half a minute it was over. He had lost.

The winner nodded gravely. 'The luck is with me still,' he said, keeping his eyes on the table that the light of triumph which had leapt into them might not be seen. 'When do you go back to your command, Vicomte?'

The unhappy man sat, as one stunned, his eyes on the painted cards which had cost him so dearly. 'The day after tomorrow,' he muttered at last, striving to collect himself.

'Then shall we say—the following evening?' the Captain asked, courteously.

The young man shivered. 'As you will,' he muttered.

'We quite understand one another,' continued the winner, eyeing his man watchfully, and speaking with more urgency. 'I may depend on you, M. le Vicomte, I presume—to keep your word?'

'The Noirterres have never been wanting to their word,' the young nobleman answered stung into passing passion. 'If I live I will put Lusigny into your hands, M. le Capitaine. Afterwards I will do my best to recover it—in another way.'

'I shall be most happy to meet you in that way,' replied the Captain, bowing lightly. And in one more minute, the door of his lodging had closed on the other; and he was alone—alone with his triumph, his ambition, his hopes for the future—alone with the greatness to which his capture of Lusigny was to be the first step. He would enjoy that greatness not a whit the less because fortune had hitherto dealt out to him more blows than caresses, and he was still at forty, after a score of years of roughest service, the governor of a paltry country town.

Meanwhile, in the darkness of the narrow streets, the Vicomte was making his way to his lodgings in a state of despair difficult to describe, impossible to exaggerate. Chilled, sobered, and affrighted he looked back and saw how he had thrown for all and lost all, how he had saved the dregs of his fortune at the expense of his loyalty, how he had seen a way of escape—and lost it for ever! No wonder that as he trudged through the mud and darkness of the sleeping town his breath came quickly and his chest heaved, and he looked from side to side as a hunted animal might look, uttering great sighs. Ah, if he could have retraced the last three hours! If he could have undone that he had done!

In a fever, he entered his lodging, and securing the door behind him stumbled up the stone stairs and entered his room. The impulse to confide his misfortunes to some one was so strong upon him that he was glad to see a dark form half sitting, half lying in a chair before the dying embers of a wood fire. In those days a man's natural confidant was his valet, the follower, half friend, half servant, who had been born on his estate, who lay on a pallet at the foot of his bed, who carried his *billets-doux* and held his cloak at the duello, who rode near his stirrup in fight and nursed him in illness, who not seldom advised him in the choice of a wife, and lied in support of his suit.

The young Vicomte flung his cloak over a chair. 'Get up, you rascal!' he cried impatiently. 'You pig, you dog!' he continued, with increasing anger. 'Sleeping there as though your master were not ruined by that scoundrel of a Breton! Bah!' he added, gazing bitterly at his follower, 'you are of the *canaille*, and have neither honour to lose nor a town to betray!'

The sleeping man moved in his chair but did not awake. The Vicomte, his patience exhausted, snatched the bonnet from his head, and threw it on the ground. 'Will you listen?' he said. 'Or go, if you choose look for another master. I am ruined! Do you hear? Ruined, Gil! I have lost all—money, land, Lusigny itself—at the cards!'

The man, roused at last, stooped with a sleepy movement, and picking up his hat dusted it with his hand, then rose with a yawn to his feet.

'I am afraid, Vicomte,' he said, in tones that, quiet as they were, sounded like thunder in the young man's astonished and bewildered ears, 'I am afraid that if you have lost Lusigny—you have lost something which was not yours to lose!'

As he spoke he struck the embers with his boot, and the fire, blazing up, shone on his face. The Vicomte saw, with stupor, that the man before him was not Gil at all—was indeed the last person in the world to whom he should have betrayed himself. The astute smiling eyes, the aquiline nose, the high forehead, and projecting chin, which the short beard and moustache scarcely concealed, were only too well known to him. He stepped back with a cry of despair. 'Sir!' he said, and then his tongue failed him. His arms dropped by his sides. He stood silent, pale, convicted, his chin on his breast. The man to whom he had confessed his treachery was the master whom he had agreed to betray.

'I had suspected something of this,' Henry of Navarre continued, after a lengthy pause, and with a tinge of irony in his tone. 'Rosny told me that that old fox, the Captain of Créance, was affecting your company somewhat too much, M. le Vicomte, and I find that, as usual, his suspicions were well-founded. What with a gentleman who shall be nameless, who has bartered a ford and a castle for the favour of Mademoiselle de Luynes, and yourself, and another I know of—I am blest with some faithful followers, it seems! For shame! for shame, sir!' he continued, seating himself with dignity in the chair from which he had risen, but turning it so that he confronted his host, 'have you nothing to say for yourself?'

The young noble stood with bowed head, his face white. This was ruin, indeed, absolute irremediable ruin. 'Sir,' he said at last, 'your Majesty has a right to my life, not to my honour.'

'Your honour!' Henry exclaimed, biting contempt in his tone.

The young man started, and for a second his cheek flamed under the well-deserved reproach; but he recovered himself. 'My debt to your Majesty,' he said, 'I am willing to pay.'

'Since pay you must,' Henry muttered softly.

'But I claim to pay also my debt to the Captain of Créance.'

The King of Navarre stared. 'Oh,' he said. 'So you would have me take your worthless life, and give up Lusigny?'

'I am in your hands, sire.'

'Pish, sir!' Henry replied in angry astonishment. 'You talk like a child. Such an offer, M. de Noirterre, is folly, and you know it. Now listen to me. It was lucky for you that I came in tonight, intending to question you. Your madness is known to me only, and I am willing to overlook it.

Do you hear? I am willing to pardon. Cheer up, therefore, and be a man. You are young; I forgive you. This shall be between you and me only,' the young prince continued, his eyes softening as the other's head sank lower, 'and you need think no more of it until the day when I shall say to you, "Now, M. de Noirterre, for Navarre and for Henry, strike!" '

He rose as the last words passed his lips, and held out his hand. The Vicomte fell on one knee, and kissed it reverently, then sprang to his feet again. 'Sire,' he said, his eyes shining, 'you have punished me heavily, more heavily than was needful. There is only one way in which I can show my gratitude, and that is by ridding you of a servant who can never again look your enemies in the face.'

'What new folly is this?' Henry asked sternly. 'Do you not understand that I have forgiven you?'

'Therefore I cannot give up Lusigny to the enemy, and I must acquit myself of my debt to the Captain of Créance in the only way which remains,' the young man replied firmly. 'Death is not so hard that I would not meet it twice over rather than again betray my trust.'

'This is midsummer madness!' said the King, hotly.

'Possibly,' replied the Vicomte, without emotion; 'yet of a kind to which your Grace is not altogether a stranger.'

The words appealed to that love of the fanciful and the chivalrous which formed part of the young King's nature, and was one cause alike of his weakness and his strength. In its more extravagant flights it gave opportunity after opportunity to his enemies, in its nobler and saner expressions it won victories which all his astuteness and diplomacy could not have compassed. He stood now, looking with half-hidden admiration at the man whom two minutes before he had despised.

'I think you are in jest,' he said presently and with some scorn.

'No, sir,' the young man answered, gravely. 'In my country they have a proverb about us. "The Noirterres," say they, "have ever been bad players but good payers." I will not be the first to be worse than my name!'

He spoke with so quiet a determination that the King was staggered, and for a minute or two paced the room in silence, inwardly reviling the obstinacy of this weak-kneed supporter, yet unable to withhold his admiration from it. At length he stopped, with a low exclamation.

'Wait!' he cried. 'I have it! *Ventre Saint Gris*, man, I have it!' His eyes sparkled, and, with a gentle laugh, he hit the table a sounding blow. 'Ha! ha! I have it!' he repeated gaily.

The young noble gazed at him in surprise, half suspicious, half in-credulous. But when Henry in low, rapid tones had expounded his plan, the young man's face underwent a change. Hope and life sprang into it.

The blood flew to his cheeks. His whole aspect softened. In a moment he was on his knee, mumbling the prince's hand, his eyes moist with gratitude. Nor was that all; the two talked long, the murmur of their voices broken more than once by the ripple of laughter. When they at length separated, and Henry, his face hidden by the folds of his cloak, had stolen to his lodgings, where, no doubt, more than one watcher was awaiting him with a mind full of anxious fears, the Vicomte threw open his window and looked out on the night. The moon had set, but the stars still shone peacefully in the dark canopy above. He remembered, his throat choking with silent emotion, that he was looking towards his home—the round towers among the walnut woods of Navarre which had been in his family since the days of St Louis, and which he had so lightly risked. And he registered a vow in his heart that of all Henry's servants he would henceforth be the most faithful.

Meanwhile the Captain of Créance was enjoying the sweets of his coming triumph. He did not look out into the night, it is true—he was over old for sentiment—but pacing up and down the room he planned and calculated, considering how he might make the most of his success. He was still comparatively young. He had years of strength before him. He would rise high and higher. He would not easily be satisfied. The times were troubled, opportunities were many, fools not few; bold men with brains and hands were rare.

At the same time he knew that he could be sure of nothing until Lusigny was actually in his possession; and he spent the next few days in painful suspense. But no hitch occurred nor seemed likely. The Vicomte made him the necessary communications; and men in his own pay informed him of dispositions ordered by the governor of Lusigny which left him in no doubt that the loser intended to pay his debt.

It was, therefore, with a heart already gay with anticipation that the Captain rode out of Créance two hours before midnight on an evening eight days later. The night was dark, but he knew his road well. He had with him a powerful force, composed in part of thirty of his own garrison, bold hardy fellows, and in part of sixscore horsemen, lent him by the governor of Montauban. As the Vicomte had undertaken to withdraw, under some pretence or other, one-half of his command and to have one of the gates opened by a trusty hand, the Captain foresaw no difficulty. He trotted along in excellent spirits, now stopping to scan with approval the dark line of his troopers, now to bid them muffle the jingle of their swords and corselets that nevertheless rang sweet music in his ears. He looked for an easy victory; but it was not any slight misadventure that would rob him of his prey. If necessary he would fight and fight hard. Still, as his company wound along the riverside or passed into the black shadow of the oak grove, which stands a mile to the east of Lusigny, he did not expect that there would be much fighting.

Treachery alone, he thought, could thwart him; and of treachery there was no sign. The troopers had scarcely halted under the last clump of trees before a figure detached itself from one of the largest trunks, and advanced to the Captain's rein. The Captain saw with surprise that it was the Vicomte himself. For a second he thought that something had gone wrong, but the young noble's first words reassured him. 'It is arranged,' M. de Noirterre whispered, as the Captain bent down to him. 'I have kept my word, and I think that there will be no resistance. The planks for crossing the moat lie opposite the gate. Knock thrice at the latter, and it will be opened. There are not fifty armed men in the place.'

'Good!' the Captain answered, in the same cautious tone. 'But you——'

'I am believed to be elsewhere, and must be gone. I have far to ride tonight. Farewell.'

'Till we meet again,' the Captain answered; and without more he saw his ally glide away and disappear in the darkness. A cautious word set the troop in motion, and a very few minutes saw them standing on the edge of the moat, the outline of the gateway tower looming above them, a shade darker than the wrack of clouds which overhead raced silently across the sky. A moment of suspense while one and another shivered— for there is that in a night attack which touches the nerves of the stoutest— and the planks were found, and as quietly as possible laid across the moat. This was so skilfully done that it evoked no challenge, and the Captain crossing quickly with a few picked men, stood in the twinkling of an eye under the shadow of the gateway. Still no sound was heard save the hurried breathing of those at his elbow, the stealthy tread of others crossing, the persistent voices of the frogs in the water beneath. Cautiously he knocked three times and waited. The third rap had scarcely sounded before the gate rolled silently open, and he sprang in, followed by his men.

So far so good. A glance at the empty street and the porter's pale face told him at once that the Vicomte had kept his word. But he was too old a soldier to take anything for granted, and forming up his men as quickly as they entered, he allowed no one to advance until all were inside, and then, his trumpet sounding a wild note of defiance, two-thirds of his force sprang forward in a compact body while the other third remained to hold the gate. In a moment the town awoke to find itself in the hands of the enemy.

As the Vicomte had promised, there was no resistance. In the small keep a score of men did indeed run to arms, but only to lay their weapons down without striking a blow when they became aware of the force opposed to them. Their leader, sullenly acquiescing, gave up his sword and the keys of the town to the victorious Captain; who, as he sat his horse in the middle of the market-place, giving his orders and sending off riders with the news, already saw himself in fancy Governor of Angoulême and Knight of the Holy Ghost.

As the red light of the torches fell on steel caps and polished hauberks, on the serried ranks of pikemen, and the circle of whitefaced townsfolks, the picturesque old square looked doubly picturesque and he who sat in the midst, its master, doubly a hero. Every five minutes, with a clatter of iron on the rough pavement and a shower of sparks, a horseman sprang away to tell the news at Montauban or Cahors; and every time that this occurred, the Captain, astride on his charger, felt a new sense of power and triumph.

Suddenly the low murmur of voices about him was broken by a new sound, the distant beat of hoofs, not departing but arriving, and coming each moment nearer. It was but the tramp of a single horse, but there was something in the sound which made the Captain prick his ears, and secured for the arriving messenger a speedy passage through the crowd. Even at the last the man did not spare his horse, but spurred through the ranks to the Captain's very side, and then and then only sprang to the ground. His face was pale, his eyes were bloodshot. His right arm was bound up in bloodstained cloths. With an oath of amazement, the Captain recognized the officer whom he had left in charge of Créance, and he thundered, 'What is this? What is it?'

'They have got Créance!' the man gasped, reeling as he spoke. 'They have got—Créance!'

'Who?' the Captain shrieked, his face purple with rage.

'The little man of Béarn! The King of Navarre! He assaulted it five hundred strong an hour after you left, and had the gate down before we could fire a dozen shots. We did what we could, but we were but one to seven. I swear, Captain, that we did all we could. Look at this!'

Almost black in the face, the Captain swore another oath. It was not only that he saw governorship and honours vanish like Will-o'-the-wisps, but that he saw even more quickly that he had made himself the laughing-stock of a kingdom! And that was the truth. To this day, among the stories which the southern French love to tell of the prowess and astute-ness of their great Henry, there is none more frequently told, none more frequently made the subject of mirth, than that of the famous exchange of Créance for Lusigny; the tradition of the move by which, between dawn and sunrise, without warning, without a word, he gave his opponents mate.

ANNE MANNING

The Masque at Ludlow

ℭℨ

I

FROM MRS LANFEAR TO FRANCES, COUNTESS OF BRIDGEWATER

This 30th of June, 1634

Madam,

For your exceptions to ye Lady Alice's silence, she through me, humbly craves your Ladyship's pardon; having at present a whitlow on her fore-finger, which hindereth her use of the pen. I am happy to report well in all other respects of my gracious young pupill. What shall I say, or rather, not say, of her most sweet conditions? I think Clotho and Lachesis must have been under the influence of Aglaia and her smiling sister graces when they twined her thread of life. Her memory is wonderful: at once or twice reading she repeats long passages, whether of prose or verse, perfectly by rote, with the most graceful gesture and charming utterance. Her progress in the classics is commendable; the justness of her remarks improving what she reads, surprises me; but I am careful not to betray my opinion of her by word or look, rather leaning to a grave severity, making light of her performances, and averring that they fall below the mark. And now, praying that your Ladyship, my Lord, and all your noble family, may enjoy felicity in this world and in that which is to come, I remain your Ladyship's poor servant to command,

MARY LANFEAR

II

FROM FRANCES, COUNTESS OF BRIDGEWATER, TO MRS LANFEAR

Lanfear,

Alice's finger had best be poulticed with bread and milk, applied warm but not hot; and afterward, when healing, ye shall apply ye following:—

Put a quarter of an oz. of benjamin, storax, and spermaceti, two-pennyworth of alkanet root, a large juicy apple chopt, a bunch of black grapes bruised, a quarter of a lb. of unsalted butter, and two oz. of bees' wax, into a stewing-pan. Simmer gently till the wax, &c., be dissolved, then strain thro' a linnen. When cold, melt it again, and put into small pots, and applie freely to the wownd.

Your commendation of Alice, my good Lanfear, joys my heart; but ye must not imperill her humbleness and simpleness by bewraying aught of ye same. Let the child be a child as long as she can; especially with so many elder sisters, some of 'em not so pretty as herself. Sneap and snub her when she needs it. Eating little and speaking little will never do a girl hurt. Young folks put themselves forward now-a-day; more fools the mothers that let them!

My Lord comes down to Ludlow shortly, to take his state in the presidency, and we shall be accompanied by a great concourse of nobility and gentry to grace his entry. We shall pick you and the children up by the way, or else send you on before. Since the death of my Lord's Countroller of the Household we are all at sixes and sevens.

> He that provides not for his kin,
> The Lord will not provide for him,

and so cousin Dick is preferred to the office; but I doubt how he will fill it. My Lord has no fears.

Tell Alice though she cannot practise her lute till her finger is heal'd, she may exercise her voice to your playing. Harry Lawes says he has no pupils of better promise and performance than my Penelope, Mary, and Alice; but he thinks Alice, as she advances in years, will pass her sisters. Ye need not to tell her this. We shall bring down Lawes; and I have desired him to compose something for ye occasion that shall exercise the children's wits and memories. Let Alice drink freely of tamarind-water if she is feverish.

Thine, my good Lanfear,
FRANCES BRIDGEWATER

III

HENRY LAWES TO JOHN MILTON

My Friend,

Lady Bridgewater hath commissioned me to prepare a Masque, or Interlude, to grace my Lord President's entry into Wales; and does me to wit that I may employ some passable good poetaster to supply verses tolerable enough for her children to learn by rote, and for me to hang music upon. In a word, will you write them? The credit will be little or none, the obligation great to

Your friend and servant,
HENRY LAWES

IV

JOHN MILTON TO HENRY LAWES

My Friend,

No office of your imposition can be without credit; yet am I not so greedy of the voice of those who in the general extol or censure amiss, as to care to be known for her Ladyship's poetaster.

So you may command me, good Mr Harry Lawes, in this matter, since you think it worth the asking, on the sole but express condition that my incognito shall be inviolably preserved.

Granted the Mask, then, you have the Masque. Let me know the number and ages of the intended enactors, and any other matter that may assist

Your sincere friend and servant to command,

JOHN MILTON

V

HENRY LAWES TO JOHN MILTON

My very worthy and esteemed good Friend,

Your manner of granting my request enhances its graciousness and my gratitude; and since you insist on the Mask, it shall be scrupulously respected. You know I am not a novice at this work; last year I set to music a Masque presented at Whitehall on Candlemas night by the gentlemen of the four Inns of Court; and the music was worthy of the versing, and the versing of the music, for both, sooth to say, were indifferent; yet Ives and I received each one hundred pounds well counted. And though I know not whether our Countess's liberality will extend to that mark, doubtless she will requite us proportionably to her own state and our deservings.

Set your wit nimbly to work, I pray you, for till I get copy from you I can do nothing. The enactors are to be my Lord Brackly, Mr Thomas Egerton, the Lady Alice, and your humble servant. You may powder freely with classical conceits (the more the better to my Lady's taste), for Lord Brackly, now in his twelfth year, is of rare endowments and acquirements; and Mr Thomas, turned of eleven, comes not behind him in stature or gifts.

For the Lady Alice, how shall I describe her? She, the youngest and fairest of eleven fair sisters, hath numbered but thirteen summers, but might pass for fifteen. The justness of her stature, comeliness of her features, symmetry of her proportions, and pure tincture of her complexion, might make her supposed one of Calypso's youngest nymphs; yet her outward graces are surpassed by those of her mind. I think Celia and Cornelia at the same tender years must have been of her mould; or you might deem her Virginia with her satchel, tripping to school. She hath the same turn for study as had Lady Jane Grey and the daughters of More; yet with infinite cheerfulness and pleasantness of humour, frank with all, without the least pride.

For music, her voice and ear are most excellent; and she hath that rare felicity so much insisted on by my good old master Giovanni Coperario (whose real appellation, as I think you know, was Johnny Cooper), that the management of her voice adds such an agreeableness to her countenance, without any constraint or effort, as that her singing is as lovely to the eye as to the ear.

Now, lest you should think I am writing of a goddess rather than of a woman scarce beyond a child, and in the vein of an inamorato rather than of a sober, prudent, music-master, I will break off and conclude with an abstrackt of what lately befel these three noble scions of the house of Egerton.

You must know that they, their tutor, and governess, have lately been on a visit at a house of one of the Egerton family in Herefordshire; and in passing through Heywood Forest they were benighted, and the Lady Alice was even lost for a time. Her brothers, I think, had gone too far into the wood to find her wild strawberries, and she, hunting for them, came upon some wild, uncouth people, who affrighted her sadly by their rough abord and outlandish jargon; but happily, her brothers, with a country servant they call Cherrycheeks, came up to her in a little while, and no harm resulted beyond the fright.

And now no more from your faithful, grateful friend and servant,

HENRY LAWES

VI

JOHN MILTON TO HENRY LAWES

Good Mr Henry Lawes, you have told me too much and yet too little; more than enough to set my fancy to work, and too little to satisfy it. The young lady you speak of must be the rarest creature that ever tripped on the green sward, and will needs be a marvel of perfect womanhood when she attains riper age. The incident of the losing in the forest will aptly form the theme of our piece; and you, as Cherrycheeks, elevated for the nonce into Thyrsis, shall speak a prologue, telling how that—

> A noble peer, of mickle trust and power,
> Has in his charge, with tempered awe to guide,
> An old and haughty nation proud in arms;

and so forth:

> Where his fair offspring
> Are coming to attend their father's state,—

but get lost in the wood; whereon you, Thyrsis (in seeming, that is, but a guardian angel in reality) have flown swiftly to the rescue; knowing, by reason of your supernal intuition, that great danger impends from a fell enchanter haunting the aforesaid wood, and offering to every belated traveller he encounters a crystal cup drugged with a potion that, once tasted, turns men into beasts.

Having set all this forth, you step aside and make way for this same enchanter, hight COMUS (Intemperance), who entereth with a charming-rod in one hand and the fatal glass in the other; and with him a rout of monsters, headed like sundry sorts of wild beasts, but otherwise like men and women, well apparelled. These make antic gestures and noises; and then Comus (I know not who shall fill the part) recites what you will find here inclosed. The lines should run off the tongue glibly; and then all knit hands and dance a fantastic measure, the accompaniment to which may have clashing of cymbals, blowing of horns, &c., as at a harvest-home or Whitsun-ale. In the midst, Comus, perceiving that something of virtue and purity beyond common, altogether foreign and adverse to the scene, draweth nigh, makes his riotous rabblement break off and hide among the trees (whence their elvish, satyric faces may from time to time peep out) while he awaits the intruder.

Then (to softest music, mark you, which ceases before her balmy lips unclose) enters our Lady bright; not in any terror or consternation, but in some natural anxiety at her belated state, and looking wistfully about for those whose voices and music had drawn her to the spot. She enlargeth on her strait in soliloquy, and sings a hymn to Echo (which I send you in the rough) in the hope to make her voice the farther heard. Note, that echo-songs always win favour of the audience: but there is no absolute need to repeat the usual trick of surprising by repetitions of sound, as in Browne's Inner Temple Masque; since this is simply an invocation.

To her enters Comus in the guise of a shepherd; and with his artful, specious tongue pretends passionate admiration of her singing, which she, though so young and unsuspecting, values at no more than it is worth; but yet addresses him for counsel how to find her brothers, whom he pretends to have seen. And here ensues a dialogue, terse and epigrammatic, which speedily issues in her submitting herself to his guidance. And so they pass out.

Then enter the two brothers—

As smooth as Hebe's their unrazored lips—

(I have seen the young gentlemen—Lord Brackly is black-haired, ruddy, and comely; the younger brother hath blue eyes and yellow locks). What they say you shall know in my next, if this jejune sketch seems to you of promise.

Your friend, as ever,
JOHN MILTON

VII

HENRY LAWES TO JOHN MILTON

The fragments you have sent me are transcendant. Lose no time, I beseech you, in supplying the rest; and hold yourself in readiness to obey a summons to Ludlow Castle. We are quite at a loss for a suitable Comus, for my Lady says she won't have bearded men play with her children; solely excepting your highly honoured friend and servant,

HENRY LAWES

VIII

JOHANNES MILTON PATRI S.P.D.

Paucis diebus abhinc, Comitissa de Bridgewater fabellam, vel, ut ita dicam, δρᾶμα Σατυρικόν, a me postulavit. Cujus consilii auctor fuit musicus ille peritissimus, noster Henricus Lawes. Opus autem ea ratione conficiendum est, ut ego carminibus, ille chordis apte congruentibus, personas in scenam inducamus apud Castellum de Ludlow, quò, rei hujus ordiendæ causa, si mihi quidam per te liceat, a nobis est eundum. Cura, igitur, de voluntate tua certiorem me facias, mi pater, atque hoc tibi persuasum habe, te filium habere ut semper amoris et officii erga te plenissimum.

IX

MR MILTON TO HIS SON JOHN

Horton, Bucks,
This 20th July, 1634.

Son John,

I have received your Latin letter, and am well content therewith, despite of its shortness, which is not generally the fault you run into, as your sentences are apt to get long-winded so as that a man scant of breath can scarce fail to draw it at the wrong place. I suppose I need not to remind you of the fable, whether of Esop or Babrius, of the two pots, one of iron, the other earth, that came into collision. My Lord of Bridgewater is the iron pot and you the other; take care that you be none the worse for him. We have never yet curried favour with the great man, and though he and I live, as it were, next door, I obliged him about the fishing, and he sent me, unaskt, a quarter of doe venison, when I would as lief have had a buck, or else my own mutton. You have not sought him, however, in this matter, but he, or rather his lady, you. We will talk it over next time you come home, but I don't think thy mother much affects it. If you do go to Ludlow, vie with none of them; rather affect an honest plainness. Neither flatter nor be flattered; be courteous and pleasant with all, but ever with a comely sobriety. Remember, my son, thine own descent and thine own deservings; so shalt thou less need that others always remember them. Money you shall have, for use, not for show or waste; still less for the pernicious games idle retainers kill time with, but in proportion to my means.

With regard to your Masque, I can have no doubt it will be well done, since you have the doing it. Many traditions still linger among us, especially in country places, wherewith poetry may be aptly enriched. I need not remind you of the precious ore, along with mica, in the old chronicles of Bede, Geoffrey of Monmouth, and others. Whatever you touch you will turn to gold; but, if you would please your father, let it have a flavour of the old Falernian, a smack of the old classicality with it. Nor yet go too far afield for a subject—

Nunquam aliud natura, aliud sapientia dixit.

To say more would be to teach him who needs not my teaching. Your mother would send you her blessing, but she knows not I am writing, being busy in her still-room, wasting ever so many bushels of charcoal to make as many pottles of waters, borage, fumitory, and all the rest, that you and I shall have to dodge the drinking of. Your loving father, so long as you live in the love of God,

JOHN MILTON

X

JOHN MILTON TO EDWARD KING

Since you say (though I scarce believe) you have so villainous a cold as to be blind of your Latin eye and deaf of your Latin ear, I will thank you for your letter in the vernacular. You question me as to what I am about, and will perhaps smile in high scorn when you are told. *Neque semper arcum tendit Apollo*; and I the rather indulge in a transient sport of fancy, that it is to serve the behest of a tried and dear friend.

The only time you graced my father's modest roof at Horton, I took you, as you may remember, to Ashridge, anciently a royal palace, in the parish of Gaddesden, in Herts, though bordering upon Buckinghamshire. Do you remember my telling you it was the seat of Sir John Egerton, Earl of Bridgewater, and relating several traits of his character? And while we sate discoursing, astride a gate overlooking a meadow pied with daisies and fragrant with cows' breath, were we not, as thou rememberest ('tis not so long agone), o' sudden startled by the sharp, snapping bark of dogs, evidently engaged in some petty warfare, which drew us to the scene of it? Two noble imps, with vermeil-tinted cheeks and ardent eyes, were inciting a couple of hounds to attack an urchin or hedge-pig that was setting up its bristling quills in self-defence. Those lads were Lord Brackly, and his brother, Mr Thomas Egerton, sons of the Earl of Bridgewater. Upon urgent request I am writing a little play, or satyric drama, in which these young gentlemen and their youngest sister shall welcome Lord Bridgewater to Ludlow. The commission jumps with my humour, and I fulfil it with a peaceful conscience; for I hold that when God did enlarge the universal diet of man's body, and add to every wholesome fruit and root the beast of the field, the bird of the air, and the fish of the river and sea, He then also, as before, left arbitrary the dieting and repasting of our minds, without particular law or prescription, wholly to the demeanour of each several man. So that one shall apply to divine philosophy, another to the Fathers, another to Aristotle, another to poetry, and another even to the drama, so it be pure and chaste.

Misdoubt me not, then, nor contemn me, for writing a Masque, wherein may be introduced pagan and pastoral conceits with no unnatural strangeness. It may be, notwithstanding, that you had heard nought of it, but that I find myself in want of a book you have in your study, of which I would gladly ask the loan. I mean Lavaterus, *De Spectris et Lemuribus*, which lies in your window-seat, along with Olaus Magnus and Georgius Agricola. I found some curious reading in it one day when you kept me waiting. You may send me the others too, if you have a mind, for I shall use them and you will not. I believe you will think yourself well repaid for them if I bring you a great beaupot of roses, coronations, columbines, sunflowers, and daffadowndillies; or, perhaps a pound of my mother's best butter, yellow as a cowslip, and printed with a cow in high relief.

Your loving friend,
JOHN MILTON

XI

EDWARD KING TO JOHN MILTON

No, that I deny! A pat of butter is not an equivalent for a posy, nor a posy for a rare book, especially if sunflowers and dandelions be cheek by jowl with roses and sops-in-wine. I send you herewith not one book but all three; but let my guerdon be a sight of the manuscript when finished. I am hugely curious to see your management of a Masque, encrusted, as it doubtless will be, with dainty allusions. It is a form of composition which neither Ben Jonson nor Carew have despised. Where lies the scene? In Arcadia, no doubt. Who are the persons? Many or few? Shepherds and shepherdesses, of course. A love-story running through the web like a golden thread. Lovers made happy at last. Then all come forward hand in hand, group in attitudes, sing a chorus, lout low to the noble company, and so an end.

EDWARD KING

XII

JOHN MILTON TO EDWARD KING

There may be a higher interest than that of love. My argument is a grave Temperance and holy Chastity. The characters will be few. I am not going to drag in

Glaucumque, Medontaque, Thersilochumque,

merely to knock them on the head. Two brothers will hold an amicable contest between fact and philosophy. The younger draws his arguments from the obvious appearance of things; the elder proceeds on a profounder knowledge, and argues from abstract principles. These interlocutors give place to a lady and an enchanter in a stately palace set out with all manner of deliciousness; and here the interest turns on the enchanter's keeping the lady spell-bound in her chair, while her mind, not consenting to his temptations, rises superior to the vain sophistry of his arguments, that would make Pleasure the Supreme Fair.

I cannot send you the manuscript; first, because it is not finished, and next, because it is impatiently awaited by the composer who is to set it to music; but hereafter you shall have a fair copy.

JOHN MILTON

XIII

HENRY LAWES TO JOHN MILTON

Last night I was spell-bound to my seat like the lovely lady, till I had finished your divine drama. I never met with anything to match it. The action, perhaps, is slow, but the poetry transcendant. The greater part must be rehearsed in measured speech or plain tune, with here and there an introit. The poet hath done so much as to leave the musician little work, for which I sincerely thank you. There is the echo-song, the invocation to Sabrina, a round when the satyrs dance in the forest, and a chorus at the end—that's all you have given me to do; so, that I may not eat the bread of idleness, I am to play Thyrsis. The young gentlemen are already forward in their parts, and Lady Alice pores over hers all day and repeats it on her pillow, I fancy, after her prayers. What we shall do for a Comus is now the problem, which you, the arch-magician, may help us to solve. Who could take the part better than yourself? You have played the Lady, ere now, in college rehearsals. Do come, 'beseech you!

Locus est, et pluribus umbris.

In default of you, we have none to look to but a gawky seventh cousin of her Ladyship's, a florid, saffron-haired youth, with cheeks as round as pippins, and somewhat too thick a tongue. He will maul your verses horribly. Think twice of it.

H. LAWES

XIV

JOHN MILTON TO HENRY LAWES

Instead of thinking twice, I dare not think once of your ensnaring proposal. I should never hear the last of it! The tide hath turned here, and is now altogether adverse to my going to Ludlow, for reasons wherewith you need not be troubled. Already I in some sort despise myself for having taken any share in such toys, which are now somewhat roughly cast at my head. In some sort I have brought it on myself: the hedgehog, if he did not patrol the orchards after dark, could never have been accused of carrying away pears and apples on his quills. After all, men come into the world for better objects than sounding of brass and tinkling of cymbals; and have too manifold temptations already to hinder and disinure them, without needlessly going astray after will o' th' wisps dancing over miry and darksome places: so have I been told within the last few hours. As for plays, masques, interludes, and all such gauds, they do but lure us from our severe but wholesome schoolmistress Truth, at whose feet meekly to sit and aptly to learn is the best docility.

In brief, that young carrotty-poll may play Comus an' he will—and can. But O, the sin and shame of committing my glib verses to his thick tongue! Prompt him, pinch him, persecute him into some conception of his part; else shall the whole piece be marred through the loutishness of one lubber. You may cudgel him into some parrot repetition, but unless the spirit of the character be in him, it will never come out of him.

To conclude, send me news of the Castle and of the Masque.

Thine,

JOHN MILTON

XV

HENRY LAWES TO JOHN MILTON

We are progressing amain. The cub who takes Comus hath learnt his part not amiss as for perfectness; but for any beauty of action or accent, we insist on them in vain. Lady Alice openly laughs at him, which only makes him open his gooseberry eyes the wider, and grin like a dog in return; but he will take no raillery from any other quarter without going into the dead sulks. Had you condescended to the part (alas, that you cannot!), then indeed we might have made another guess affair of the Masque; but since you will not or must not, we must e'en do as we can.

Lady Bridgewater sends some of us in advance of her to Ludlow next week, whence you may expect to hear from me.

H. LAWES

XVI

HENRY LAWES TO JOHN MILTON

Ludlow Castle

Since you desire an account of this place, and are unlikely to see it, I will not delay to gratify your wish by wasting time on the mischances of the journey, but take up my parable at the journey's end. In Shakspearian phrase,

> This castle hath a pleasant seat; the air
> Nimbly and sweetly recommends itself
> Unto the senses.

It is founded on a rock overlooking the river Corve, and was built by Roger Montgomery in the reign of Henry the First. Many of the original towers still remain, overhanging a great fosse, but some of them are ruinated, and even certain of the royal apartments lie open to the weather, overrun with ivy, and prankt with stonecrop and snap-dragon.

The habitable part of the Castle was chiefly rebuilt by Sir Henry Sidney when lord president, including twelve goodly rooms and a wardrobe. He also repaired Mortimer's Tower, and made it a receptacle of the ancient records, perishing for lack of safeguard. All the new building over the gate is owing to Sir Henry Sidney; and he also restored a fair room under the court-house, and made a great wall round the wood-yard, and a brave condyt within the inner court.

All this have I learnt of the old grey-haired warden who had the place in charge.

The great Hall has a groined roof, and richly painted windows, flinging or, azure, and gules on the ancient stone floor. In one of them is an impalement of St Andrew's Cross, with Prince Arthur's arms. Also there's a great stone escutcheon of Prince Arthur's arms. On the left side of this hall, which indeed is exceeding magnifical, are the arms of my lord Earl of Warwick, the Earl of Derby, Earl of Worcester, Earl of Pembroke, and Sir Henry Sidney. On the other side, the arms of North and South Wales; two red lions and two golden lyons for Prince Arthur. At the end of the hall is a pretty device, how the hedgehog broke his chain and came to Ludlow from Ireland. (Query, may not another hedgehog break his chain and come to Ludlow from Horton?)

There is at the entrance of the hall a great iron portcullis of huge weight, also a music gallery, and oaken screen. At the upper end a dais and long oaken table running across; two inferior tables running down the hall. In my Lord's private dining-room, cupboards for plate, and a table to play shuffle-board.

Now, for the Chapel, it is so trim and bravely painted with the arms of many English kings and of the lords of the castle, that you need not wish a fairer place for devotion. The bell rings (or will ring when we all come together) for prayers a quarter of an hour before breakfast, during which time the retinue make ready for covering the tables, but do not cover.

Before the breakfast, which is at eight o'clock, the castle keys are brought in; also during dinner and supper.

I need not remind you that Prince Arthur, son of Harry Seventh, died in this castle; it being the palace of the Prince of Wales, appendent on his principality. But many of the state chambers are altogether unusable, given over to the bats and owls, yet with frequent tokens of ancient pomp peeping out from amidst the rubbish of their mouldering fragments. They are reported haunted; so as that my guide averred the maids would not approach that wing of the building after dark, save in couples.

Over the stable-doors are the arms of Queen Elizabeth, Lord Pembroke, &c. Something less than twenty years ago, the creation of Prince Charles to the principality was solemnized here with uncommon splendour.

I spare you the various courts, staircases, nests of turret-chambers, galleries, offices, &c. The landscape on all sides is agreeably diversified; the prospect in some directions ranging over woods, in others over large and beautiful meadows, dotted with sheep and cattle. The pleasance, or garden-plot within the castle, has been greatly neglected; and the hangings, where there are any, are dropping from the walls; but large provision of furniture is hourly arriving in baggage-carts, and the damp old walls echo to many and cheerful voices.

Yester-even, what time the crescent moon peered with pale visage through the silvered clouds, I strayed about the precincts and lone deserted courts,—buttresses and battlements casting strange unearthly shadows on the weedy ground,—and anon I conjured up for myself the old place in all its bravery when Prince Arthur and his six-months' bride, Doña Catalina, arrived here in all the glory of youth and love, to take their state upon them—she being mounted on a palfrey with velvet housings, attended by eleven ladies, and received at the castle gates with loud blaring trumpets and deafening huzzas.

Methought how soon all this bravery paled, and the fair young Prince, plague-smitten, lay in the pangs of death, while the young Spanish girl wept over him; and mysterious shades seemed to float past me, sighing as they went. All at once these sad shadows were dispersed by the lively din of young voices and mocking laughter; and Lady Alice and Lord Brackly darted into the court, he clutching at her skirts and trying to get from her something which she held beyond his reach, high above her head.

This morning, I was straying round the Castle, marking well her bulwarks, and considering her towers, when, just over against Mortimer's Tower, the Lady Alice came up to me, bearing a fair white lily, accompanied by Lord Brackly and a great stag-hound; and quoth she, 'We are to have Scambling-day, Master Lawes, because that the carts have brought plenty of tables and stools, but no mugs nor trenchers, pots nor pans; and the clerk of the kitchen was ready to hang himself, till we told him we would dine on bread and cheese, spread on the grass.'

'That will be mighty pleasant,' quoth I. 'And pray, where may be the spot selected for your Ladyship's refection?'

'Under yonder hawthorn,' said she, 'where the ground is carpeted with euphrasy and wild thyme. Come and see it.'

So we all walked together to the spot, and when we reached it, my Lord Brackly exclaimed, 'O, what a goodly stage to play "Love's Labour's Lost," without need of scene-shifting! Sister, you shall be the Princess, and I'll be Boyet, and carve the capon.'

'What part will your Lordship fit upon me?' quoth I.

'You,' said he, 'shall be Costard, and cry, "O, sweet guerdon! elevenpence farthing better than remuneration!" '

'Nay,' said Lady Alice, 'Master Lawes shall not be condemned to so rough a part, but play Biron.'

On which I fell a posturing, and mouthed,

> What! I? *I* love? I sue? I seek a wife?

till they tired 'emselves with laughing. Said I, 'Why not stick to the matter in hand, and apply ourselves to a right good serious rehearsal of our Masque, which I am fain to think we are not too forward in?'

'Aye, that is well thought of,' said Lady Alice. 'Go, brother, and call Tom.'

'I will save your Lordship that trouble,' said I, 'and also hunt up Cuthbert Curlypate, who is no doubt slumbering in some sunny nook.'

At this point Madam Lanfear came sailing out, starched as buckram, crying, 'What are you all about?'

'We are going to rehearse the Masque,' says Lord Brackly. 'Here's a marvellous convenient place for our rehearsal. "This green plot shall be our stage, this hawthorn-brake our tyring-house; and we will do it in action as we will do it before the Duke." '

'Oh then, I will play audience,' says she, settling herself on the bank.

'Aye, goody, or peer through the bushes for one of Comus's court,' said my Lord, lightly, which I could see she liked not.

In brief, we severally returned to the Castle for what we wanted; and Cuddy, being summoned to the rehearsal, made a Jack-in-the-Green of himself, and, encouraged by his disguising, gave forth his versings lustily, but altogether broke down in the second scene; so that the brothers rushed in before their time, and belaboured him with their foils. Too much of this.

Thy friend,
H. LAWES

XVII

JOHN MILTON TO HENRY LAWES

Not a word too much. Continue your diurnall, I beseech you.

I despise myself somewhat for being so concerned in this toy; but one would not willingly let die the fruitage of one's brain. Can this Lady Alice be but thirteen? I happen never to have seen her, and am concerned to know whether she be indeed a just representative of my Innominata, or whether, indeed, I have not made my lady too old and sententious for so young a player.

Your picture of the Castle on the Border attracts me: a rugged shell for so fair a nut.

> Thy friend,
> JOHN MILTON

XVIII

THE LADY ALICE EGERTON TO THE COUNTESS
OF BRIDGEWATER

Madam,

I trust that this may find my honoured parents in good health. May it please your Ladyship to grant your poor daughter's humble request, that for the ensuing pageant I may have a white silk gown shot with silver threads, and pearls for my hair. Your Ladyship knoweth I do not in usual affect much finery, but this once I hope you will grant the humble request of

> Your Ladyship's unworthy child,
> ALICE EGERTON

XIX

THE COUNTESS OF BRIDGEWATER TO LADY ALICE EGERTON

Daughter Alice,

At your time of life, I never once had anything finer than dimity or cloth, according to the season, and two silk suits for festivals, one red, t'other green, which lasted me till I grew out of 'em. And I think taffeta quite fine enough for a child of your years, for though you're tall, you're but a chit; and moreover, Alice, you droop your head too much, which lessens your height and will make your nose red. 'Tis not the clothes but the carriage that makes the lady. However, you shall have the gown, on condition you learn to bridle. And I trust you will keep your heart in all humbleness and lowliness, not puffed up as the manner of some is, but submitting yourself to those that have the rule over you. And so long as you do that, be sure of the love and blessing of your mother,

FRANCES BRIDGEWATER

Ye ought to make much of your golden opportunities now, Alice, for they will never occur again. For me, I cannot command a quarter of an hour's leisure.

XX

THE COUNTESS OF BRIDGEWATER TO RICHARD EGERTON

Cousin Dick,

You shall please direct the marshals and ushers of the hall to order the carriages and conveyances for the stuff, when my Lord changeth house; how they shall be occupyde, how appointed; and that matters shall not be left undone, or but half done, to the last, so as that on the day of my Lord's departour, only the hanging-stuff and beds shall remain to be removed. As to say, the stuff in my Lord's chamber and mine, the stuff in the great chamber, and in the chamber wherein my Lord maketh ready, &c. And my Lord's stuff shall be removed in three carts, and mine in three carts; and one cart for stuff over and above remaining. And two carts for the children's stuff; and one cart for the officers following, to wit, the gentlemen carvers, sewers, cup-bearers, and waiters, for carrying their five beds, and none other to be allowed them.

For all the stuff pertaining to the sellar, pantry, and buttery, one cart; and none other to be allowed 'em.

For the kitchen, squillery, larder, and pastry, two carriages. And everything belonging to the bakehouse and brewhouse, one carriage.

Item, the workmen of the household, which is to say, the joiner, smith, painter, minestrels, and huntsmen, to have one cart for their stuff; and none other to be allowed 'em.

Number of all the carriages employed in the removal, besides the chariot, seventeen.

Cousin Dick, the removal of this great household with all its stuff is a great charge. Thou art in office to see that other men mind their offices; see that thou mindest thine own office, so shall thy Lord have pleasure in thy duty. Be not given to wine, wherein is excess, and let not thine hand know a bribe. Suffer not the catour to make any entries till the things are brought in. Let the clerk of the kitchen take account betimes what shall need to be provided for so great a company as will soon be at Ludlow; as, so many bucks, so many swans, so many beeves, muttons, veals, so many baskets of fish, so many partridges, cranes, herns, wypes, woodcocks, ruffs and reys, &c. And that no herbs be wastefully bought, seeing that the cooks may have enough and to spare out of my Lord's gardens. Also that the clerks of the kitchen cast up daily the cheque-roll, and see that the accounts tally with the empcions in the catour's journal-book. If he defraud but a peppercorn a day, 'tis three hundred, sixty, and five peppercorns by the year.

Note, that the clerks of the brevements breve every stranger by name that cometh to my Lord's house, else shall tag, rag, and bobtail be walking in upon us; and what is to hinder them of the spoons?

And at all odd times, Cousin Dick, I charge you to peruse over the Booke of Riwles of the Household, till ye have them at your fingers' ends, which shall in some sort serve in stead of long experience. And so rest you merry, and keep you faithful to my Lord.

Your kinswoman and well-wisher,
FRANCES BRIDGEWATER

XXI

TO FRANCES, COUNTESS OF BRIDGEWATER

Madame, may it lik your Ladyship, seeing that my Lord's clerk coump-troller being deceast, another is put in his place, I would hint in your ear, that ye shall bid the new clerk coumptroller dailly to have an ey to the slaighter house when any viaunds shall be slain there; and there to see the suitt clean taken out *without any bribe*, and there weighed and then brought into the stor-house, and from thens delivered by the clerks to the chaundler at dew time or times: and so shall the slaighter-man make the vailles no larger than he ought to do; and his horable Lordship and your Ladyship not be defrauded of your owne.

Your Ladyship's servant to command (except in name),
ROBIN NAMELESS

XXII

THE COUNTESS OF BRIDGEWATER TO RICHARD EGERTON

Cousin Dick,

Read what's writ overleaf, and see how your predecessor was befooled. One had need have as many eyes as Argus to spy out all the cheats in an overgrown house. I am not so sure that Master Nameless himself may not heretofore have had a hand in the peculation; else, how comes he to be so knowing? I depend much on you, Cousin Dick; see you requite my reliance.

FRANCES BRIDGEWATER

XXIII

THE COUNTESS OF BRIDGEWATER TO MRS LANFEAR

Lanfear,

I think the young gentleman that writ the Masque must have meant to make us all water-drinkers—however, that may pass. It will do no harm to warn the children against winebibbing and gluttonishness. I have looked through the fair copy Lawes sent me, and find no fault, except prolixity. I hope we may not fall asleep, and vex the children, especially just after supper.

You will get the beasts' heads when we come; some of them are excellently put out of hand. You must make the best you can of Cuddy; he's a head higher than Alice, and won't look the part amiss. There could not be a better Thyrsis than Lawes. 'Tis a pity he cannot take both parts. Mary would like hugely to be Sabrina, if it were not playing second to her younger sister, but she cannot be spared from court; so Lettice can take the part, having a passable voice; and it may get her a husband, for her face is her fortune; but don't let her step out of her place. My good Lanfear, I write to you as if you were an *alter ego*. State brings weight; never wish yourself other than you are. I have promised Alice the white and silver gown; doubtless the child will look a very angel in it. For the nymph Sabrina, a pretty habiliment might be copied that was devised for a water-nymph in Ben Jonson's masque, enacted when King James's eldest son was created Prince of Wales. Her head-dress was a murex-shell, ornamented with coral; a veil of silver gossamer depending therefrom; a bodice of water-blue silk, branched with silver sea-weed, a half-tunic of silver gauze, brocaded with gold sea-weed; and a train of river-blue silk, figured with columns of white lace, sea-weed pattern. I'm afraid if we dress Lettice in these fallalls, we shall turn her head for life. Inferior water-nymphs in satin tunics of palest river-blue, with silver flowers; their long hair hanging loose in waving curls, crowned with water-lilies. Girls are glad enough to be fantastically dressed.

FRANCES BRIDGEWATER

XXIV

MRS FYTTON TO MRS LANFEAR

Esteemed good Friend,

When you acknowledge the buckram suit, sent herewith, prithee let me have news of the state entry, and all thereto belonging. I remember Mr Harry Lawes when he was a chubby-cheeked chorister in Salisbury cathedral, with a pipe as sweet as any lark's.

Your assured good friend,
HENRIETTA FYTTON

XXV

MRS LANFEAR TO MRS FYTTON

Ludlow Castle,
Sept. 30, 1634.

My good and dear Friend,

The entry was very fine. All the country assembled to do it honour, as well as others from distant parts. My Lord President, in his collar and mantle, on his managed steed, with velvet housings. My Lady in wondrous good liking, on her white horse, richly caparisoned. She was in purple velvet, fringed and buttoned with gold, and a velvet hat laced with gold, and surmounted by three feathers. Six drums and six trumpets at the gate sounded at their approach. Also guns were fired. Four-and-twenty young ladies of the principality, who afterwards danced in the Masque, with Lady Alice at their head, as fair as May, received them with flowers. The cornets, sackbuts, and drums, could hardly be heard for the huzzas; and flags were streaming in all quarters. Then the deputy-lieutenant did make a fair speech to my Lord, who made a fair speech in reply. My Lord and Lady, and all the fine company, then alighted, on scarlet cloth laid over the stones; and a herald made proclamation; and two gentlemen-ushers did lead the way into the castle with their rods of office, crying, 'By your leaves, gentlemen, stand by!' My Lord and Lady following after, hand in hand, with train-bearers, my Lord Brackly and Mr Egerton, and the goodly bevy of daughters (no finer family can England boast), and money being scattered among the people, for whom also casks of ale were set running, and sheep roasted.

Then we all went to prayers in the chapel, which was full to overflowing; and then in great order they all went in procession to make the circuit of all the castle, the music still sounding, beginning with the great hall, where supper was being spread, and ending with the kitchen, where cooks in white nightcaps were fuming over the fires, that the company might see the preparations making for the sack-posset, in a vessel nearly as big as Guy of Warwick's porridge-pot.

I had had thrust upon me the charge of Master William Egerton, a most rebellious, disobedient young gentleman, aged five, whom I much regretted his grandmamma had given me in charge; for he rollicked and rantipoled to that degree that it was misery to look after him; and he dragged my gown apart at the gathers.

When the company perambulated the castle, he must needs rush down a dark passage, bawling 'Whoop! whoop!' and I must therefore follow after him, hallooing him to come back, because of a ruinated doorway at the end. He, bouncing through it, went clean down a gap where had once been a stair-landing, and so into a water-butt with an awful splash. 'Twas no good to shriek, though I did it, nor could I jump after him, but must needs go round. I tremble to think how it might have gone with the child had he lain there top-side-'tother-way till I reached him. However, providential to relate, a young man chancing to be in the court and see the babe project himself in that way through the ruined doorway and fall into the butt, pulled him out by the leg, roaring pretty lustily, but more frightened than hurt. I was wetted to the skin by merely taking him in my arms; however, I carried him off to his bed, after thanking the young man, who was one of the many strangers drawn to Ludlow by the shew. I think I lookt into twenty bedrooms on my way to the nursery, without finding a woman-servant at her post—all of 'em off duty, scaring about and gaping after the shew, leaving their mistresses' caskets, trinkets, and maybe money, open to any that had a fancy to help themselves. I was mad with 'em for it, because Master William was no regular charge of mine, only made over to me to keep out of the Countess's way; and I could not leave him till I had pulled off his wet things and laid him in bed, where he insisted on holding my hand till he fell asleep, because the maids had told him Ludlow Castle was full of ghosts.

Meantime, the window being open, I heard the sounds of far-off merriment, that to a younger woman had been tantalizing enough; and whenever I tried to draw my finger out of his little hot hand, he would open his eyes wide and say, 'Lanfear! are you there? Oh!'

Presently, a voice under the window said, in surprise, 'What, Jack! you here after all?'

To which another responded, 'Indeed, you lured me hither, but I have done myself wrong, have I not?'

The first rejoined, 'How should that be? You can have done yourself no wrong that I know of, and for certain you come to do us good; though so late in the day that no lodging has been assigned for you. I will speak immediately to the chamberlain.'

Then the other interrupts: 'Trouble not thyself, good Harry, for I come but for a single night, somewhat to the discontent of elder people. Let me but have a corner of thy chamber for this night, and it sufficeth.'

'Right welcome shall you be to it, Jack,' said the first, whose trick of speech I now felt assured I knew. 'But now come along, for I am wanted, and may not tarry. I will just get you a cup of wine, and then carry you to our green-room.'

'You may spare the wine,' says the other, 'for I never care to drink it; so let us repair to the green-room at once.'

All this was spoken rapidly and with jocundity, as between familiars. I loosened my finger from the hold of Master William, who was now fast asleep, and spying from the lattice, saw Master Lawes quitting the court, along with the young gentleman who had pulled the child out of the water-butt.

I now pinned in my gathers, and then rummaged out an idle baggage named Audrey, who, like all the others, had been gadding. Her I well rated, as representing the rest; then sent her, in spite of the pouts, to sit by Master William till Mrs Nurse returned.

After this I posted off at my best speed to the chamber we call the green-room, where all the preparations were made for the Masque that was to be enacted as soon as the supper-tables were withdrawn.

Here I found a scene of confusion impossible to describe; people capped with heads of horses, asses, oxen, wolves, wild boars, pigs, and geese, slipped on over their own heads, and amusing themselves by howling, braying, and grunting in character. These were selected from the mixed company, many of 'em persons of quality, who had no regular part set down for them, but were to leap about, and glare through the trees, as the rabble rout of Comus.

Also the nymph Sabrina, in river-blue silk and silver, with water-lilies in her hair, very fantastical and pretty, with inferior nymphs attending.

I passed all these to get at my Lady Alice, seated in her magic chair, arrayed in her gown of white silk shot with silver, feeling assured she would eclipse 'em all; but, to my consternation, I perceived her shedding of tears, while Lord Brackly was red with anger, and Mr Thomas looking full of mischief.

Cuthbert, habited as Comus, garlanded and smeared with berries, and his eyes set in his head, was absolutely sottish and drunken, he having found his way to the sack-posset, and taken so much of it on pretence of getting up his courage to play his part, that he was now absolutely disguised with drinking.

I heard broken utterances of 'the masque must be given up,'—'he will disgrace us all,'—'he hath lost all memory of his part;'—and meanwhile an usher, with his 'by your leave, gentlemen,' comes from the great hall, to say my Lord and Lady, and all the noble company, are waiting.

'Go and tell them, Brackly,' says Lady Alice, crying, 'the masque is spoilt.'

'In faith, then, I will not,' says my lord. 'I'll try first whether I cannot thresh this numskull till I've beat the sack out and the sense in.'

'There's no sense to beat in,' says Mr Thomas, turning on his heel. 'What! is there no one of all this goodly company that will play Comus for us at a pinch?'

Oh, they would any or all have been so glad to do it (I believe 'em!) if they had known it in time! only it was too late now to con the part.

'But you shall be prompted all through,' urges Mr Thomas, looking from one to another, 'prompted in an audible voice.'

But no, not one of 'em was bold enough (and in faith I think they were right); they could roar, cackle, bray, as long as you would, but not recite a long part at one reading. Meanwhile Cuddy was beginning to blubber and rub his head (my lord having rapped it pretty smartly with the flat of his sword, a real one), and Mr Lawes, taking him by the arm, pulled him away a little, to where the strange young gentleman, whom I shall call Innominato, stood apart, calm, quiet, attent, and frowning a little, but most beautiful of aspect, strangely contrasting with the monsters about him.

'Recover yourself, Cuddy,' says Mr Lawes, not unkindly, 'for the conjuncture is pressing. All this large company will be disgraced, and the assemblage in the hall disappointed, if you collect not yourself sufficiently to repeat your part without blunder. Compose yourself, and try at it once more; you shall have enow of prompting.'

Cuddy, after one more ubbaboo, bursts forth in stentorian stuttering:—

> Welcome, welcome, Joy and Feast,
> Welcome—(plague it) Bird and Beast—
> Midnight shout and Revelry,
> Tipsy rout and . . .

'O, murder, murder!' cries Innominato, clapping his hands to his ears; while Cuddy, quite beside himself, begins to prance like a Mohawk in the war-dance, bawling out 'Hip, hip, hurra,' and then suddenly falling all along on the ground.

I never saw a creature look more disgusted than Lady Alice; while some openly laughed, and others cried shame. All the while, Mr Lawes kept plucking his friend's sleeve, and saying, '*You* know the part—*you* do it!'—and the other, 'No, no;' and putting him off.

Of a sudden, Mr Lawes goes up and whispers Lady Alice and Lord Brackly, who start, and look towards the Innominato with curiosity and pleasing awe; then, after a word or two among themselves, my lord takes Lady Alice by the hand and leads her with gracefullnesse to the stranger, to whom she, once lifting her humid eyes and no more, makes request, in what words I know not, being only near enough to catch the prettiness of the action; but he, I could see, bent low, yet with the mien of a prince, and said in a rich, most harmonious utterance, as if every word had a value and beauty of its own,—

'Your ladyship's behest has my obedience.'

And without another word, he picked Cuddy's ivy-wreath from the ground, set it on his own graceful, redundant locks, and, throwing on a kind of domino which lay at hand, took up the charming-rod, and stood looking calmly around with the austere majesty of Prospero rather than the tipsy jollity of Comus.

An irrepressible smile of triumph shone on Mr Lawes's features. I never saw creatures look more relieved or grateful than Lady Alice and her brothers. She said:

'Sir, I can never forget this goodness. You have nobly come forward to help us in our strait.'

'O, name it not, madam!' said he. 'The only requital I sue for is, that my share in the Masque, and my proper name, may remain unmentioned.'

'You may rely on us,' said she, softly.

Meanwhile, a whisper ran round, 'Who is he?' And Mr Thomas, who loves fun, put it about, 'A duke's son in disguise!'

Mr Lawes, having now gone on the stage as Thyrsis, was reciting his prologue; and the prompter, now coming in to say Comus would be wanted immediately, Innominato, who had been standing with folded arms in deep reverie, probably recalling his part, which, it appeared, he knew by rote already, instantly went forward and stepped upon the stage; we crowding every coign of vantage where we might see and hear.

For my part, I stood on a table behind one of the great screens, and could see about three-fourths of the stage pretty well, over the heads of others; but the audience on the daïs I saw not. What a clapping of hands there was when he went on, followed by such a silence, you might have heard a pin drop.

Nothing could be more charming than his declamation, which yet had a certain measuredness and severity that the theme, methought, demanded not. However, all were mighty attent, with a kind of strained attention to a very grave exercise of the intellect.

They relieved themselves by dancing the Measure with due jocundity; and then all the inferior players fell back among the manufactured

trees, which were real greens tied on frames, to make way for the lovely
Lady.

I never saw Lady Alice look more sweet! She had already been listening
with concentrated attention to Innominato's recitation; and, reaching up
to my hand where I stood, squeezed it and whispered, 'Now, Goodie, I
have full confidence. Nothing can put me out now!' I was so glad to hear
the dear Child say this; and to see her go forward with such wonderful
and beautiful composedness; articulating every word so distinctly and
sweetly as to be heard in every part of the hall. In the echo-song, she
outdid herself, though at first her voice trembled a little; but I loudly
whispered, 'Let my Lord President hear the words,' which fixed her
thought on him and not on the company.

Innominato, with his grave, harmonious voice, then resumed; she, with
more vivacity, replicated; he, bending his serious regards on her with
kindliness, rejoined; she responded with artless brevity. Thus one, then
the other, alternated in graceful question and reply; she looking full at
him with her honest blue eyes, and with an amused smile now and then
playing about her honey-sweet lips—he never once laying aside the mien
and delivery of a young Sage; and, when she reached

> Shepherd, lead on,

conducting her away by the hand amid continued applause.

This scene gave the key-note to all that followed. The noble Brothers
gave their dialogue with marvellous spirit; and when Mr Lawes, who in
the rehearsals had been quite the master actor, re-appeared, he seemed in
comparison tame; being in fact only up to his usual level, whereas they had
transcended theirs. In young people, stimulated by unwonted applause,
this is easily accounted for.

But then came the touch-stone; to wit, the scene in Comus's stately
palace, set out with all manner of deliciousness; which was simply represented by withdrawing the frameworks with greens and discovering the
lower end of the hall, hung with scarlet and gilt leather, and having a
corner cupboard and sideboard of plate, with a table spread as for feasting.
This scene, I knew, would be the great test of Lady Alice. As for Innominato, we already had a dim perception of what was in him, for he
had given the tone and colour to the whole piece; a classic gravity, more
consistent with a Greek play than a mumming interlude; but though this
would have less savour for the mixed crowd about the doors, there were
those on the daïs could appreciate it.

My Lady Alice, then, is discovered in the enchanted chair, from which
she essays to rise; but the magician, in dumb show, puts her back, offering
her the Cup of Intemperance, which she refuseth.

That noble colloquy then ensued, in which I no more seemed to see and hear a girl of thirteen, but a woman ripe in all virtue, eloquence, and wisdom; for the words put into her mouth seemed by the justice and force of their delivery, her own.

As again and again he tempted her, there was something of passionateness in her utterance of:

> 'Twill *not*, false traitor!
> 'Twill *not* restore the truth and honesty
> That thou hast banished from thy tongue with lies!
> Was *this*—(*looking about her with flashing eyes*)
> Was this the cottage and the safe abode
> Thou told'st me of? What grim aspects are these?
> These ugly-headed monsters? Mercy guard me!

She almost shrieked out this ejaculation, and the rest was lost in a murmur of applause.

Then the deep, earnest voice, with its bitter sarcasm, rejoined,—

> O foolishness of men! that lend their ears
> To those budge doctors of the Stoic school,
> And fetch their precepts from the Cynic tub!
> Praising the lean and sallow Abstinence.

Never was Intemperance preached in a way so unlikely to win disciples. Lady Alice seemed scarce able to hear the sophistical argument to an end —again and again essaying to leave her chair, yet unable; and then, when she burst in with her withering rebuke, a pause ensued; and in yet deeper, more soul-subduing accents, he went on with,—

> She fables not! I feel that I do fear
> Her words, set off by some supernal power.

Suddenly, a clash of swords, and disturbance without. The brothers, beautiful as Castor and Pollux, rushed in with drawn swords—dashed the cup out of his hand and shivered it to atoms. The rabble rout closed them round and a grand mêlée ensued; but finally they and Comus were driven out. Here was a fine, moving scene. There remained a striking picture. The brothers, lowering their swords, approached their sister—she remained charmed to stone in her chair, her senses all bound up in alabaster!

This, though little in the telling, was as effective as anything in the whole piece; so rarely did Lady Alice personate the character.

The attendant Spirit comes in. 'What!' cries he, 'have you let the foul Enchanter 'scape? Oh, you mistook! you should have snatched his wand, and bound *him* fast!'

Too late! The brothers, in grief, seek to know what is to be done. The Spirit tells them there is a certain nymph not far from thence, by name Sabrina, who is goddess of the river Severn, and who has power to reverse the spells of malign enchanters, if duly invoked in song.

On this, Mr Lawes gave us a charming cantata, invoking Sabrina to 'rise and heave her rosy head from her coral-paven bed;' and then Sabrina, attended by water-nymphs, did enter in most graceful guise and reverse the spell. And so the Lady is disenchanted, the brothers embrace her, and the Spirit leads them, amid singing and dancing, up to the daïs, to present them to my Lord and Lady, who embrace them all round, every one pressing upon them with compliments and felicitations. Then, the musicians playing loudly in the gallery, there ensues a general dance.

But amidst this, my Lady Alice, being wondrously flushed and over-wrought, obtains in a whisper, leave of her lady-mother to slip away to her own chamber and go to bed; and the dear Child puts me off from following her, saying, 'No, Goody, no—I need to be alone—you must stay for the posset.'

Meanwhile my Lord President calls aloud for Comus, that he may compliment him handsomely. But Comus, strange to relate, is nowhere to be found, he has vanished into thin air, as an enchanter should. This causes wonder, laughter, and is presently forgotten; for the sack-posset is being wheeled in, on a huge stand, by ever so many cooks and butlers, with cornets flourishing; and every one is anxious to have a taste.

Sure, never was so inconsequent a sequel to a Masque preaching Temperance! The healths began, first in spoons, then in silver cups; and though the healths were many, and a great variety of names given to them, it was observed after one hour's hot service that the posset had sunk in the vessel only one inch! So then my Lord called in all the household and hangers on, including many strangers, and they upon their knees did drink my Lord President's health in brimful tankards. This lasted till midnight.

Oh, what a parody on the Masque! And there lay Cuddy in his remote turret, disgraced by drinking a pint of sack, while all these were swilling quarts! However, his default had procured us the rarest Actor ever seen, so nobody could owe him a grudge.

My Lady, having made me drink her health and my Lord President's, and the healths of all the noble family, sent me with a cup of the posset to my Lady Alice, to warm her and do her good. But I was checked, on going in, by seeing the good young creature at her prayers; and when she rose from her knees and I offered her the drink, she said,

'Feel how my hands burn already, Goody! I'm sure I want no hot drinks. And do you know, the gentleman that played Comus, said aside to me ere he went, "Too much acting will be to you the poisoned cup—shun it, like wine!"—Oh! how well it went off, after all! but what a feverish dream it has been! I don't think I shall want to act again.'

Her pretty lip quivered as she spoke, but I feigned not to perceive it; and, since she protested against even a mortice of wax, I left her, as she desired, in quiet and darkness.

There was no quiet in any other quarter of the Castle! Even in the far-off nursery, Master William had been woke up and was roaring. Few men went sober to bed. Oh, what a parody!—And now, it has all faded away—and

> E'en like the baseless fabric of a vision,
> Left not a wrack behind.

Lord Jerningham

∞

I had some knowledge of George Vestries five years before he doffed the Stuart colours and joined us, when there was a stir in London over his betrothal to Anne Wilding, maid-of-honour to Henrietta Maria. Half England talked of this lady, of her beauty, gaiety, wit, and of her charities. Those were sharp days. For my part, I was employed in dangerous fomentations, but, like the town, I found time to hear of this alliance. Then the embers flamed up betwixt Court and Parliament. Vestries's father, Lord Hampton, went with the Sovereign; but the young man remained ready to range himself with the opposing forces. Thus, this prudent pair made out to save their goods; for, when the King was brought to book, Hampton must flee the country, but the noble estates lay fallow for his son, and were not confiscated. Old Sir Henry Wilding followed Charles Stuart to his own mortal cost; his two sons were shot at Naseby; and, himself being taken in Scotland, he was declared a traitor, and done to death in Edinburgh by Argyll.

His daughter, the maid-of-honour, was fled, none knew whither; some said to the Queen in France; while Vestries (thoughtful of his fields in Hampshire) allowed it to be said that his betrothal was set aside; and, lamenting ostentatiously the course his own father had pursued, rarely spoke of Sir Henry, and did not lift his hand when the old man came to trial.

The common rumour had it that Vestries was too stanch a Parliament man and Mistress Anne too bitter (she had reason!) on the other side to speak kindly with each other this side of Judgement Day; yet it was believed that she had loved him dear, and that he, for his part, had greater affection for her than for all else in this world save the keeping of the great estates. For this latter was the first principle with him; it was the marrow of his heart, and he regarded it preciously as a point of honour with posterity.

Time was when I liked him very well; for upon the happy fight at Naseby, which made me a Colonel of Horse, he joined us of the Parliament side, in the field, and soon after became a captain in my regiment. He was a man of parts, and, bearing a lively sword in action, he could

show a most engaging front to his companions in arms at the board. He
was drawn to select service with my command by hearing that among us
a man was not like to be shot for an oath or damned for a bottle of wine,
the officers enjoying some acquaintance beyond the pale of the Psalms
and the peregrinations of Joshua. For this (when they had whispers of it)
the Levellers mouthed and spat sulphur at us whiles, but our work—and
we had enough—was our reply. I boast there were never soldiers more
precise in their duty, nor, at that time, more devoted to Oliver; but we
were concerned in the political issue, not the religious. We nourished
the Republic in its cradle with our glad blood, and faith! a divine, or a
ranting corporal in our camp might only curse us into sweeter drowsing.
Vestries was, in seeming, a man of the temper of those who fought about
me. Sooth, we all liked him, 'tis the truth; he became my near friend and
familiar, and so I came to know the true case between him and Anne
Wilding.

The King had done with cheating, with tricking us, and with life, and
the block was set away. It was an October morning, and we lay in a wood,
near Oxford, encamped. I was seated on a fallen log, writing, when a
ragged fellow darted by me and lost himself in the underbrush. Vestries
came flying in pursuit, passed me; then, after I had helped him to beat up
the bush, in vain, he came and flung himself down, groaning, in my
bivouac. He reached up and handed me a scribbled paper.

'Read it,' he said. 'It is from a lady you must know of.'

It was a letter from Mistress Wilding. He knew not where she hid
herself, he told me, but two months past had found means to get a letter
to her through a Scots prisoner, who was very secret with him, and would
only tell him how to set the missive upon the first stage of its journey.
He had written this letter, he groaned, in an hour of acute remembrance
of her, an hour of agony when it seemed he must have her, could not
face the years without her, and now this tattered lackey had come out of
nowhere, had squeezed the answering script into his hand, and, stealing
off while Vestries read it, vanished lightly, as I saw.

'Read it,' said Vestries again. 'It's a word from the only woman ever I
had eyes for! And be sure she writes to the most unhappy wretch in
England!'

The letter was short; there was none of the fripperies or elegance and
shepherdizing that now find vogue:

Disloyal to the King, couldst thou be loyal to me? Thou needst not protest and
filly-fally with a lady that is not for bearing with thee much further, yet likes thee
dear. Ah, sir, my love has gone a great way; thine shrank back at the first
hardship—but now comes, still reaching timid hands after me! Would have me—
and fears to have me! Thy father almost lost thee Hampshire. Thy sweetheart

might tip it out of the scale and set it wasting under the butcher. Even in this anguished missive of thine there is hinting and hold-back a-plenty. Even so, I wear it on my heart! Was it conscience set thee under Cromwell? I may prate of conscience a little, for loyalty hath rewarded our folk sadly, making it a sore task to write to one under such arms as you bear.

Yet do I put love over conscience—nay! conscience is the finer part of love! Couldst thou leave all to follow either? 'Tis the only way thou shalt ever have me, and if thou canst take me so, thou shalt have me. And though that time last but one little moment, it shall be worth all. If I could see thee of my way of thinking, death would be naught. So, I will put thee to the test. Thou hast sworn much and pled much, but ever words—words! I will put thee to the test: see thou to it that thou be ready, dear one. Leave all and follow love when I come. Look to see me in thy camp before a year is gone. Beware my coming with fire and sword, for, know you, dear enemy, I bear arms—I bear arms, and fight on the border for the Martyr's memory, and I plot through the realm for the blessed young Majesty, King Charles the Second!

Shalt hear no more till I come; then be ready, love, be ready! If thou wilt be true for that space, one little moment shall be worth all.

'A mighty seditious document!' quoth I.

'Heaven help me!' cried the Captain from the ground. 'For God knows what she may do! She is sweet as May, with May's soft humours, but there was always a wildness in her. . . . A strange woman, very beautiful, but . . . strange—strange! If I could know what she means!' He rubbed his brow with his hands. 'If I could get her off safely to the Continent——'

'If you could, it might be well if it were done soon, Captain Vestries,' I said. 'I make sure she is in a perilous mix-up; these are evil days for enemies to the Commonwealth, such as she confesses—nay, professes herself. If I were in your place——'

'*You!*' he broke out loudly. 'You would clap her in jail to be tried! But you shall not come near her!'

'I hope not,' I answered earnestly.

'*I* am no machine of war!' he cried. 'I am a man of heart, do you understand? Lord, Lord, she has endured such miseries! You do not know her, nor what she might do in her desperation, and having no one in the world but me—and I against her!' He struck his hands together like one in despair. 'My cousin Jerningham should have had her; he would have haled her out of all this, and kept her safely to a woman's work, if it took a beating, and he *would* have beat her if it needed!—for he is as tempestuous a harebrain as herself. He loved her very well, long ago. He went to sea, and when he came back I had won her; so he sailed again and forgot her. Yes, I won her—woe worth us both! What is to be done? Not that I dream you can tell me!'

'I can tell you nothing but to come at some means to get her to the Dutch or to France. The plain truth is best, Vestries: friends of the Stuarts are to grow very few.' I spoke slowly, that he might weigh the words as I did. 'I say these are bad times for such; even the slip of a word or two is a chancy thing wherever it happens. Get her away if you can.'

'Get her away!' he groaned. 'How can I? If I knew how to reach her, could I move in the matter? You say the truth; a word or two goes a long way these times. What could I do if I knew? She would not go! You are a sweet comforter, Colonel Thomas Breed! Oh, mercy! She bears arms, she says—and on a lost side! And she so gently reared!'

I left him lying there in that strange folly, and returned to my writing. He was a man in love; in spite of all that came, I affirm it. Moreover, he was capable of a marvellous loyalty, but it was neither to man nor woman, but to the senseless stretch of earth of which some scribbled sheets declared him master. And if this faith he kept with his estates could have gone with his love to Anne Wilding, I think it would have made that love she sought so eagerly, what sort she held to be worth all—even in one moment.

From that day of her last letter to him, a wan apprehension sat upon the soul of George Vestries, and was visible to me. He was full of a loud gaiety in company, but the mood beneath his manner mocked him. And though the autumn waned and the winter wore away and his mistress had not descended upon our lines, yet she was a lady of her word, and, to most ways of thinking, desperate. He was a whiter-faced man each month, in the waiting for her coming.

On March 1, I lay overnight with three squadrons of horse at Rochester. That evening I and Major Hatton, Captain Vestries, Captain Merrifield, Lieutenant Chesley, of my regiment, and Colonel Lorrimer, and a half-score of other officers of a division of foot, which was quartered at Rochester, supped at The Parliament Man, an inn, late The King's Arms, redubbed again since that day. 'Tis the honest hosts of England and their signs that are the politicians' almanac.

Vestries was our host, making the feast in honour of his kinsman, Jerningham, who had just come up from the sea (where he had served our cause) to join us at Rochester.

The men who sat at the board were such as I would have picked from the whole world for battle or bout; and how warmly they come before my mind's eye today, as they sang and laughed that night! A burly group in carelessness, cuirasses doffed, long swords depending from chair-backs, great limbs at ease, great hearts in cheer; Hatton, Chesley, Merrifield —could I evoke those stalwarts from the past to ride at my side once more, I might not need to lay exiled old bones among the Dutch! Poor

Merrifield!—a soldier, not too quick to do his own thinking, but unquestioning and fateful in his strokes for the right cause; as kindly a companion at the table as a squire might wish, a hearty drinker and a hearty doer—he had ill-luck!

The Cavaliers will tell you that our troops and their officers were one fanatical crew; but we had some who could do as noble trencher-service as any gartered lackey about the court today, and we proved it in our first toast to Jerningham, for it showed the clean bottom of the punch-bowl.

John Lawrence, Viscount Jerningham, was a big young man, long-faced and high-featured, and something singular to look upon, for his hair was very light-coloured, almost white, but his eyes were dark, beneath black brows. His reputation (which had lately gone over the country) led us to expect another man, something of the rough blade, for he had commanded a sloop whose doings cost Parliament no little stress of imagination to acquit of piracy. She had fought wildly on many coasts, ever with such stirring abandon as brought her off safely; so we looked to discover in her master a spice of the bravo, but we found such flavour lacking in him. He carried about him, rather, the air of the ante-chamber. When he bowed, we felt the soldier, but the soldier at court; there were rustling of gowns and murmur of silken whisperings; yet he had the free, strong voice of a sailor, and his was a laugh to win your liking at the first sound of it. But I thought him the lighter kind of cynic, one with a humorous condescension toward life, and little care of it; and deep in the dark of his eyes there glinted a spark that might flame up in the right wind—a glow hinting suddenness of temper, strange capacities, and not altogether sane—so that I could understand his cousin's having called him 'tempestuous' and 'harebrained'. Such was he, the man of most spirit I ever saw.

But the guest of honour carried the gaiety of the evening upon his broad shoulders very winningly. After a number of general choruses, he answered our pressure for a song with a Cavalier ballad, which he said he offered as a curiosity for our inspection. He sang well, and was heartily applauded, though I felt it needful to question the propriety of repeating Cavalier sentiments except as evidence before judges. My courtesy was not equal to Lord Jerningham's.

'I hope, sir,' quoth I, growing red, 'you own to no feeling of disaffection towards our leaders?'

'Nay,' said he, with his rich laugh, 'if I did, I trust you would not believe me so foolhardy as to let it be known to Colonel Breed! I have often heard of you, sir, and I esteem the honour I assume for myself in pledging you.'

With that we sent the healths around, giving throat to many a jovial song, pledging every member of the company in turn, and when the echo of the last shout was lost among the rafters and we were all reseated, my lord remained upon his feet.

'Let us annul these cups,' said he. 'Let us begin again' (whereupon there went up a great cheer), 'and for the first toast I propose something sweeter and less rusty than ourselves. Does ever a soldier fall without a name upon his lips as his soul is spent? It is in that final test he finds which name he loved the best—for a soldier does not live long enough to discover it before!'

'There spake the sailor!' shouted Chesley.

'Pledge the name,' Jerningham pursued, laughing. 'The name you dream might come upon your lips if you fell tonight, though God forbid the thought! Drink to something better and worse than war or men. And I call the officer who defends the fidelity of landsmen!'

Thus challenged, Chesley rose.

'To Phœbe!' he said gallantly. 'I have no need to die to learn the name! To Phœbe, peerless among her sex, unversed in the arms of Mars, but wise in the arts of Venus!'

'Nevertheless,' quoth Lord Jerningham, with a bow to Chesley, 'we hope that though unversed in the arms of Mars, the lady may not long prove entirely unacquainted with those of a valiant son of his.'

And Chesley, in confusion, called Captain Vestries.

Those of us who were acquainted with his story were curious to see what he would do. I knew him. I thought he would not toast his love in any company of Parliament men who might recognize 'Anne' as the Royalist Mistress Wilding; yet he had won the girl from Jerningham, and must feel a bitter shame if, under his cousin's eyes, he dared not name her; and yet—Hampshire haunted him. 'Twas a dangerous time, and a little thing might divorce old lands from old names. The man's hesitation was palpable. He uttered a half-incoherent plea for excuse, not rising, while many eyes were bent upon him curiously. Then Jerningham himself came to his relief.

'Cousin,' he said, as Vestries fidgeted, 'there is a rare comfit of the low countries in a little silver box in my chamber above. I should like our friends to judge of it, if you will oblige me by fetching it hither. You know how loutish lazy I am, and you were always the quicker. I believe when we taste the comfits we shall accept your service in fetching them to acquit you of a toast.'

There was some noisy protest at this, but Vestries, with a quick look of gratitude towards his kinsman, hastily quitted the room, and Jerningham suggesting that we fill the time with a song, we roared forth 'Fill the Bowl' with good will. At the conclusion we discovered the landlord bobbing in the doorway.

'Is your business pressing!' asked my lord, as the song ceased and we fell silent. 'Have we drunk the house dry?'

The landlord explained that a gentleman had arrived at the inn; that the place, being full, he could obtain no lodging. He had travelled a great distance since morning, was chilled with the wind; the night was now far advanced, and he refused to go further. He craved our permission to sit quietly by our fire, since we occupied the main room of the inn. He wished not to intrude, but the kitchen, where the menials sat, was his only other shelter, and he disliked turning the poor hinds out of doors to the sheds on such a night.

This last set me a-thinking: it was high-tainted with Cavalier; moreover, it seemed to my quick suspicion that the gentleman misspoke himself, and that to intrude must be precisely his desire. I had my mouth open to say so, but Jerningham at once bade the landlord to entreat the gentleman's presence, and, almost as he spoke, the stranger appeared upon the threshold and bowed with grace, though not uncovering.

The picture he made, framed in the low doorway, I can see well today, here in my exile among the Dutch. Cavalier from head to foot! Not tall, delicately fashioned, a proud, slender figure wrapped in a long riding-cloak, with muddy boots of buff, a long rapier in a frayed velvet scabbard, a great, plumed, flapping hat, a strange, thin face, wan in its shadows, and brown hair a-tangle about it, blown by the winds of the March night, violet depths under the large eyes, the features sharpened by either thickness or hunger; nevertheless, there was, in the mien and looks of the storm-tossed apparition, a certain wild brightness, the like of which I never saw, and, flecked with mud and weary-white as he was—even the deep lace at his throat and wrists torn, wet, and splashed with the road—his bearing was at once dainty and magnificent.

Jerningham bowed to the ground.

'We shall be honoured to have you join us,' he said.

The stranger bowed again, and begged our excuses in a low voice, which methought something too high-keyed for the length of his rapier.

'Nay, sir,' said Jerningham heartily, as the stranger moved towards a settee by the hearth. 'We beg you, do us the honour to join us at the board, and you may sit upon the fire side of it, and dry yourself at your ease. I am John Lawrence; pray, let me make these gentlemen known to you.'

'My lord, he looks like a bloody-hearted Stuart man,' whispered Merrifield, at his elbow.

'Let us have no politics, in Heaven's name,' laughed the Viscount softly, laying his hand on Merrifield's shoulder, 'over a bowl of punch!'

The stranger let his hand fall negligently upon the back of the settee, turned with some languor towards Jerningham, and said:

'For this courtesy I thank my lord Jerningham, who, if I mistake not, has but lately returned from the high seas where he has been serving against the King?'

Merrifield uttered a great oath; angry and amazed faces were turned towards the young man from every part of the table. I half rose to my feet, but the Viscount laughed aloud.

' "Against the King of Spain," the gentleman would say if you gave him time to finish,' he cried. 'Is it not so, sir?'

The newcomer looked at him, and then at us, with a sudden gay look, strangely like that of a teasing girl.

'Ay, sirs,' and he laughed with Jerningham. 'Against the King of Spain!'

Upon this Merrifield humbly craved pardon for his blunder, and the laugh went round the company—until it reached me; for I did not join it.

'You cannot refuse us, sir,' said my lord gaily. 'In the name of my cousin, Captain Vestries, whose entertainment this is, I make you welcome, and answer for his eager entreaty. Come, sir, we shall be glad of your assistance with a fresh bowl which should not be ill known to you.'

The guest bowed, said quietly:

'Indeed, I take it as a great honour that I may join such loyal hearts— and my name is Charlton,' and accepted the chair which the Viscount himself placed for him at the curve of the table near the fire.

As he took his seat, George Vestries re-entered the room. I saw a faint look of surprise in the Captain's face as he beheld a new figure at the table, and then Jerningham's broad back cut off my view of him, as he went to take the box of comfits. With that there rose the roistering clamour of a drinking song; and I remember that I saw (without noting it) the Viscount introducing Vestries to Mr Charlton. The latter had risen, and the three gentlemen were bowing very low.

When I next glanced at Vestries he was again in his chair (across from Charlton), sitting sidewise, in a heaped-up fashion, with a multitude of fine drops on his brow and cheeks, which were so sickly pale that it startled me, until I thought the liquor might have gone far with him—it seemed he would be first under the table.

'Gentlemen,' said our new guest, as the song came to a finish, 'I think you were passing the toasts before I came in, and I would not interrupt you. May I ask how far they had been honoured?'

'Up to Captain Vestries,' said Fanshawe of the foot, a good man, but heavy. 'We were drinking to our Ladies of Heart, sir.'

'Much honoured, they!'

'Vestries!' cried Fanshawe. ''Tis your turn, Captain!'

'I seem to recall he had given his,' interposed Jerningham, 'before he went for the comfits.'

'He had not,' shouted Fanshawe, laughing, and pounding on the table. 'Vestries! Vestries!'

'Nay,' said my lord, still striving to spare his relative, 'I do remember we excused him for fetching the comfits.'

'We did, indeed,' seconded Chesley, who repented his thoughtlessness.

But 'Vestries! Vestries!' shouted the leaden-headed Fanshawe, while the others were obliged to take it up. 'I appeal to Mr Charlton as, perforce, an impartial judge!' he cried.

'Escape appears difficult for you, Captain Vestries,' said Charlton. 'To the lady of your heart—if, mayhap, there be one?'

I marked that my lord, who was regarding his relative with some sympathy, turned sharply towards the stranger as though struck with something in his tone. Charlton looked up, their eyes met; then Charlton's gaze fell slowly till it rested upon George Vestries.

The Captain unsteadily got to his feet. He supported himself by the back of his chair, and several, considering him far gone, nudged their neighbours.

'I drink,' he said thickly, 'to loyalty!'

'Loyalty to your lady?' It was Charlton's high, clear voice.

'Loyalty—to country,' answered the Captain, with a catch in the throat. 'Fidelity to honour! I call Colonel Breed,' he ended, half audibly.

'Faith!' quoth I, rising quickly, as a buzz went round, 'I've no sweetheart. I have no love, no lady wife, nor, please Heaven, in no sense shall I ever own a widow. I can sigh out no toast to a whining damsel, and, being a bad hand at carpet-dancing, let me say for you, "Lord save any lady from my drinking!" But since it needs be a toast gallant, I pledge not a woman, but Woman, the Creator's last gift to man—and would He had been less generous! To Woman, then! A hearty health, drink deep; much happiness to her, say I! I call the guest of honour.'

Amid a great outcry against me, his lordship rose and bowed. I have been all my life a hard republican, but I like to speak of John Lawrence by his title; for he was a lordly man, and it suits him.

He inclined twice toward the assembly, addressing himself especially to Mr Charlton, as if desirous of causing him to feel one of us, and apparently much taken, as were we all, with the odd charm of the stranger's presence. The latter gave him a quick smile, but I observed that his eyes turned sharply from Jerningham to the Captain, who was bent low over the table, his head in his hands.

'I would not offend against your patience,' said my lord. 'Reproach me, but not her I honour, if the toast be long. Brothers of the board, Colonel Breed has given us a right word. Roving on many shores, I have ever found women to be alike: tender jades, harsh as flint; leal and true, fickle as the wind of this night; sweet as honey, bitter as aloes!'

'But this is a mass of contradictions!' objected Lorrimer.

Charlton, who sat next, turned to him with a little laugh. 'No more nor no less!'

'No man may come nearer the truth,' said my lord, with a fine, dry smile. 'If I were a painter, I could never decide whether to depict a woman pointing down and leading up, or pointing up and leading down!'

'Do both, my lord,' said Charlton. 'Both—for different men.'

'True, and wisely understood,' answered Jerningham. 'And such as she is, divine or too human, sweet or too sweet, we offer her our heart— sometimes from a distance! It may be better to send it, than to stand close and give it in her hand. Friends, we love her better far than near, more in one moment than in eternity!

> Oh, love of mine, what sorry years
> I'd spend beside thee! Strange it seems,
> But though thy presence, sweet, endears,
> I loved thee better in my dreams!

And thereupon there was a general protest and great remonstrance.

'You cannot prove me ungallant!' my lord responded, 'for I exclude from such bad lines all true mistresses of true lovers, and, above all, the lady to whom I shall drain this glass, though she be no sweetheart of mine.'

Now, with this there was a pricking up of ears.

'A sailor roams in divers parts,' he continued. 'Demoiselles, mesdames, señoritas, signorinas, signoras, señoras, all flicker before his glass; I have the honour to swear that none of these is worth the outermost tucker of a true English maiden's petticoat! I have dangled a trifle at the heels of some ladies of Spain and other lands, but I never found true love in my heart except once. Then I was not more than a lad, with my cousin, our gentle host there, and our choice fell upon the same sweetling.

'Old loves—first loves—they are the best of all! Out of the springtime past I evoke her lively image. When I first saw her—knee-high she was— her head was crowned with the treasure of a Spanish galleon, gold, like out-of-doors in June—nor did my heart grow cold to her as it darkened! Her I evoke, the one child in my boyish mind! To her I drain this goblet and summon your deep quaffing! First love is best!'

'A name!—a name!' was the cry, as we rose to honour the toast— 'a name!'

'Most willingly,' he answered, 'and without encroachment on any man's prerogative. My dear cousin and old friend, George Vestries, will take only as I give—in friendship to him and his—this toast from the memory of the one innocent time in a life not too innocent. And I will not show myself so churlish as to mistrust your gentleness, gentlemen. I hold no surnames from this company.'

He bent his head slightly, and then, 'To Mistress Anne Wilding!' he cried, and drained his goblet to the dregs.

George Vestries stood uncertainly, leaning this way and that; he grew of a yellow pallor, and returned his cousin's bow with a tremulous head, while my lord's glance rested upon him kindly, but with a veiled pity.

The toast was drunk almost in silence, and then Merrifield began to bellow, 'Charlton!—Charlton! We call the latest guest! Charlton!'

'Will you so honour us?' said I.

The newcomer rose to his feet.

George Vestries uttered a sharp exclamation, and lifted both hands in a singular movement of protest. Lord Jerningham, breathing deeply, leaned across the table with an excitement that came near agitation; but, before either could speak, the stranger lifted a hand in a slight gesture which was imperial. 'Gladly I propose a health,' he said, 'I should take shame not to give you that one which should have been first of all.'

And then I heard George Vestries make a sort of moan, I saw his cousin's eyes flash again, and I understood who had come among us. 'I will put thee to the test,' she had written, 'Be ready, love—be ready.' No man stood before us, but there, uplifting her glass, girt with a soldier's arms, stained with a soldier's riding, stood the maid of Charles Stuart's court, the tragic waif, Anne Wilding. Vestries had known her at once, and Jerningham had recognized her—not upon her entrance, I think, but soon after (though I never knew)—and now that I understood, I marvelled at the dullness of the others, who looked upon only a handsome, haggard, storm-tossed lad.

So she stood and fronted us, elect to her hour, her face limned with light on the outlines, as a cunning painter draws his saints, her eyes in shadow, melancholy, wistful, fixed upon George Vestries with a serious sweetness that bore yet a tinge of fond, wild mirth, her light figure poised in gallant dignity; and for all the Cavalier's dress and ruffling sword she wore out of loyalty to the rotten cause that all her kin had died for, there was that in her look which made modesty to drape about her like a scarf.

I knew what she would do to test her lover, and I believe she thought he would fail her. He was to have the chance. And now, I still say this, as I have always said: Had I been in his place—had it been a mere question of groves, meadows, villages, and a stone house with me, and had I been that lady's love, I would have seen the tawdry things at the furthermost corners of hell before I would have let her come to where she now stood!

But for me the Republic was all: my youth was a dream of what it was to be; my old age is an anguish for what it might have been. I poured most of my blood in the dust for it, and now am a pariah near a Dutch grave because I gave my blood for its sake. I would have abated not one jot of my duty to the Republic though Gabriel came to bid me. But had I been George Vestries—ah well! the old must strive for mildness. He dribbled at his fate, and the land owned him.

A party of Royalists at supper would have pledged the King in the first glass; and I saw Mistress Wilding's purpose. My duty was plain: she had borne arms against the Commonwealth, and, by her own confession, engaged in plots against it; but I was not dog enough to interrupt her before she finished with her lover and had done with what she intended. And the water came to my hard eyes as I saw the brave figure she made—a girl alone among us, who were her enemies and her King's.

Vestries raised his head piteously; but her steadfast gaze held his and detained his purpose to speak.

'Gentlemen,' she said, 'soldiers of the Commonwealth, as you would be called, you drank to loyalty at the behest of Captain Vestries. Are there two kinds? Is loyalty to your country to be divided from loyalty to your King?'

There was a stir among us, an oath, a sharp cry of 'Treason!' and George Vestries staggered back from the table. But she lifted the long glass above her head with a joyous gesture.

'To King Charles!—to King Charles! A speedy end to Oliver Cromwell and his traitors! Who loves me drinks the toast!'

There was an instant of incredulous silence; then, with a shout and a crash of cups on the table, came turmoil and chaos. I moved towards the girl.

'You are under arrest!' I cried.

'I will not be taken alive,' she said. 'Who loves me drinks my toast!'

Vestries fell in his chair with a groan and dropped his head on his arms. Never have I felt a more consuming pity than for that woman then. For one instant, for the merest flicker of time, she lost the bravery of her bearing, and I saw her lip quivering, her hand shaking wildly. I stepped towards her, my hand uplifted, when Jerningham thrust himself between us and caught my wrist as she gave one last look at her pitiful lover.

My lord released me instantly. He had not even glanced at me; his eyes were fixed on her, and in them I saw the glow of that fanatical spark I had marked shoot up into flame.

'Look no more at him!' he cried, tossing his head back proudly, so that he seemed the tallest man I had ever seen—'look no more at him, nor ever away from me!'

He plunged a goblet in the wine, drew his sword and crossed it against the glass on high.

'I care no more,' he cried, 'who shall call himself Protector or King than I care to know what Frenchman's lap-dog died last night. But were there death in the cup, here's a health to King Charles! A speedy end to Oliver Cromwell and his traitors!'

Then he stood at her side and faced us. There was a great shouting, greater cursing, the crash of glass and overturning of chairs, and a clattering of steel as rapiers came out of their sheaths; but I moved towards Jerningham in advance of the others.

'My lord,' I cried wildly, 'the wine is in your head! We know you for a loyal servitor of the Government. We can forget your words. But—this——'

'Do not say more!' exclaimed my lord imperiously.

'If any gentleman knows aught that would misbecome me, were it spoken,' said the girl, 'he will die before he speaks it!' And so my lips were sealed, and those of George Vestries.

'We have heard treason here tonight,' cried Merrifield, 'and, by Heaven, you both stand trial for it!'

'I will not be taken!' whispered Anne Wilding.

'No,' said my lord.

George Vestries got to his feet and came toward us with some incoherent babbling. Two officers caught him by the arms and held him. An instant's silence fell upon us—the hatefullest pause I ever endured. Then Viscount Jerningham made a low bow to me.

'Colonel Breed,' he said, 'am I to have the distinction of engaging you?'

'An honour, sir, to me,' I replied sadly. I drew and saluted him. 'But your companion, my lord, is not to fight.'

He thanked me with a look.

'Friend,' he said to her, 'you must leave your quarrel to me. John Lawrence never drew sword in a cause he liked so well!'

He whispered something to her rapidly, imperiously, and I made sure it was a command to escape whilst he kept our attention to the struggle. She shook her head, smiling upon him gloriously, and took his left hand and pressed it quickly to her lips. Then he thrust her suddenly behind him, leaped forward, and engaged me. And even as he came at me, though that strange light, not wholly sane, played in his eyes, his face was so finely commanded by the soul of him that he seemed to be on the point of making me a friendly speech.

He was far cooler than I, and I immediately received a wound of my own making. I saw his point come searching my breast and parried wildly, and, in that quick necessity, upwards. The parry came too late, and the end of his blade tore through my cheek and across my forehead. The blood flew into my eyes, and I made out that several of my companions ran forward to help me. I shouted to them savagely to fall back, but could not check them. Jerningham's companion sprang to meet them.

I was half blinded. I had begun for my life, thrusting again and again to the full stretch of arm and weapon; but ours was not the only steel that sang in the room. Close by—so close that the four rapiers seemed one series of sparkles from a common centre, like a star—the girl had suddenly engaged Merrifield, and they fought at our side.

'For God's sake, Merrifield,' I cried, 'have done! Do him no hurt!'

'Why?' he panted, for the rapier was no toy in her hands.

I would have answered, but at that instant he beat down her guard and thrust behind it. I was thrusting at the same time; my lord saw Merrifield's stroke, and, with a cry, he turned and struck at the Captain's blade with his bare left hand—too late; it had gone home. His own parry was lost; I saw my sword enter his body before I could withhold it. I tried to recoil, but the first impulse had been fulfilled ere my hand obeyed its orders not to strike. So it was I killed John Lawrence, and that great pair fell together.

I stumbled back with the red sword in my hand; then I threw myself upon the floor, and took his head upon my knees. 'God help me, my lord, what have I done?' I said.

'Here is nothing for regret, Colonel Breed,' he answered.

He turned his eyes to where Henrietta's maid-of-honour lay. Merrifield, white-lipped and horror-stricken, was bending over her. From the other side of the room still came the prattle of George Vestries, where they held him. I knew my lord's wish; I moved him gently, very gently, close to Anne Wilding.

'I would I had been a better fighter,' he whispered, as the haze stole over his bright eyes. 'Still, I cherish it a matchless honour that—I—did fight—in—your cause.'

She raised herself on one arm and leaned over him.

'Sweet!' she said, and bent down and kissed him on the lips.

'First love is best,' he murmured, and he gave that half-sigh that comes last of all.

Her supporting arm wavered, and she fell across his body with that arm about his neck.

So he won her from his cousin.

Iconoclasts

Sergeant Salathiel Christ-With-Us Roper
Begins a Letter to His Friend AD 1651

∞

To my friend and fellow-worker in God, greeting. You bade me let you know how we fared and, as the Captain in charge hath tonight divers papers to dispatch to my Lord, the time seemeth opportune for the sending of this poor screed to you. The work proceeds apace; and if I say that oftentimes the sight of the destruction that is inseparable from the cause giveth me pain, you will understand that it is not because I favour the idolatrous and ungodly images, windows, and such baubles, but rather because I grieve that the Devil hath entrenched himself so firmly behind that which is pleasing to the eye. In this district there have been closed one hundred and four alehouses—and, except in the case of which more hereafter, with but little opposition and no bloodshed. At the great church at Welland were found divers Popish vestments, prick-song books and a Book of Common Prayer. All of these were destroyed with great zeal. At Cawley a Babylonish font, after the troopers had defiled the waters thereof, was used to baptize their horses. I am not, however, of the opinion that the work is put forward by such measures, which savour to me of the old mummery; nevertheless some are of the opinion that play of this kind offers the men encouragement and thereby strengthens their zeal.

At Welland we met opposition from a strange quarter. There we found an ale-house cunningly placed by the sea and away from the highroad. A sign over the door, which was at once cut down together with a great part of the thatch, bore the effigy of King Henry VIII. A body of us went within to empty the barrels and break down the copper in which the Devil's brew is made. Men for this work have, as you know, to be chosen with discretion; for many a one having put his hand to the plough hath looked backward, to be like Lot's wife confounded. Within were but three people, an old woman of an alien race, wearing an immodest gown of scarlet cloth, a man whom we took for her son, and a young girl. Yet did these three bolt and lock the door of the cellar and refuse to give up the key. Then the Captain, like the thunderbolt of Jehovah that he is, gave order to bind the people and break down the door of the cellar by force.

This was done and within was found, not ale only, but divers other liquors of which Scripture saith: 'Wine is a mocker, strong drink is raging.' All these were emptied out save only the brandy, which the Captain bade spare, it being a great virtue in sickness, and therewith had set aside to take to his lodging, where he keepeth sundry simples and medicines for the relief of sickness. At that the man calls out: 'I have seen robbers of many kinds and colours, but never before have I seen those that rob do it in the name of God.' Upon this the Captain says that unless he stills his tongue it shall be tied for him. 'As you will,' says the man, 'I would as soon be gagged with a piece of filthy cloth as with ill-quoted Scripture texts, and the good liquor like Abel's blood crying from the ground.' At this evidence of stiffneckedness the Captain begins to look around for some reason for offence, for here doubtless was an enemy that should be destroyed, root and branch. And there upon the wall was one of those images of Mary, the like of which I should not have thought would have been in England since the Queen left. There were flowers on a ledge before it. We had stumbled by chance upon one of Baal's strongholds. The Captain, with his sword, cuts off the plaster head and scatters the flowers, whereat the old woman gives a great cry, and said that it belonged to the holy woman who was dead. Much else she said, but her speech was broken. And the man began to struggle with the trooper who was holding him.

'I care nothing for the image,' he shouted, 'but that you should come into a private place to give rein to your whim—why, Charles himself, whom you dogs called tyrant, would have knocked on the door and doffed his hat to the lady of the house.' And he indicated the snuffling old coloured woman, who was still muttering about a holy woman. And I saw our Captain signal with his eyes to the man who was holding the tavern-keeper, and he took him by the throat as if to shake him into silence, but he bent back his head. And there was a deep silence in the kitchen. The girl began to cry: 'You have killed him.'

'It is not killing, to destroy the enemies of the Lord. Read the story of Samuel and Agag, else. And ye who remain had better look to your ways.' And with that we came out of the tavern, for the Captain's voice was final. One man after another added his word—'Even so perish all Thine enemies, Lord,' and other such. For a moment I regretted the death of one who, meseemed, was a brave man. Think not from this, my friend, that I waver in the good work . . . but hadst thou seen him, speaking his mind freely, in the face of us all.

Methinks that overmuch of my time hath been devoured by this account of an affair of little moment; you will wish to know more of the revelation of faith. . . .

A. M. BURRAGE

Mr Codesby's Behaviour

ວວ

I

Mr Codesby, the latest victim of that public benefactor Mr Titus Oates, went on his way from Newgate to Tyburn in the manner becoming to the times. Part of his sentence was 'to be drawn on an hurdle' to the place of execution, and this was being done. Plaguey uncomfortable it was. The cobbles on Ludgate Hill were never too smooth even under wheels and springs, and just now Ludgate Hill was badly in need of repairs.

London was, at the time, maintaining an expensive court. The dimensions of the royal harem were a matter merely for vulgar speculation. But Charles had owned to ten children, all by different mothers, and each of these ladies was now titled and dowered. Little money was left for road repairs. Thus was the lot of the wicked made harder by the profligacy of the court.

Not that Mr Codesby was a wicked man—from now on there shall be no secret as regards his innocence—but the charge of high treason had been found a fact by judge and jury. Certainly, Mr Codesby could not account for a paper, laboriously setting forth a vile plot to murder His Majesty, which was found in his chimney—where it had been placed by a friend of Mr Oates.

Fleet Street turned out apathetically to see the soldiers and the doomed man on the sledge. The sight was common enough, although, to be sure, there would be a great crowd around the gallows, and much trade done by the vendors of ballads and oranges—the sometime friends of Mistress Nelly Gwynn, now the mother of a little earl.

Mr Codesby was himself supplied with oranges, at his own expense, and he sucked at one from time to time—in a delicate and gentlemanly fashion behind his handkerchief—reserving the pulp and peel to fling into the face of any of the rabble who might venture near in order to insult a dying man. He was sleepy, but otherwise not in the mood for death. He had been awake all night, or most of it, conversing with the Ordinary of Newgate. This was not because he required the services of the Ordinary, but because the reverend gentleman—who was with him even now—declined to leave him.

No man had more respect for a minister of religion than had Mr Codesby; but a minister who repeatedly urges repentance for an uncommitted crime is apt to be a little tiresome during a man's last hours. And, to say truth, Mr Codesby had been concerned more with the chances of his reprieve, the odds against which were lengthening as the hours dwindled. He was not an irreligious man, but he thought, without worrying too much about it, that there must be justice and mercy elsewhere for one who had found neither here.

All that concerned him now was to put as bold a face on this last phase as a man might show. For it was not merely a matter of hanging by the neck until he was dead. While still living he was to be cut down, mutilated on the quartering-block beside the brazier, and in fact endure treatment —the common treatment of undistinguished traitors—which it were not seemly to discuss in detail.

Not until now had he given up all hope. Friends, he knew, had been at Whitehall storming the King's mercy for a reprieve. But the King was often enough not to be found during the day, and not the Queen's own self knew where to seek him at night. Save for the ordeal before him, Mr Codesby counted himself already as dead as the drowned cat an unkind hand had flung into the sledge.

Thus, physically uncomfortable and suffering mentally, he continued the Traitor's Ride.

II

In his youth Mr Codesby had been told of the serene happiness which is the companion of conscious innocence. This sensation he now knew to be vastly overrated. And if eternal peace were to requite him for man's injustice, he would not have minded borrowing a little on account.

It was a long ride for the condemned man, apart from its attendant discomforts. Sufferers want to be done with suffering—even by escape through the Gate of Death. But there was another matter to be considered. He knew that he would be permitted to speak to the crowd from the cart which would await him under the gallows. It would be expected of him; for he knew the rules of this ghastliest game of all. He must say nothing to excuse himself, nothing to vilify his accusers, nothing in criticism of the law or of His Sovereign Majesty. The best type of scaffold speech was that which urged the audience to take note, to repent in time of its sins, to witness his, the prisoner's, own repentance, and to remark what small and unchecked failings had grown to bring him to his present pass. And the mass of the audience was invariably so impressed that, after the head and limbs were severed, and the other work done, the bulk of them crowded the neighbouring taverns and got drunk on gin.

The procession reached the gallows at last. Pikes scattered and flung back the pressing crowd. Mr Codesby found himself beside the cart, on which was already standing one of the executioners, paring the trailing-rope to a convenient length.

In the midst of an open half-circle, which soldiers had formed about the cart, was a man kneeling before a brazier full of black coals, bellows in both hands and puffing with his mouth. This second executioner was a grimy fellow, his hands black and his cheeks stained with coal. He had endured much raillery from the crowd, and still the brazier emitted only a breath or two of smoke.

Mr Codesby's legs were freed, and he was pushed up a short ladder on to the cart, the Jack Ketch who waited there bending to seize his shoulders.

'Pardon me for what I must do,' he muttered in Mr Codesby's ear.

'God may do so,' answered Mr Codesby, and added contemptuously: 'but you have your guinea for this work.'

III

The voice of the crowd had risen and grew higher yet as Mr Codesby faced the throng with the noose about his neck. Vaguely he understood the meaning of this babbling clamour. They were asking for a speech. Mr Codesby, still ignorant of what he was to say, held up his hand, and the noise dwindled and fell almost to silence, as if a door had been closed upon it.

The Sheriff gave him the usual leave and the customary warning. A word of self-defence, a word against King or Justice, and the cart would be drawn, leaving him to strangle without further speech or the last ghostly counsel of the Ordinary.

Mr Codesby was silent for some seconds, for more reasons than one. First, the thoughts which he had collected were scattered again by the sight of faces he knew in the crowd. He marked the features of acquaintances and even friends, eight, ten, a round dozen of them. And, since he knew why they were there, a great rage and hatred blazed up in his heart.

Had they borne him any ill-will, he had forgiven them, as—he had been pitifully whispering to God—he forgave his enemies. These were no enemies of his until that moment. Enemies now, because they had come to watch him die for the mere sight's sake, and so that they might have a ring about them in the tavern when they came to tell the tale.

In sudden unchristian rage the man forgot his plight. But he held his gall. Did he shriek insults and epithets at them, the cart were moved to leave him strangling—even as the sobs would have choked him could he have thought that they had come in sympathy to this hard death-bed.

Speak! Yes, he would speak, and say that for which no sheriff would wish to silence him.

His gaze, taking in the half-circle before him, met suddenly the eyes of the grimy fellow kneeling by the brazier. He knew that face for all the dirt upon it. This—this was surely young Job Otter, the idlest of all idle apprentices, the fellow he had sent home as worthless. So young Job Otter had found other work, and thus, through no plan of his, was to take vengeance on his former master?

Recognition checked Mr Codesby before he could utter a word. Then thoughts came billowing upon him. Young Otter's eyes were sad and friendly, and yet there was in them a smile, and, strangest of all, a message. Yes, surely a message, and as that thought leaped upon him the blackened lips of the youth parted, as if upon a word.

Yes, surely he was saying something silently. The lips writhed. It was as if the youth were teaching a child the lip-shape for the utterance of a word. And Job Otter, seeing that he had caught his old master's gaze, laboured desperately to give him the reading.

Inspiration comes to men who stand on death's threshold. With a sudden leap and tingle of understanding Mr Codesby read the word on young Otter's lips. Nay, he read two words, 'Time' and 'Reprieve'.

He understood. His friends were still seeking the King, combing the lodgings of my lady Castlemaine and the lesser sluts. There was still hope, could he be found. All such could be read in two words by a dying man. . . .

IV

The Sheriff was addressing him. 'See you say nothing to the people in excuse. . . .'

Mr Codesby scarcely heard him. He was preparing and shuffling the fragments of his speech. It came from him in a great voice, beginning 'Friends and Fellow Sinners.'

Our Mr Codesby found himself armed at first with the blessed weapon of tact. Inspiration had laid this in his hand like a magic sword. He did not proclaim his innocence, but ascribed his present pass to folly—which none could gainsay.

Of the sins and follies of his youth he made much more than he had been wont to make in the family circle, and ten good minutes slid by without the Sheriff glancing at a clock. For the Sheriff heard of all manner of delightful peccadillos which he too might have committed, and found himself most impiously regretting his own wasted opportunities to sin.

But when a pause came in the speech he looked askance at Job Otter, puffing over the scarcely smouldering coals, and roared at him: 'In God's name, fellow, what are you at? Go get a flint that will strike and tinder that will catch.'

'The wood and coals are damp, your honour.'

'Damp! They must not be damp! Hurry, man, hurry! Else we shall be here all morning, and myself scarce breakfasted!'

Job Otter bent to his work with the bellows, but turned his head a little and lowered an eyelid towards the man who stood noosed upon the cart.

'I see old and familiar faces before me,' Mr Codesby cried, his voice rising with his courage, 'and for the last time I exhort ye. Some of ye might have stood where I stand now, and may yet so stand. John Dillop! Have ye repented yet of the shillings and pence ye lost at sell-the-goose, which ye were used to play in the church porch?—aye, and on the Lord's Day, too!'

There was a mild sensation in the crowd, only checked by Mr Codesby's voice.

'And you, Samuel Watchet! Did conscience bring ye to make recompense to the alderman whose dog ye stole?'

'It's a lie!' shouted Samuel Watchet, blushing deeply; but his voice was drowned.

'And who laughed when another took Sal Cook to church, three weeks before her twins were born? Are ye repentant now, David Hall? The Day of Judgement is not far off!'

More protests were lost in another roar. Mr Codesby took fire from his own rhetoric.

'My masters, I tell you that a dying man has more than earthly vision. Some here that have come to see me die had been better under their own roof-tree today. Aye, by Heaven, 'tis so! Charles Claff! Your wife is at home and I have seen another's shadow cross the threshold.

'Henry Melling! Your daughter has gone forth to meet the young cheapjack. I hope good will come of it, but I cannot read the future.'

There was a stir in the crowd and a sudden shout of laughter mingled with awe.

'Aye, there are many among you who will curse this their holiday and my death's day. Methinks I am a prophet new inspired. Richard Smith! You had done well to lock your till. That prodigal of yours has long fingers and an itching palm.'

A struggle ensued, as of somebody trying to force a way through the press of people. Voices rose to a tumult, and Mr Codesby had once more to hold up a hand for quiet.

Kneeling beside the brazier and biting upon his lip, Job Otter worked the bellows, but forced only a little smoke to rise and mingle with the breath of the multitude.

'Od's sake, give him heed!' roared the Sheriff. ''Tis sure he speaks inspired!'

Mr Codesby continued in this wise, inspired indeed. Words poured from him, while the minute-hand of the clock went round on its eternal pilgrimage. Not since Job turned and fell upon his comforters had such a scene been witnessed, until suddenly the prisoner's voice was drowned.

A thudding of hoofs coming nearer and halting on the fringe of the crowd produced the diversion, and the cry of a newcomer was immediately taken up, flaming to heaven like fire in summer grass.

'Way in the King's name! Way in the King's name! Reprieve! Pardon! Reprieve!'

Mr Codesby waited until a dusty horseman had forced a passage and thrust a sealed paper into the Sheriff's hands. Then he sat down upon the cart and hid his face between his hands, while the roar dinned in his ears.

The soldiers still held the crowd, but one within the circle laid a hand on Mr Codesby's knee. The reprieved man looked up through sudden tears at his late 'prentice.

'Lad,' he said, 'lad, how shall I thank thee? But not now, not now. The Nag's Head. I go there as soon as I may be conveyed. Lord, how they yell—they who had come to see me die! I'll be there anon. The Nag's Head. As soon as they give me passage.'

Then he stood upright, bowed to the Sheriff, and bowed to the huzzas which threatened him with deafness. The mob, cheated of the sight for which it had gathered, yet knew how to be generous. For hangings were common enough, but last-moment reprieves a rarity.

Not for nigh upon two hours could Mr Codesby reach the Nag's Head. He went in one of London's four hundred hackney-coaches, through a lane forced for him by the soldiers, and ultimately reached his destination, where he was unknown by name and where there was yet no rumour of a reprieve.

There he waited quietly in the sand-strewn parlour until Job Otter, pale and shiftily smiling, came in and took him by the hand, and, over glasses of brown madeira, Mr Codesby blurted out what was in his heart.

Job Otter smiled again. 'Master,' he said, 'you are well quit. There are those who came to see you die who have gone home with heavy hearts and scores to settle.'

'Ah!' agreed Mr Codesby thankfully, 'belike you have the right of it there. I am no mischief-merchant, I. But I had to eke the time. And now, lad, I will not have you a Jack Ketch's nark. Come back with me and continue learning the trade where ye left off. I give you your 'dentures—I can do no less. D' you mind why I sent you home before? D' you mind that, lad?'

A twinkle shot the bright hazel of Job Otter's eyes. 'That I do!' he chuckled. 'Od's life, I do! Ye told me I was too poor a hand at lighting the fires!'

A. E. W. MASON

Mr Mitchelbourne's Last Escapade

✿

It was in the kitchen of the inn at Framlingham that Mr Mitchelbourne came across the man who was afraid, and during the Christmas week of the year 1681. Lewis Mitchelbourne was young in those days, and esteemed as a gentleman of refinement and sensibility, with a queer taste for escapades pardonable by reason of his youth. It was his pride to bear his part in the graceful tactics of a minuet, while a saddled horse waited for him at the door. He delighted to vanish of a sudden from the lighted circle of his friends into the byways where none knew him, or held him of account; not that it was all vanity with Mitchelbourne, though no doubt the knowledge that his associates in London town were speculating upon his whereabouts tickled him pleasurably through many a solitary day. But he was possessed both of courage and resource, qualities for which he found too infrequent an exercise in his ordinary life; and so he felt it good to be free for awhile, not from the restraints but from the safeguards with which his social circumstances surrounded him. He had his spice of philosophy too, and discovered that these sharp contrasts—luxury and hardship—treading hard upon each other, and the new strange people with whom he fell in, kept fresh his zest of life.

Thus it happened that at a time when families were gathering cheerily each about a single fireside, Mr Mitchelbourne was riding alone through the muddy and desolate lanes of Suffolk. The winter was not seasonable; men were not tempted out of doors. There was neither briskness nor sunlight in the air, and there was no snow upon the ground. It was a December of dripping branches, and mists and steady pouring rains, with a raw sluggish cold which crept into one's marrow.

The man who was afraid, a large corpulent man, of a loose and heavy build, with a flaccid face and bright little inexpressive eyes like a bird's, sat on a bench within the glow of the fire.

'You travel far tonight?' he asked nervously, shuffling his feet.

'Tonight!' exclaimed Mitchelbourne, as he stood with his legs apart taking the comfortable warmth into his bones. 'No further than from this fire to my bed,' and he listened with enjoyment to the rain which cracked upon the window like a shower of gravel flung by some mischievous urchin. He was not suffered to listen long, for the corpulent man began again.

'I am an observer, sir. I pride myself upon it, but I have so much humility as to wish to put my observations to the test of fact. Now, from your carriage, I should judge you to serve His Majesty.'

'A civilian may be straight. There is no law against it,' returned Mitchelbourne, and he perceived that the ambiguity of his reply threw his questioner into a great alarm. He was at once interested. Here, it seemed, was one of those encounters which were the spice of his journeyings.

'You will pardon me,' continued the stranger with a great assumption of heartiness, 'but I am curious, sir, curious as Socrates, though I thank God I am no heathen. Here is Christmas, when a sensible gentleman, as upon my word I take you to be, sits to his table and drinks more than is good for him in honour of the season. Yet here are you upon the roads of Suffolk which have nothing to recommend them. I wonder at it, sir.'

'You may do that,' replied Mitchelbourne, 'though, to be sure, there are two of us in the like case.'

'Oh, as for me,' said his companion shrugging his shoulders, 'I am on my way to be married. My name is Lance,' and he blurted it out with a suddenness as though to catch Mitchelbourne off his guard. Mitchelbourne bowed politely.

'And my name is Mitchelbourne, and I travel for my pleasure, though my pleasure is mere gypsying, and has nothing to do with marriage. I take comfort from thinking that I have no friend from one rim of this county to the other, and that my closest intimates have not an inkling of my whereabouts.'

Mr Lance received the explanation with undisguised suspicion, and at supper, which the two men took together, he would be forever laying traps. Now he slipped some outlandish name or oath unexpectedly into his talk, and watched with a forward bend of his body to mark whether the word struck home; or again he mentioned some person with whom Mitchelbourne was quite unfamiliar. At length, however, he seemed satisfied, and drawing up his chair to the fire, he showed himself at once in his true character, a loud and gusty boaster.

'An exchange of sentiments, Mr Mitchelbourne, with a chance acquaintance over a pipe and a glass—upon my word I think you are in the right of it, and there's no pleasanter way of passing an evening. I could tell you stories, sir; I served the King in his wars, but I scorn a braggart, and all those glories are over. I am now a man of peace, and, as I told you, on my way to be married. Am I wise? I do not know, but I sometimes think it preposterous that a man who has been here and there about the world, and could, if he were so meanly-minded, tell a tale or so of success in gallantry, should hamper himself with connubial fetters. But a man must settle, to be sure, and since the lady is young, and not wanting in looks or breeding or station, as I am told——'

'As you are told?' interrupted Mitchelbourne.

'Yes, for I have never see her. No, not so much as her miniature. Nor have I seen her mother either, or any of the family, except the father, from whom I carry letters to introduce me. She lives in a house called "The Porch" some miles from here. There is another house hard by to it, I understand, which has long stood empty and I have a mind to buy it. I bring a fortune, the lady a standing in the county.'

'And what has the lady to say to it?' asked Mitchelbourne.

'The lady!' replied Lance, with a stare. 'Nothing but what is dutiful, I'll be bound. The father is under obligations to me.' He stopped suddenly, and Mitchelbourne, looking up, saw that his mouth had fallen. He sat with his eyes starting from his head and a face grey as lead, an image of panic pitiful to behold. Mitchelbourne spoke but got no answer. It seemed Lance could not answer—he was so arrested by a paralysis of terror. He sat staring straight in front of him, and apparently at the mantelpiece, which was just on a level with his eyes. The mantelpiece, however, had nothing to distinguish it from a score of others. Its counterpart might be found to this day in the parlour of any inn. A couple of china figures disfigured it, to be sure, but Mitchelbourne could not bring himself to believe that even their barbaric crudity had power to produce so visible a discomposure. He inclined to the notion that his companion was struck by a physical disease, perhaps some recrudescence of a malady contracted in those foreign lands of which he vaguely spoke.

'Sir, you are ill,' said Mitchelbourne. 'I will have a doctor, if there is one hereabouts to be found, brought to your relief.'

He sprang up as he spoke, and that action of his roused Lance out of his paralysis. 'Have a care,' he cried almost in a shriek. 'Do not move! For pity, sir, do not move,' and he in his turn rose from his chair. He rose trembling, and swept the dust off a corner of the mantelpiece into the palm of his hand. Then he held his palm to the lamp.

'Have you seen the like of this before?' he asked in a low shaking voice.

Mitchelbourne looked over Lance's shoulder. The dust was in reality a very fine grain of a greenish tinge.

'Never!' said Mitchelbourne.

'No, nor I,' said Lance, with a sudden cunning look at his companion, and opening his fingers he let the grain run between them. But he could not remove as easily from Mitchelbourne's memories that picture he had shown him of a shaking and shaken man.

Mitchelbourne went to bed divided in his feelings between pity for the lady Lance was to marry, and curiosity as to Lance's apprehensions. He lay awake for a long time speculating upon that mysterious green seed which could produce so extraordinary a panic, and in the morning his curiosity predominated. Since, therefore, he had no particular destination, he was easily persuaded to ride to Saxmundham with Mr Lance, who, for his part, was most earnest for a companion. On the journey Lance gave further evidence of his fears. He had a trick of looking backwards whenever they came to a corner of the road—a habitual trick, it seemed, acquired by a continued condition of fear. When they stopped at midday to eat of an ordinary, he inspected the guests through the chink at the hinges of the door before he would enter the room; and this, too, he did as though it had long been natural to him. He kept a bridle in his mouth, however; that little pile of grain upon the mantelshelf had somehow warned him into reticence, so that Mitchelbourne, had he not been addicted to his tobacco, would have learnt no more of the business, and would have escaped the extraordinary peril which he was subsequently called upon to face.

But he *was* addicted to his tobacco, and no sooner had he finished his supper that night at Saxmundham, than he called for a pipe. The maid-servant fetched a handful from a cupboard, and spread them upon the table, and amongst them was one plainly of Barbary manufacture. It had a straight wooden stem painted with hieroglyphics in red and green, and a small reddish bowl of baked earth. Nine men out of ten would no doubt have overlooked it, but Mitchelbourne was the tenth man. His fancies were quick to kindle, and taking up the pipe he said in a musing voice—

'Now, how in the world comes a Barbary pipe to travel so far over seas, and herd in the end with common clays in a little Suffolk village?'

He heard behind him the grating of a chair violently pushed back. The pipe seemingly made its appeal to Mr Lance also.

'Has it been smoked?' he asked in a grave low voice.

'The inside of the bowl is stained,' said Mitchelbourne.

Mitchelbourne had been inclined to believe that he had seen last evening the extremity of fear expressed in a man's face; he had now to admit that he had been wrong. Mr Lance's terror was a Circe to him, and sunk him into something grotesque and inhuman; he ran once or twice in a little tripping, silly run backwards and forwards, like an animal trapped, and out of its wits; his face had the look of a man suffering from nausea; so that Mitchelbourne seeing him, was ashamed and hurt for their common nature.

'I must go,' said Lance, babbling his words. 'I cannot stay. I must go.'

'Tonight?' exclaimed Mitchelbourne. 'Six yards from the door you will be soaked!'

'Then there will be the fewer men abroad. I cannot sleep here! No, though it rained pistols and bullets I must go.' He went out into the passage, and secretly calling his host, asked for his score. Mitchelhourne made a further effort to detain him.

'Make an enquiry of the landlord first. It may be a mere shadow that frightens you.'

'Not a word, not a question,' Lance implored. The mere suggestion increased a panic which seemed incapable of increase. 'And as for the shadow, why, that's true. The pipe's the shadow, and the shadow frightens me. A shadow! Yes! A shadow's a horrible threatening thing! Show me a shadow cast by nothing, and I am with you. But you might as easily hold that this Barbary pipe floated hither across the seas of its own will. No! 'Ware shadows, I say.' And so he continued harping on the word till the landlord fetched in the bill.

The landlord had his dissuasions too, but they availed not a jot more than Mr Mitchelbourne's.

'The road is as black as a pauper's coffin,' said he, 'and damnable with ruts.'

'So much the better,' said Lance.

'There is no house where you can sleep nearer than Glemham, and no man would sleep there could he kennel elsewhere.'

'So much the better,' said Lance. 'Besides, I am expected tomorrow evening at "The Porch", and Glemham is on the way.' He paid his bill, slipped over to the stables, and lent a hand to the saddling of his horse. Mitchelbourne, though for once in his life he regretted the precipitancy with which he welcomed strangers, was still sufficiently provoked to see the business to its end. His imagination was seized by the thought of this fat and vulgar person fleeing in terror through English lanes from a Barbary Moor. He had now a conjecture in his mind as to the nature of that greenish seed. He accordingly rode out with Lance towards Glemham.

It was a night of extraordinary blackness; you could not distinguish a hedge until the twigs stung across your face; the road was narrow, great tree-trunks with bulging roots lined it; at times it was very steep—and, besides and beyond every other discomfort, there was the rain. It fell pitilessly straight over the face of the country with a continuous roar as though the earth was a hollow drum. Both travellers were drenched to the skin before they were free of Saxmundham, and one of them, when after midnight they stumbled into the poor tumbledown parody of a tavern at Glemham, was in an extreme exhaustion. It was no more than an ague, said Lance, from which he periodically suffered, but the two men slept in the same bare room, and towards morning Mitchelbourne was awakened from a deep slumber by an unfamiliar voice talking at an incredible speed through the darkness in an uncouth tongue. He started up upon his elbow; the voice came from Lance's bed. He struck a light. Lance was in a high fever, which increased as the morning grew.

Now, whether he had the sickness latent within him when he came from Barbary, or whether his anxieties and corpulent habit made him an easy victim to disease, neither the doctor nor any one else could determine. But at twelve o'clock that day Lance was seized with an attack of cholera, and by three in the afternoon he was dead. The suddenness of the catastrophe shocked Mr Mitchelbourne inexpressibly. He stood gazing at the still features of the man whom fear had, during these last days, so grievously tormented, and was solemnly aware of the vanity of those fears. He could not pretend to any great esteem for his companion, but he made many suitable reflections upon the shears of the Fates and the tenuity of life, in which melancholy occupation he was interrupted by the doctor, who pointed out the necessity of immediate burial. Seven o'clock the next morning was the hour agreed upon, and Mitchelbourne at once searched in Lance's coat pockets for the letters which he carried. There were only two, superscribed respectively to Mrs Ufford at 'The Porch', near Glemham, and to her daughter Drusilla. At 'The Porch', Mitchelbourne remembered, Lance was expected this very evening and he thought it right at once to ride thither with his gloomy news.

Having, therefore, sprinkled the letters plentifully with vinegar and taken such rough precautions as were possible to remove the taint of infection from the letters, he started about four o'clock. The evening was most melancholy. For, though no rain any longer fell, there was a continual pattering of drops from the trees and a ghostly creaking of branches in a light and almost imperceptible wind. The day, too, was falling; the grey overhang of cloud was changing to black, except for one wide space in the west, where a pale spectral light shone without radiance; and the last of that was fading when he pulled up at a parting of the roads and enquired of a man who chanced to be standing there his way to 'The Porch'. He was directed to ride down the road upon his left hand until he came to the second house, which he could not mistake, for there was a dyke or moat about the garden wall. He passed the first house a mile further on, and perhaps half a mile beyond that he came to the dyke and the high garden wall, and saw the gables of the second house loom up behind it black against the sky. A wooden bridge spanned the dyke and led to a wide gate. Mitchelbourne stopped his horse at the bridge. The gate stood open and he looked down an avenue of trees into a square of which three sides were made by the high garden wall, and the fourth and innermost by the house. Thus the whole length of the house fronted him, and it struck him as very singular that neither in the lower nor the upper windows was there anywhere a spark of light; nor was there any sound but the tossing of the branches and the wail of the wind among the chimneys. Not even a dog barked or rattled a chain, and from no chimney breathed a wisp of smoke. The house in the gloom of that melancholy evening had a singularly eerie and tenantless look; an oppressive silence reigned there; and Mitchelbourne was unaccountably conscious of a growing aversion to it, as to something inimical and sinister.

He had crossed the mouth of a lane, he remembered, just at the first corner of the wall. The lane ran backwards from the road parallel with the side wall of the garden. Mitchelbourne had a strong desire to ride down that lane and inspect the back of the house before he crossed the bridge into the garden. He was restrained for a moment by the thought that such a proceeding must savour of cowardice. But only for a moment. There had been no doubting the genuine nature of Lance's fears, and those fears were very close to Mr Mitchelbourne now. They were feeling like cold fingers about his heart. He was almost in the icy grip of them.

He turned and rode down the lane until he came to the end of the wall. A meadow stretched behind the house. Mitchelbourne unfastened the catch of a gate with his riding-whip and entered it. He found himself upon the edge of a pool, which on the opposite side wetted the house wall. About the pool some elder trees and elms grew and overhung, and their boughs tapped like fingers upon the window-panes. Mitchelbourne was assured that the house was inhabited, since from one of the windows a strong yellow light blazed, and whenever a sharper gust blew the branches aside, swept across the face of the pool like a flaw of wind.

The lighted window was in the lowest storey, and Mitchelbourne, from the back of his horse, could see into the room. He was mystified beyond expression by what he saw. A deal table, three wooden chairs, some ragged curtains drawn back from the windows, and a single lamp, made up the furniture. The boards of the floor were bare and unswept; the paint peeled in strips from the panels of the walls; the discoloured ceiling was hung with cobwebs; the room in a word matched the outward aspect of the house in its look of long disuse. Yet it had occupants. Three men were seated at the table in the scarlet coats and boots of the King's officers. Their faces, though it was wintertime, were brown with the sun, and thin and drawn as with long privation and anxiety. They had little to say to one another, it seemed. Each man sat stiffly in a sort of suspense and expectation, with now and then a restless movement or a curt word as curtly answered.

Mitchelbourne rode back again, crossed the bridge, fastened his horse to a tree in the garden, and walked down the avenue to the door. As he mounted the steps, he perceived with something of a shock, that the door was wide open and that the void of the hall yawned black before him. It was a fresh surprise, but in this night of surprises, one more or less, he assured himself, was of little account. He stepped into the hall and walked forwards, feeling with his hands in front of him. As he advanced, he saw a thin line of yellow upon the floor ahead of him. The line of yellow was a line of light, and it came, no doubt, from underneath a door, and the door, no doubt, was that behind which the three men waited. Mitchelbourne stopped. After all, he reflected, the three men were English officers wearing His Majesty's uniform, and, moreover, wearing it stained with their country's service. He walked forward and tapped upon the door. At once the light within the room was extinguished.

It needed just that swift and silent obliteration of the slip of light upon the floor to make Mitchelbourne afraid. He had been upon the brink of fear ever since he had seen that lonely and disquieting house; he was now caught in the full stream. He turned back. Through the open doorway he saw the avenue of leafless trees tossing against a leaden sky. He took a step or two and then came suddenly to a halt. For all around him in the darkness he seemed to hear voices breathing and soft footsteps. He realized that his fear had overstepped his reason; he forced himself to remember the contempt he had felt for Lance's manifestations of terror; and swinging round again he flung open the door and entered the room.

'Good evening, gentlemen,' said he airily, and he got no answer whatsoever. In front of him was the grey panel of dim twilight where the window stood. The rest was black night and an absolute silence. A map of the room was quite clear in his recollections. The three men were seated he knew at the table on his right hand. The faint light from the window did not reach them, and they made no noise. Yet they were there. Why had they not answered him? he asked himself. He could not even hear them breathing, though he strained his ears. He could only hear his heart drumming at his breast, the blood pulsing in his temples. Why did they hold their breath? He crossed the room, not knowing what he did, bereft of his wits. He had a confused ridiculous picture of himself wearing the flaccid, panic-stricken face of Mr Lance, like an ass' head, Fear holding the wand of Titania. He reached the window and stood in its embrasure, and there one definite practical thought crept into his mind. He was visible to these men who were invisible to him. The thought suggested a precaution, and with the trembling haste of a man afraid, he tore at the curtains and dragged them till they met across the window, so that even the faint grey glimmer of the night no longer had entrance. The next moment he heard the door behind him latch, and a key turn in the lock. He crouched beneath the window and did not stand up again until a light was struck, and the lamp relit.

The lighting of the lamp restored Mr Mitchelbourne, if not to the full measure of his confidence, at all events to an appreciation that the chief warrant for his trepidation was removed. What he had with some appearance of reason feared was a sudden attack in the dark. With the lamp lit, he could surely stand in no danger of any violence at the hands of three king's officers whom he had never come across in all his life. He took, therefore, an easy look at them. One, the youngest, now leaned against the door, a youth of a frank honest face, unremarkable but for a firm set of the jaws. A youth of no great intellect, thought Mitchelbourne, but tenacious, a youth marked out for subordinate command, but never likely for all his sterling qualities to kindle a woman to a world-forgetting passion, or to tread with her the fiery heights where life throbs at its fullest. Mr Mitchelbourne began to feel quite sorry for this young officer of the limited capacities, and he was still in the sympathetic mood when one of the two men at the table spoke to him. Mitchelbourne turned at once. The officers were sitting with a certain air of the theatre in their attitudes, one a little dark man and the other a stiff, light-complexioned fellow with a bony, barren face, unmistakably a stupid man and the oldest of the three. It was he who was speaking, and he spoke with a sort of aggravated courtesy like a man of no breeding counterfeiting a gentleman upon the stage.

'You will pardon us for receiving you with so little ceremony. But while we expected you, you on the other hand were not expecting us, and we feared that you might hesitate to come in if the lamp was burning when you opened the door.'

Mitchelbourne was now entirely at his ease. He perceived that there was some mistake and made haste to put it right.

'On the contrary,' said he, 'for I knew very well you were here. Indeed, I knocked at the door to make a necessary enquiry. You did not extinguish the lamp so quickly but that I saw the light beneath the door, and besides I watched you some five minutes through the window from the opposite bank of the pool at the back of the house.'

The officers were plainly disconcerted by the affability of Mr Mitchelbourne's reply. They had evidently expected to carry off a triumph, not to be taken up in an argument. They had planned a stroke of the theatre, final and convincing, and behold the dialogue went on! There was a riposte to their thrust.

The spokesman made some gruff noises in his throat. Then his face cleared.

'These are dialectics,' he said superbly, with a wave of the hand.

'Good,' said the little dark fellow at his elbow, 'very good!'

The youth at the door nodded superciliously towards Mitchelbourne.

'True, these are dialectics,' said he with a smack of the lips upon the word. It was a good, cunning, scholarly word, and the man who could produce it so aptly worthy of much admiration.

'You make a further error, gentlemen,' continued Mitchelbourne, 'you no doubt are expecting someone, but you were most certainly not expecting me. For I am here by the purest mistake, having been misdirected on the way.'

Here the three men smiled to each other, and their spokesman retorted with a chuckle.

'Misdirected indeed you were. We took precautions that you should be. A servant of mine was stationed at the parting of the roads. But we are forgetting our manners,' he added, rising from his chair. 'You should know our names. The gentleman at the door is Cornet Lashley, this is Captain Bassett, and I am Major Chantrell. We are all three of Trevelyan's regiment.'

'And my name,' said Mitchelbourne, not to be outdone in politeness, 'is Lewis Mitchelbourne, a gentleman of the county of Middlesex.'

At this each of the officers was seized with a fit of silent laughter; but before Mitchelbourne had time to resent their behaviour, Major Chantrell said indulgently—

'Well, well, we shall not quarrel about names. At all events we all four are lately come from Tangier.'

'Oh, from Tangier,' cried Mitchelbourne. The riddle was becoming clear. That extraordinary siege where a handful of English redcoats, unpaid and ill-fed, fought a breached and broken town against countless hordes for the honour of their king during twenty years, had not yet become the property of the historian. It was still an actual war in 1681. Mitchelbourne understood whence came the sunburn on his antagonists' faces, whence the stains and the worn seams of their clothes. He advanced to the table and spoke with a greater respect than he had used.

'Did one of you,' he asked, 'leave a Moorish pipe behind you at an inn of Saxmundham?'

'Ah,' said the Major with a reproachful glance at Captain Bassett. The captain answered with some discomfort—

'Yes. I made that mistake. But what does it matter? You are here none the less.'

'You have with you some of the Moorish tobacco?' continued Mitchelbourne.

Captain Bassett fetched out of his pocket a little canvas bag, and handed it Mitchelbourne, who untied the string about the neck, and poured some of the contents into the palm of his hand. The tobacco was a fine, greenish seed.

'I thought as much,' said Mitchelbourne. 'You expected Mr Lance tonight. It is Mr Lance whom you thought to misdirect to this solitary house. Indeed Mr Lance spoke of such a place in this neighbourhood, and had a mind to buy it.'

Captain Bassett suddenly raised his hand to his mouth, not so quickly however but Mitchelbourne saw the grim amused smile upon his lips. 'It is Mr Lance for whom you now mistake me,' he said abruptly.

The young man at the door uttered a short contemptuous laugh; Major Chantrell only smiled.

'I am aware,' said he, 'that we meet for the first time tonight, but you presume upon that fact too far. What have you to say to this?' And dragging a big and battered pistol from his pocket, he tossed it upon the table, and folded his arms in the best transpontine manner.

'And to this?' said Captain Bassett. He laid a worn leather powder flask beside the pistol, and tapped upon the table triumphantly.

Mr Mitchelbourne recognized clearly that villainy was somehow checkmated by these proceedings and virtue restored, but how he could not for the life of him determine. He took up the pistol.

'It appears to have seen some honourable service,' said he. This casual remark had a most startling effect upon his auditors. It was the spark to the gunpowder of their passions. Their affectations vanished in a trice.

'Service, yes, but honourable! Use that lie again, Mr Lance, and I will ram the butt of the pistol down your throat!' cried Major Chantrell. He leaned forward over the table in a blaze of fury. Yet his face did no more than match the faces of his comrades.

Mitchelbourne began to understand. These simple soldier-men had endeavoured to conduct their proceedings with great dignity and a judicial calmness; they had mapped out for themselves certain parts which they were to play as upon a stage; they were to be three stern imposing figures of justice; and so they had become simply absurd and ridiculous. Now, however, that passion had the upper hand of them, Mitchelbourne saw at once that he stood in deadly peril. These were men.

'Understand me, Mr Lance,' and the Major's voice rang out firm, the voice of a man accustomed to obedience. 'Three years ago I was in command of Devil's Drop, a little makeshift fort upon the sands outside Tangier. In front the Moors lay about us in a semicircle. Sir, the diameter of the circle was the line of the sea at our backs. We could not retire six yards without wetting our feet, not twenty without drowning. One night the Moors pushed their trenches up to our palisades; in the dusk of the morning I ordered a sortie. Nine officers went out with me and three came back—we three. Of the six we left behind five fell, by my orders to be sure, for I led them out; but, by the living God, you killed them. There's the pistol that shot my best friend down, an English pistol. There's the powder flask which charged the pistol, an English flask filled with English powder. And who sold the pistol and the powder to the Moors, England's enemies? You, an Englishman. But you have come to the end of your lane tonight. Turn and turn as you will you have come to the end of it.'

The truth was out now, and Mitchelbourne was chilled with apprehension. Here were three men very desperately set upon what they considered a mere act of justice. How was he to dissuade them? By argument? They would not listen to it. By proofs? He had none to offer them. By excuses? Of all unsupported excuses which can match for futility the excuse of a mistaken identity? It springs immediate to the criminal's lips. Its mere utterance is almost a condemnation.

'You persist in error, Major Chantrell,' he nevertheless began.

'Show him the proof, Bassett,' Chantrell interrupted with a shrug of the shoulders, and Captain Bassett drew from his pocket a folded sheet of paper.

'Nine officers went out,' continued Chantrell, 'five were killed, three are here. The ninth was taken a prisoner into Barbary. The Moors brought him down to their port of Marmora to interpret. At Marmora your ship unloaded its stores of powder and guns. God knows how often it had unloaded the like cargo during these twenty years—often enough it seems to give you a fancy for figuring as a gentleman in the county. But the one occasion of its unloading is enough. Our brother-officer was your interpreter with the Moors, Mr Lance. You may very likely know that, but this you do not know, Mr Lance. He escaped, he crept into Tangier with this your bill of lading in his hand,' and Bassett tossed the sheet of paper towards Mitchelbourne. It fell upon the floor before him but he did not trouble to pick it up.

'Is it Lance's death that you require?' he asked.

'Yes! yes! yes!' came from each mouth.

'Then already you have your wish. I do not question one word of your charges against Lance. I have reason to believe them true. But I am not Lance. Lance lies at this moment dead at Great Glemham. He died this afternoon of cholera. Here are his letters,' and he laid the letters on the table. 'I rode in with them at once. You do not believe me, but you can put my words to the test. Let one of you ride to Great Glemham and satisfy himself. He will be back before the morning.'

The three officers listened so far with impassive faces, or barely listened, for they were as indifferent to the words as to the passion with which they were spoken.

'We have had enough of the gentleman's ingenuities, I think,' said Chantrell, and he made a movement towards his companions.

'One moment,' exclaimed Mitchelbourne. 'Answer me a question! These letters are to the address of Mrs Ufford at a house called "The Porch". Is it near to here?'

'It is the first house you passed,' answered the Major, and as he noticed a momentary satisfaction flicker upon his victim's face, he added, 'But you will not do well to expect help from "The Porch"—at all events in time to be of much service to you. You hardly appreciate that we have been at some pains to come up with you. We are not likely again to find so many circumstances agreeing to favour us, a dismantled house, yourself travelling alone and off your guard in a county with which you are unfamiliar and where none know you, and just outside the window a convenient pool. Besides—besides,' he broke out passionately, 'there are the little mounds about Tangier, under which my friends lie,' and he covered his face with his hands. 'My friends!' he cried in a hoarse and broken voice, 'my soldier-men! Come, let's make an end. Basset, the rope is in the corner. There's a noose to it. The beam across the window will serve'—and Bassett rose to obey.

But Mitchelbourne gave them no time. His fears had altogether vanished before his indignation at the stupidity of these officers. He was boiling with anger at the thought that he must lose his life in this futile ignominious way for the crime of another man, who was not even his friend and who besides was already dead. There was just one chance to escape it seemed to him. And even as Bassett stooped to lift the coil of rope in the corner he took it.

'So that's the way of it,' he cried, stepping forward. 'I am to be hung up to a beam till I kick to death, am I? I am to be buried decently in that stagnant pool, am I? And you to be miles away before sunrise, and no one the wiser! No, Major Chantrell, I am not come to the end of my lane,' and before either of the three could guess what he was at, he had snatched up the pistol from the table and dashed the lamp into a thousand fragments.

The flame shot up blue and high, and then came darkness.

Mitchelbourne jumped lightly back from his position to the centre of the room. The men he had to deal with were men who would follow their instincts. They would feel along the walls; of so much he could be certain. He heard the coil of rope drop down in a corner to his left; so that he knew where Captain Bassett was. He heard a chair upset in front of him, and a man staggered against his chest. Mitchelbourne had the pistol still in his hand, and struck hard, and the man dropped with a crash. The fall followed so closely upon the upsetting of the chair that it seemed part of the same movement and accident. It seemed so clearly part that a voice spoke on Mitchelbourne's left, just where the empty hearth would be.

'Get up! Be quick!'

The voice was Major Chantrell's, and Mitchelbourne had a throb of hope. For since it was not the Major who had fallen nor Captain Bassett, it must be Lashley. And Lashley had been guarding the door, of which the key remained in the lock. If only he could reach the door and turn the key! He heard Chantrell moving stealthily along the wall upon his left hand, and he suffered a moment's agony, for in the darkness he could not surely tell which way the Major moved. For if he moved to the window, if he had the sense to move to the window and tear aside those drawn curtains, the grey twilight would show the shadowy moving figures. Mitchelbourne's chance would be gone. And then something totally unexpected and unhoped-for occurred. The god of the machine was in a freakish mood that evening. He had a mind for pranks and absurdities. Mitchelbourne was strung to so high a pitch that the ridiculous aspect of the occurrence came home to him before all else, and he could barely keep himself from laughing aloud. For he heard two men grappling and struggling silently together. Captain Bassett and Major Chantrell had each other by the throat, and neither of them had the wit to speak. They reserved their strength for their struggle. Mitchelbourne stepped on tiptoe to the door, felt for the key, grasped it without so much as a click, and then suddenly turned it, flung open the door and sprang out. He sprang against a fourth man—the servant no doubt who had misdirected him—and both tumbled on to the floor. Mitchelbourne however tumbled on the top. He was again upon his feet, while Major Chantrell was explaining matters to Captain Bassett; he was flying down the avenue of trees before the explanation was finished. He did not stop to untie his horse; he ran, conscious that there was only one place of safety for him—the interior of Mrs Ufford's house. He ran along the road till he felt that his heart was cracking within him, expecting every moment that a hand would be laid upon his shoulder or that a pistol shot would ring out upon the night. He reached the house and knocked loudly at the door. He was admitted breathless by a man who said to him at once, with the smile and familiarity of an old servant—

'You are expected, Mr Lance.'

Mitchelbourne plumped down upon a chair and burst into uncontrollable laughter. He gave up all attempt for that night to establish his identity. The fates were too heavily against him. Besides he was now quite hysterical.

The manservant threw open a door.

'I will tell my mistress you have come, sir,' said he.

'No, it would never do,' cried Mitchelbourne. 'You see, I died at three o'clock this afternoon. I have merely come to leave my letters of presentation. So much I think a proper etiquette may allow. But it would never do for me to be paying visits upon ladies so soon after an affair of so deplorable a gravity. Besides I have to be buried at seven in the morning, and if I chanced not to be back in time I should certainly acquire a reputation for levity which, since I am unknown in the county, I am unwilling to incur'—and leaving the butler stupefied in the hall, he ran out into the road. He heard no sound of pursuit.

M. R. JAMES

Martin's Close

∞

Some few years back I was staying with the rector of a parish in the West, where the society to which I belong owns property. I was to go over some of this land: and, on the first morning of my visit, soon after breakfast, the estate carpenter and general handyman, John Hill, was announced as in readiness to accompany us. The rector asked which part of the parish we were to visit that morning. The estate map was produced, and when we had showed him our round, he put his finger on a particular spot. 'Don't forget,' he said, 'to ask John Hill about Martin's Close when you get there. I should like to hear what he tells you.' 'What ought he to tell us?' I said. 'I haven't the slightest idea,' said the rector, 'or, if that is not exactly true, it will do till lunch-time.' And here he was called away.

We set out; John Hill is not a man to withhold such information as he possesses on any point, and you may gather from him much that is of interest about the people of the place and their talk. An unfamiliar word, or one that he thinks ought to be unfamiliar to you, he will usually spell— as c-o-b cob, and the like. It is not, however, relevant to my purpose to record his conversation before the moment when we reached Martin's Close. The bit of land is noticeable, for it is one of the smallest enclosures you are likely to see—a very few square yards, hedged in with quickset on all sides, and without any gate or gap leading into it. You might take it for a small cottage garden long deserted, but that it lies away from the village and bears no trace of cultivation. It is at no great distance from the road, and is part of what is there called a moor, in other words, a rough upland pasture cut up into largish fields.

'Why is this little bit hedged off so?' I asked, and John Hill (whose answer I cannot represent as perfectly as I should like) was not at fault. 'That's what we call Martin's Close, sir: 'tes a curious thing 'bout that bit of land, sir: goes by the name of Martin's Close, sir. M-a-r-t-i-n Martin. Beg pardon, sir, did Rector tell you to make enquiry of me 'bout that, sir?' 'Yes, he did,' 'Ah, I thought so much, sir. I was tell'n Rector 'bout that last week, and he was very much interested. It 'pears there's a murderer buried there, sir, by the name of Martin. Old Mr Samuel Saunders, that formerly lived yurr at what we call South-town, sir, he had a long tale 'bout that, sir: terrible murder done 'pon a young woman, sir. Cut her throat and cast her in the water down yurr.' 'Was he hung for it?' 'Yes, sir, he was hung just up yurr on the roadway, by what I've 'eard, on the Holy Innocents' Day, many 'undred years ago, by the man that went by the name of the bloody judge: terrible red and bloody, I've 'eard.' 'Was his name Jefferies, do you think?' 'Might be possible 'twas— Jefferies—J-e-f—Jefferies. I reckon 'twas, and the tale I've 'eard many times from Mr Saunders—how this young man Martin—George Martin— was troubled before his crule action come to light by the young woman's sperit.' 'How was that, do you know?' 'No, sir, I don't exactly know how 'twas with it: but by what I've 'eard he was fairly tormented: and rightly tu. Old Mr Saunders, he told a history regarding a cupboard down yurr in the New Inn. According to what he related, this young woman's sperit come out of this cupboard: but I don't racollact the matter.'

This was the sum of John Hill's information. We passed on, and in due time I reported what I had heard to the Rector. He was able to show me from the parish account-books that a gibbet had been paid for in 1684, and a grave dug in the following year, both for the benefit of George Martin; but he was unable to suggest any one in the parish, Saunders being now gone, who was likely to throw any further light on the story.

Naturally, upon my return to the neighbourhood of libraries, I made search in the more obvious places. The trial seemed to be nowhere reported. A newspaper of the time, and one or more newsletters, however, had some short notices, from which I learnt that, on the ground of local prejudice against the prisoner (he was described as a young gentleman of a good estate), the venue had been moved from Exeter to London; that Jefferies had been the judge, and death the sentence, and that there had been some 'singular passages' in the evidence. Nothing further transpired till September of this year. A friend who knew me to be interested in Jefferies then sent me a leaf torn out of a secondhand bookseller's catalogue with the entry: JEFFERIES, JUDGE: *Interesting old MS trial for murder*, and so forth, from which I gathered, to my delight, that I could become possessed, for a very few shillings, of what seemed to be a verbatim report, in shorthand, of the Martin trial. I telegraphed for the manuscript and got it. It was a thin bound volume, provided with a title written in longhand by someone in the eighteenth century, who had also added this note: 'My father, who took these notes in court, told me that the prisoner's friends had made interest with Judge Jefferies that no report should be put out: he had intended doing this himself when times were better, and had shew'd it to the Revd Mr Glanvil, who incourag'd his design very warmly, but death surpriz'd them both before it could be brought to an accomplishment.'

The initials W. G. are appended; I am advised that the original reporter may have been T. Gurney, who appears in that capacity in more than one State trial.

This was all that I could read for myself. After no long delay I heard of some one who was capable of deciphering the shorthand of the seventeenth century, and a little time ago the typewritten copy of the whole manuscript was laid before me. The portions which I shall communicate here help to fill in the very imperfect outline which subsists in the memories of John Hill and, I suppose, one or two others who live on the scene of the events.

The report begins with a species of preface, the general effect of which is that the copy is not that actually taken in court, though it is a true copy in regard of the notes of what was said; but that the writer has added to it some 'remarkable passages' that took place during the trial, and has made this present fair copy of the whole, intending at some favourable time to publish it; but has not put it into longhand, lest it should fall into the possession of unauthorized persons, and he or his family be deprived of the profit.

The report then begins:

This case came on to be tried on Wednesday, the 19th of November, between our sovereign lord the King, and George Martin Esquire, of (I take leave to omit some of the place-names), at a sessions of oyer and terminer and gaol delivery, at the Old Bailey, and the prisoner, being in Newgate, was brought to the bar.

Clerk of the Crown. George Martin, hold up thy hand (which he did).

Then the indictment was read, which set forth that the prisoner 'not having the fear of God before his eyes, but being moved and seduced by the instigation of the devil, upon the 15th day of May, in the 36th year of our sovereign lord King Charles the Second, with force and arms in the parish aforesaid, in and upon Ann Clark, spinster, of the same place, in the peace of God and of our said sovereign lord the King then and there being, feloniously, wilfully, and of your malice aforethought did make an assault and with a certain knife value a penny the throat of the said Ann Clark then and there did cut, of the which wound the said Ann Clark then and there did die, and the body of the said Ann Clark did cast into a certain pond of water situate in the same parish (with more that is not material to our purpose) against the peace of our sovereign lord the King, his crown and dignity.'

Then the prisoner prayed a copy of the indictment.

LCJ (Sir George Jefferies). What is this? Sure you know that is never allowed. Besides, here is a plain indictment as ever I heard; you have nothing to do but to plead to it.

Pris. My lord, I apprehend there may be matter of law arising out of the indictment, and I would humbly beg the court to assign me counsel to consider of it. Besides, my lord, I believe it was done in another case: copy of the indictment was allowed.

LCJ. What case was that?

Pris. Truly, my lord, I have been kept close prisoner ever since I came up from Exeter Castle, and no one allowed to come at me and no one to advise with.

LCJ. But I say, what was that case you allege?

Pris. My lord, I cannot tell your lordship precisely the name of the case, but it is in my mind that there was such an one, and I would humbly desire——

LCJ. All this is nothing. Name your case, and we will tell you whether there be any matter for you in it. God forbid but you should have anything that may be allowed you by law: but this is against law, and we must keep the course of the court.

Att. Gen. (Sir Robert Sawyer). My lord, we pray for the King that he may be asked to plead.

Cl. of Ct. Are you guilty of the murder whereof you stand indicted, or not guilty?

Pris. My lord, I would humbly offer this to the court. If I plead now, shall I have an opportunity after to except against the indictment?

LCJ. Yes, yes, that comes after verdict: that will be saved to you, and counsel assigned if there be matter of law: but that which you have now to do is to plead.

Then after some little parleying with the court (which seemed strange upon such a plain indictment) the prisoner pleaded *Not Guilty*.

Cl. of Ct. Culprit. How wilt thou be tried?

Pris. By God and my country.

Cl. of Ct. God send thee a good deliverance.

LCJ. Why, how is this? Here has been a great to-do that you should not be tried at Exeter by your country, but be brought here to London, and now you ask to be tried by your country. Must we send you to Exeter again?

Pris. My lord, I understood it was the form.

LCJ. So it is, man: we spoke only in the way of pleasantness. Well, go on and swear the jury.

So they were sworn. I omit the names. There was no challenging on the prisoner's part, for, as he said, he did not know any of the persons called. Thereupon the prisoner asked for the use of pen, ink, and paper, to which the LCJ replied: 'Ay, ay, in God's name let him have it.' Then the usual charge was delivered to the jury, and the case opened by the junior counsel for the King, Mr Dolben.

The Attorney-General followed:

May it please your lordship, and you gentlemen of the jury, I am of counsel for the King against the prisoner at the bar. You have heard that he stands indicted for a murder done upon the person of a young girl. Such crimes as this you may perhaps reckon not to be uncommon, and, indeed, in these times, I am sorry to say it, there is scarce any fact so barbarous and unnatural but what we may hear almost daily instances of it. But I must confess that in this murder that is charged upon the prisoner there are some particular features that mark it out to be such as I hope has but seldom if ever been perpetrated upon English ground. For as we shall make it appear, the person murdered was a poor country girl (whereas the prisoner is a gentleman of a proper estate) and, besides that, was one to whom Providence had not given the full use of her intellects, but was what is termed among us commonly an innocent or natural: such an one, therefore, as one would have supposed a gentleman of the prisoner's quality more likely to overlook, or, if he did notice her, to be moved to compassion for her unhappy condition, than to lift up his hand against her in the very horrid and barbarous manner which we shall show you he used.

Now to begin at the beginning and open the matter to you orderly: About Christmas of last year, that is the year 1683, this gentleman, Mr Martin, having newly come back into his own country from the University of Cambridge, some of his neighbours, to show him what civility they could (for his family is one that stands in very good repute all over that country) entertained him here and there at their Christmas merrymakings, so that he was constantly riding to and fro, from one house to another, and sometimes, when the place of his destination was distant, or for other reason, as the unsafeness of the roads, he would be constrained to lie the night at an inn. In this way it happened that he came, a day or two after the Christmas, to the place where this young girl lived with her parents, and put up at the inn there, called the New Inn, which is, as I am informed, a house of good repute. Here was some dancing going on among the people of the place, and Ann Clark had been brought in, it seems, by her elder sister to look on; but being, as I have said, of weak understanding, and, besides that, very uncomely in her appearance, it was not likely she should take much part in the merriment; and accordingly was but standing by in a corner of the room. The prisoner at the bar, seeing her, one must suppose by way of a jest, asked her would she dance with him. And in spite of what her sister and others could say to prevent it and to dissuade her——

LCJ. Come, Mr Attorney, we are not set here to listen to tales of Christmas parties in taverns. I would not interrupt you, but sure you have more weighty matters than this. You will be telling us next what tune they danced to.

Att. My lord, I would not take up the time of the court with what is not material: but we reckon it to be material to show how this unlikely acquaintance begun: and as for the tune, I believe, indeed, our evidence will show that even that hath a bearing on the matter in hand.

LCJ. Go on, go on, in God's name: but give us nothing that is impertinent.

Att. Indeed, my lord, I will keep to my matter. But, gentlemen, having now shown you, as I think, enough of this first meeting between the murdered person and the prisoner, I will shorten my tale so far as to say that from then on there were frequent meetings of the two: for the young woman was greatly tickled with having got hold (as she conceived it) of so likely a sweetheart, and he being once a week at least in the habit of passing through the street where she lived, she would be always on the watch for him; and it seems they had a signal arranged: he should whistle the tune that was played at the tavern: it is a tune, as I am informed, well known in that country, and has a burden, *Madam, will you walk, will you talk with me?*

LCJ. Ay, I remember it in my own country, in Shropshire. It runs somehow thus, doth it not? [Here his lordship whistled a part of a tune, which was very observable, and seemed below the dignity of the court. And it appears he felt it so himself, for he said]: But this is by the mark, and I doubt it is the first time we have had dance-tunes in this court. The most part of the dancing we give occasion for is done at Tyburn. [Looking at the prisoner, who appeared very much disordered.] You said the tune was material to your case, Mr Attorney, and upon my life I think Mr Martin agrees with you. What ails you, man? staring like a player that sees a ghost!

Pris. My lord, I was amazed at hearing such trivial, foolish things as they bring against me.

LCJ. Well, well, it lies upon Mr Attorney to show whether they be trivial or not: but I must say, if he has nothing worse than this he has said, you have no great cause to be in amaze. Doth it not lie something deeper? But go on, Mr Attorney.

Att. My lord and gentlemen—all that I have said so far you may indeed very reasonably reckon as having an appearance of triviality. And, to be sure, had the matter gone no further than the humouring of a poor silly girl by a young gentleman of quality, it had been very well. But to proceed. We shall make it appear that after three or four weeks the prisoner became contracted to a young gentlewoman of that country, one suitable every way to his own condition, and such an arrangement was on foot that seemed to promise him a happy and a reputable living. But within no very long time it seems that this young gentlewoman, hearing of the jest that was going about that countryside with regard to the prisoner and Ann Clark, conceived that it was not only an unworthy carriage on the part of her lover, but a derogation to herself that he should suffer his name to be sport for tavern company: and so without more ado she, with the consent of her parents, signified to the prisoner that the match between them was at an end. We shall show you that upon the receipt of this intelligence the prisoner was greatly enraged against Ann Clark as being the cause of his misfortune (though indeed there was nobody answerable for it but himself), and that he made use of many outrageous expressions and threatenings against her, and subsequently upon meeting with her both abused her and struck at her with his whip: but she, being but a poor innocent, could not be persuaded to desist from her attachment to him, but would often run after him testifying with gestures and broken words the affection she had to him: until she was become, as he said, the very plague of his life. Yet, being that affairs in which he was now engaged necessarily took him by the house in which she lived, he could not (as I am willing to believe he would otherwise have done) avoid meeting with her from time to time. We shall further show you that this was the posture of things up to the 15th day of May in this present year. Upon that day the prisoner comes riding through the village, as of custom, and met with the young woman: but in place of passing her by, as he had lately done, he stopped, and said some words to her with which she appeared wonderfully pleased, and so left her; and after that day she was nowhere to be found, notwithstanding a strict search was made for her. The next time of the prisoner's passing through the place, her relations enquired of him whether he should know anything of her whereabouts; which he totally denied. They expressed to him their fears lest her weak intellects should have been upset by the attention he had showed her, and so she might have committed some rash act against her own life, calling him to witness the same time how often they had beseeched him to desist from taking notice of her, as fearing trouble might come of it: but this, too, he easily laughed away. But in spite of this light behaviour, it was noticeable in him that about this time his carriage and demeanour changed, and it was said of him that he seemed a troubled man. And here I come to a passage to which I should not dare to ask your attention, but that it appears to me to be founded in truth, and is supported by testimony deserving of credit. And, gentlemen, to my judgment it doth afford a great instance of God's revenge against murder, and that He will require the blood of the innocent.

[Here Mr Attorney made a pause, and shifted with his papers: and it was thought remarkable by me and others, because he was a man not easily dashed.]

LCJ. Well, Mr Attorney, what is your instance?

Att. My lord, it is a strange one, and the truth is that, of all the cases I have been concerned in, I cannot call to mind the like of it. But to be short, gentlemen, we shall bring you testimony that Ann Clark was seen after this 15th of May, and that, at such time as she was so seen, it was impossible she could have been a living person.

[Here the people made a hum, and a good deal of laughter, and the Court called for silence, and when it was made]——

LCJ. Why, Mr Attorney, you might save up this tale for a week; it will be Christmas by that time, and you can frighten your cook-maids with it [at which the people laughed again, and the prisoner also, as it seemed]. God, man, what are you prating of—ghosts and Christmas jigs and tavern company—and here is a man's life at stake! (To the prisoner): And you, sir, I would have you know there is not so much occasion for you to make merry neither. You were not brought here for that, and if I know Mr Attorney, he has more in his brief than he has shown yet. Go on, Mr Attorney. I need not, mayhap, have spoken so sharply, but you must confess your course is something unusual.

Att. Nobody knows it better than I, my lord: but I shall bring it to an end with a round turn. I shall show you, gentlemen, that Ann Clark's body was found in the month of June, in a pond of water, with the throat cut: that a knife belonging to the prisoner was found in the same water: that he made efforts to recover the said knife from the water: that the coroner's quest brought in a verdict against the prisoner at the bar, and that therefore he should by course have been tried at Exeter: but that, suit being made on his behalf, on account that an impartial jury could not be found to try him in his own country, he hath had that singular favour shown him that he should be tried here in London. And so we will proceed to call our evidence.

Then the facts of the acquaintance between the prisoner and Ann Clark were proved, and also the coroner's inquest. I pass over this portion of the trial, for it offers nothing of special interest.

Sarah Arscott was next called and sworn.

Att. What is your occupation?

S. I keep the New Inn at——.

Att. Do you know the prisoner at the bar?

S. Yes: he was often at our house since he come first at Christmas of last year.

Att. Did you know Ann Clark?

S. Yes, very well.

Att. Pray, what manner of person was she in her appearance?

S. She was a very short thick-made woman: I do not know what else you would have me say.

Att. Was she comely?

S. No, not by no manner of means: she was very uncomely, poor child! She had a great face and hanging chops and a very bad colour like a puddock.

LCJ. What is that, mistress? What say you she was like?

S. My lord, I ask pardon; I heard Esquire Martin say she looked like a puddock in the face; and so she did.

LCJ. Did you that? Can you interpret her, Mr Attorney?

Att. My lord, I apprehend it is the country word for a toad.

LCJ. Oh, a hop-toad! Ay, go on.

Att. Will you give an account to the jury of what passed between you and the prisoner at the bar in May last?

S. Sir, it was this. It was about nine o'clock the evening after that Ann did not come home, and I was about my work in the house; there was no company there only Thomas Snell, and it was foul weather. Esquire Martin came in and called for some drink, and I, by way of pleasantry, I said to him, "Squire, have you been looking after your sweetheart?' and he flew out at me in a passion and desired I would not use such expressions. I was amazed at that, because we were accustomed to joke with him about her.

LCJ. Who, her?

S. Ann Clark, my lord. And we had not heard the news of his being contracted to a young gentlewoman elsewhere, or I am sure I should have used better manners. So I said nothing, but being I was a little put out, I begun singing, to myself as it were, the song they danced to the first time they met, for I thought it would prick him. It was the same that he was used to sing when he come down the street; I have heard it very often: *Madam, will you walk, will you talk with me?* And it fell out that I needed something that was in the kitchen. So I went out to get it, and all the time I went on singing, something louder and more bold-like. And as I was there all of a sudden I thought I heard some one answering outside the house, but I could not be sure because of the wind blowing so high. So then I stopped singing, and now I heard it plain, saying, *Yes, sir, I will walk, I will talk with you*, and I knew the voice for Ann Clark's voice.

Att. How did you know it to be her voice?

S. It was impossible I could be mistaken. She had a dreadful voice, a kind of a squalling voice, in particular if she tried to sing. And there was nobody in the village that could counterfeit it, for they often tried. So, hearing that, I was glad, because we were all in an anxiety to know what was gone with her: for though she was a natural, she had a good disposition and was very tractable: and says I to myself, 'What, child! are you returned, then?' and I ran into the front room, and said to Squire Martin as I passed by, 'Squire, here is your sweetheart back again: shall I call her in?' and with that I went to open the door; but Squire Martin he caught hold of me, and it seemed to me he was out of his wits, or near upon. 'Hold, woman,' says he, 'in God's name!' and I know not what else: he was all of a shake. Then I was angry, and said I, 'What! are you not glad that poor child is found?' and I called to Thomas Snell and said, 'If the Squire will not let me, do you open the door and call her in.' So Thomas Snell went and opened the door, and the wind setting that way blew in and overset the two candles that was all we had lighted: and Esquire Martin fell away from holding me; I think he fell down on the floor, but we were wholly in the dark, and it was a minute or two before I got a light again: and while I was feeling for the fire-box, I am not certain but I heard someone step 'cross the floor, and I am sure I heard the door of the great cupboard that stands in the room open and shut to. Then, when I had a light again, I see Esquire Martin on the settle, all white and sweaty as if he had swounded away, and his arms hanging down; and I was going to help him; but just then it caught my eye that there was something like a bit of a dress shut into the cupboard door, and it came to my mind I had heard that door shut. So I thought it might be some person had run in when the light was quenched, and was hiding in the cupboard. So I went up closer and looked: and there was a bit of a black stuff cloak, and just below it an edge of a brown stuff dress, both sticking out of the shut of the door: and both of them was low down, as if the person that had them on might be crouched down inside.

Att. What did you take it to be?

S. I took it to be a woman's dress.

Att. Could you make any guess whom it belonged to? Did you know any one who wore such a dress?

S. It was a common stuff, by what I could see. I have seen many women wearing such a stuff in our parish.

Att. Was it like Ann Clark's dress?

S. She used to wear just such a dress: but I could not say on my oath it was hers.

Att. Did you observe anything else about it?

S. I did notice that it looked very wet: but it was foul weather outside.

LCJ. Did you feel of it, mistress?

S. No, my lord, I did not like to touch it.

LCJ. Not like? Why that? Are you so nice that you scruple to feel of a wet dress?

S. Indeed, my lord, I cannot very well tell why: only it had a nasty ugly look about it.

LCJ. Well, go on.

S. Then I called again to Thomas Snell, and bid him come to me and catch any one that come out when I should open the cupboard door, 'for,' says I, 'there is some one hiding within, and I would know what she wants'. And with that Squire Martin gave a sort of a cry or a shout and ran out of the house into the dark, and I felt the cupboard door pushed out against me while I held it, and Thomas Snell helped me: but for all we pressed to keep it shut as hard as we could, it was forced out against us, and we had to fall back.

LCJ. And pray what came out—a mouse?

S. No, my lord, it was greater than a mouse, but I could not see what it was: it fleeted very swift over the floor and out at the door.

LCJ. But come; what did it look like? Was it a person?

S. My lord, I cannot tell what it was, but it ran very low, and it was of a dark colour. We were both daunted by it, Thomas Snell and I, but we made all the haste we could after it to the door that stood open. And we looked out, but it was dark and we could see nothing.

LCJ. Was there no tracks of it on the floor? What floor have you there?

S. It is a flagged floor and sanded, my lord, and there was an appearance of a wet track on the floor, but we could make nothing of it, neither Thomas Snell nor me, and besides, as I said, it was a foul night.

LCJ. Well, for my part, I see not—though to be sure it is an odd tale she tells—what you would do with this evidence.

Att. My lord, we bring it to show the suspicious carriage of the prisoner immediately after the disappearance of the murdered person: and we ask the jury's consideration of that; and also to the matter of the voice heard without the house.

Then the prisoner asked some questions not very material, and Thomas Snell was next called, who gave evidence to the same effect as Mrs Arscott, and added the following:

Att. Did anything pass between you and the prisoner during the time Mrs Arscott was out of the room?

Th. I had a piece of twist in my pocket.

Att. Twist of what?

Th. Twist of tobacco, sir, and I felt a disposition to take a pipe of tobacco. So I found a pipe on the chimney-piece, and being it was twist, and in regard of me having by an oversight left my knife at my house, and me not having over many teeth to pluck at it, as your lordship or any one else may have a view by their own eyesight——

LCJ. What is the man talking about? Come to the matter, fellow! Do you think we sit here to look at your teeth?

Th. No, my lord, nor I would not you should do, God forbid! I know your honours have better employment, and better teeth, I would not wonder.

LCJ. Good God, what a man is this! Yes, I *have* better teeth, and that you shall find if you keep not to the purpose.

Th. I humbly ask pardon, my lord, but so it was. And I took upon me, thinking no harm, to ask Squire Martin to lend me his knife to cut my tobacco. And he felt first of one pocket and then of another and it was not there at all. And says I, 'What! have you lost your knife, Squire?' And up he gets and feels again and he sat down, and such a groan as he gave, 'Good God!' he says, 'I must have left it there.' 'But,' says I, 'Squire, by all appearance it is *not* there. Did you set a value on it,' says I, 'you might have it cried.' But he sat there and put his head between his hands and seemed to take no notice to what I said. And then it was Mistress Arscott come tracking back out of the kitchen place.

Asked if he heard the voice singing outside the house, he said 'No', but the door into the kitchen was shut, and there was a high wind: but says that no one could mistake Ann Clark's voice.

Then a boy, William Reddaway, about thirteen years of age, was called, and by the usual questions, put by the Lord Chief Justice, it was ascertained that he knew the nature of an oath. And so he was sworn. His evidence referred to a time about a week later.

Att. Now, child, don't be frighted: there is no one here will hurt you if you speak the truth.

LCJ. Ay, if he speak the truth. But remember, child, thou art in the presence of the great God of heaven and earth, that hath the keys of hell, and of us that are the king's officers, and have the keys of Newgate; and remember, too, there is a man's life in question; and if thou tellest a lie, and by that means he comes to an ill end, thou art no better than his murderer; and so speak the truth.

Att. Tell the jury what you know, and speak out. Where were you on the evening of the 23rd of May last?

LCJ. Why, what does such a boy as this know of days. Can you mark the day, boy?

W. Yes, my lord, it was the day before our feast, and I was to spend sixpence there, and that falls a month before Midsummer Day.

One of the Jury. My lord, we cannot hear what he says.

LCJ. He says he remembers the day because it was the day before the feast they had there, and he had sixpence to lay out. Set him up on the table there. Well, child, and where wast thou then?

W. Keeping cows on the moor, my lord.

But, the boy using the country speech, my lord could not well apprehend him, and so asked if there was any one that could interpret him, and it was answered the parson of the parish was there, and he was accordingly sworn and so the evidence given. The boy said—

'I was on the moor about six o'clock, and sitting behind a bush of furze near a pond of water: and the prisoner came very cautiously and looking about him, having something like a long pole in his hand, and stopped a good while as if he would be listening, and then began to feel in the water with the pole: and I being very near the water—not above five yards— heard as if the pole struck up against something that made a wallowing sound, and the prisoner dropped the pole and threw himself on the ground, and rolled himself about very strangely with his hands to his ears, and so after a while got up and went creeping away.'

Asked if he had had any communication with the prisoner, 'Yes, a day or two before, the prisoner, hearing I was used to be on the moor, he asked me if I had seen a knife laying about, and said he would give sixpence to find it. And I said I had not seen any such thing, but I would ask about. Then he said he would give me sixpence to say nothing, and so he did.'

LCJ. And was that the sixpence you were to lay out at the feast?
W. Yes, if you please, my lord.

Asked if he had observed anything particular as to the pond of water, he said, 'No, except that it begun to have a very ill smell and the cows would not drink of it for some days before.'

Asked if he had ever seen the prisoner and Ann Clark in company together, he began to cry very much, and it was a long time before they could get him to speak intelligibly. At last the parson of the parish, Mr Matthews, got him to be quiet, and the question being put to him again, he said he had seen Ann Clark waiting on the moor for the prisoner at some way off, several times since last Christmas.

Att. Did you see her close, so as to be sure it was she?

W. Yes, quite sure.

LCJ. How quite sure, child?

W. Because she would stand and jump up and down and clap her arms like a goose (which he called by some country name: but the parson explained it to be a goose). And then she was of such a shape that it could not be no one else.

Att. What was the last time that you so saw her?

Then the witness began to cry again and clung very much to Mr Matthews, who bid him not be frightened. And so at last he told this story: that on the day before their feast (being the same evening that he had before spoken of) after the prisoner had gone away, it being then twilight and he very desirous to get home, but afraid for the present to stir from where he was lest the prisoner should see him, remained some few minutes behind the bush, looking on the pond, and saw something dark come up out of the water at the edge of the pond furthest away from him, and so up the bank. And when it got to the top where he could see it plain against the sky, it stood up and flapped the arms up and down, and then run off very swiftly in the same direction the prisoner had taken: and being asked very strictly who he took it to be, he said upon his oath that it could be nobody but Ann Clark.

Thereafter his master was called, and gave evidence that the boy had come home very late that evening and been chided for it, and that he seemed very much amazed, but could give no account of the reason.

Att. My lord, we have done with our evidence for the King.

Then the Lord Chief Justice called upon the prisoner to make his defence; which he did, though at no great length, and in a very halting way, saying that he hoped the jury would not go about to take his life on the evidence of a parcel of country people and children that would believe any idle tale; and that he had been very much prejudiced in his trial; at which the LCJ interrupted him, saying that he had had singular favour shown to him in having his trial removed from Exeter, which the prisoner acknowledging, said that he meant rather that since he was brought to London there had not been care taken to keep him secured from interruption and disturbance. Upon which the LCJ ordered the Marshal to be called, and questioned him about the safe keeping of the prisoner, but could find nothing: except the Marshal said that he had been informed by the underkeeper that they had seen a person outside his door or going up the stairs to it: but there was no possibility the person should have got in. And it being enquired further what sort of person this might be, the Marshal could not speak to it save by hearsay, which was not allowed. And the prisoner, being asked if this was what he meant, said no, he knew nothing of that, but it was very hard that a man should not be suffered to be at quiet when his life stood on it. But it was observed he was very hasty in his denial. And so he said no more, and called no witnesses. Whereupon the Attorney-General spoke to the jury. [A full report of what he said is given, and, if time allowed, I would extract that portion in which he dwells on the alleged appearance of the murdered person: he quotes some authorities of ancient date, as St Augustine *de cura pro mortuis gerenda* (a favourite book of reference with the old writers on the supernatural) and also cites some cases which may be seen in Glanvil's, but more conveniently in Mr Lang's books. He does not, however, tell us more of those cases than is to be found in print.]

The Lord Chief Justice then summed up the evidence for the jury. His speech, again, contains nothing that I find worth copying out: but he was naturally impressed with the singular character of the evidence, saying that he had never heard such given in his experience; but that there was nothing in law to set it aside, and that the jury must consider whether they believed these witnesses or not.

And the jury after a very short consultation brought the prisoner in Guilty.

So he was asked whether he had anything to say in arrest of judgment, and pleaded that his name was spelt wrong in the indictment, being Martin with an I, whereas it should be with a Y. But this was overruled as not material, Mr Attorney saying, moreover, that he could bring evidence to show that the prisoner by times wrote it as it was laid in the indictment. And, the prisoner having nothing further to offer, sentence of death was passed upon him, and that he should be hanged in chains upon a gibbet near the place where the fact was committed, and that execution should take place upon the 28th December next ensuing, being Innocents' Day.

Thereafter the prisoner being to all appearance in a state of desperation, made shift to ask the LCJ that his relations might be allowed to come to him during the short time he had to live.

LCJ. Ay, with all my heart, so it be in the presence of the keeper; and Ann Clark may come to you as well, for what I care.

At which the prisoner broke out and cried to his lordship not to use such words to him, and his lordship very angrily told him he deserved no tenderness at any man's hands for a cowardly butcherly murderer that had not the stomach to take the reward of his deeds: 'and I hope to God,' said he, 'that she *will* be with you by day and by night till an end is made of you.' Then the prisoner was removed, and, so far as I saw, he was in a swound, and the court broke up.

I cannot refrain from observing that the prisoner during all the time of the trial seemed to be more uneasy than is commonly the case even in capital causes: that, for example, he was looking narrowly among the people and often turning round very sharply, as if some person might be at his ear. It was also very noticeable at this trial what a silence the people kept, and further (though this might not be otherwise than natural in that season of the year), what a darkness and obscurity there was in the court room, lights being brought in not long after two o'clock in the day, and yet no fog in the town.

* * *

It was not without interest that I heard lately from some young men who had been giving a concert in the village I speak of, that a very cold reception was accorded to the song which has been mentioned in this narrative: *Madam, will you walk?* It came out in some talk they had next morning with some of the local people that that song was regarded with an invincible repugnance—it was not so, they believed, at North Tawton— but here it was reckoned to be unlucky. However, why that view was taken no one had the shadow of an idea.

CLEMENCE DANE

The Emancipation of Mrs Morley

∞

Queen Anne was in her parlour, having dined, and the Queen was alone. The page-in-waiting at the head of the back stairs forbore to whistle. The voices of chambermaids and grooms were hushed. Even the new favourite, Abigail Masham, knew herself unwanted between two-thirty and four of a Sunday afternoon. The loyal silence was unbroken, save for the April bees buzzing in like petty treasons through the open casements, and the gardener's shears eternally clip-clipping at the late Queen's vegetable statuary.

For this was the hour when Queen Anne took her nap, and the silence was a matter of affection, not of discipline.

A queen who conspired with her bedchamber women to send comforts to prisoners condemned for treason against her stout person was not a queen to be harsh with disturbers of her Sunday nap. The prevailing peace, therefore, was a tribute paid unprompted by the only class which truly valued their Queen Anne. The lords might mock and the ladies shrug, but Anne Hyde's well-meaning daughter had the common people behind her.

'Be quiet, can't you, with your milk-pails? The Queen's asleep.'

But for once the Queen was not asleep. For spring winds whispered in the poplar by the window, and the sunlight, filtering through that eternal conversation, danced irreverently upon the Queen's dinner-flushed countenance. Then, as she lifted her one beauty, her celebrated hand, to shade her eyes, the light dance dropped suddenly to the cause of her wakefulness —the letter lying open on her lap. The feminine handwriting stood out sharp as a threat.

'I am glad your Majesty is going to Kensington to make use of the fresh air and take care of your health. I will follow you thither and wait every day till it is convenient for you to see me, as what I have to say is of such a nature as to require no answer.'

If only, thought the poor Queen, the Duchess of Marlborough would stop writing letters! But at least the former insolent intimacy of address had ceased. It was 'your Majesty' now, not 'Mrs Morley'. Then the Queen flushed as she remembered that such intimacy was her own fault.

It was she who had suggested, half a lifetime ago, that Princess Anne and Sarah Jennings should have citizen names for each other.

'Let me beg of you not to call me highness at every word, but to be as free with me as one friend ought to be with another.'

Yes, she remembered writing it. She remembered choosing her name. 'I'll be Mrs Morley. And you?'

She could still hear the young Sarah's loud voice in answer. 'With my frank open temper, highness, I'll take a free name—Freeman.'

But Mrs Freeman had grown too free.

'Whose fault?' the Queen asked herself, and meekly answered, 'Mine'. Hers, no doubt. But she had so badly needed a leader. She had loved her husband, her good, kind, dull George of Denmark, but he needed a leader as much as she did. So they had both looked up to Sarah and to Sarah's clever husband.

Sarah and he had shown her that it was right to desert her own royal father when the revolution came and brought in sister Mary and brother William of Orange. She had not meant to be a wicked daughter. One must keep out the Catholics—Sarah's husband said so; and preserve one's rights—as Sarah said. And the people—they hated King James.

Yet for a long while the people had pointed at her behind her back for deserting him. She had lived it down, but for a while she had known how they whispered and pointed. She was *Regan* and Mary was *Goneril*, the wicked daughters in a stupid play she had not been able to read. What was it called? *King Lear*. 'Greedy Regan', they called her. How cruel people were when you did your best, cruel and insolent! But none was so insolent as the writer of the letter in her lap. Insolent—for had she not refused the Duchess audience? And here was the answer—flat defiance.

'I will follow you thither and wait every day till it is convenient for you to see me.'

'It will never be convenient,' said the Queen aloud, and struck her writing table with her gouty hand, as never in her life had she struck a suppliant. For she seldom sat down to her table without remembering a certain interview held long ago with Uncle Charles. She was the suppliant then. She saw his dark face again, and his kind quizzing eyes, as he told her that he had sent away her Mulgrave, her lover, her poet, and that she was to marry George of Denmark. And when she had wept, he had said—not mocking her, for once!——

'You see, niece Anne, you may have to wear my crown one day.' And then began to quote, as usual, that tiresome playwright who wrote *Lear* and had died in his grandfather's time. 'Uneasy lies the head that wears a crown, niece!' said King Charles.

It had proved true, though how the man had guessed it she could not imagine. Even her own poet, her lost dear Mulgrave, had not guessed how hard it was to be a King's daughter. She had tried to tell him indeed, but the letter never reached him. She knew that now. And why had it never reached him? Because her own familiar Sarah Jennings, her 'Mrs Freeman', in whom she trusted, had delivered it not to Mulgrave but to King Charles.

And that, too, she had only found out thirty years later, when Masham, clearing out her Uncle Charles' desk, had happened on a secret drawer and its papers. They lay in her desk now—notes of a gaming debt, his half-written scrawl to the beloved sister in France, his own report on the dissection of a maid of honour's stillborn child. How horrid her uncle could be! And at the bottom of the bundle lay her own letter to Mulgrave, with the pencilled note on the back—'Paid Sarah Jennings, for delivery, fifty guineas.' There had been treachery even then.

It was all so long ago that the Queen's pulse could not beat as fast at remembrance of that remote treachery as it beat at the remembrance of later, pettier wrongs. Nevertheless the perpetual frown between the cloudy eyes deepened and the Queen's mouth began to droop in that sullenness which was her anger, as she read her letter.

'What I have to say is of such a nature as to require no answer.'

She should have none!

Queen Anne at that moment of decision was no ignoble figure. She was a sonsy widow, happier at cards than at the council table; but she was also, if anyone had watched her then, a queen more concerned for her broken soldiers than for their glorious victories. Not long ago she had reminded a listening parliament and the greatest soldier of the age that, 'notwithstanding the arts of those who delight in war, both time and place *are* appointed for opening a treaty of peace!'

No, Queen Anne was no longer afraid of the Duke of Marlborough. 'Will this bloodshed never cease?' she had cried when they told her of another victory. And she did not thank him in her heart for Blenheim, Oudenarde, Malplaquet and all the other crown-jewels of blood.

And if Mrs Freeman, by the grace of Anne, Duchess of Marlborough, came to Kensington as she threatened, then Mrs Morley, by the grace of God, Queen of England, Scotland, and Ireland, Defender of the Faith, would not find it convenient ever again to see her false friend.

But when she heard a familiar voice raised in expostulation outside her door, then she realized that whatever the Queen might resolve against the Duchess, Mrs Morley was still afraid of Mrs Freeman. When the page-in-waiting scratched upon the door, Mrs Morley touched the little bell which permitted his entry with a shaking hand.

The page also was a trifle breathless. His deep obeisance was accompanied by a glance over his shoulder as he delivered his message. Her Grace the Duchess of Marlborough waited without: Her Grace wished him to enquire if Her Majesty would see her then or whether she should come at some other time. Again he glanced sideways, and the Queen, although she loved etiquette, forgave him; for her spirits were rising as her slow wits at last realized that the message was a humble one. Could it be that Mrs Freeman was ceasing to despise her 'poor, unfortunate, faithful Morley'? Still—must she see her? Where was Masham, to give her good advice?

'Taking the plunge at last,' she said with piteous reluctance, 'I will see her,' and added hastily, as the scared page, much relieved, was retreating, 'And afterwards fetch me Masham!'

For she whose unroyal kindly weakness was her need for kind voices and kind hearts about her had a yearning at this moment of loss, vengeance, and farewell, for the presence of her enemy's cousin. Abigail Masham might be clumsy, but how tender were her hands! Her nose was red and her chest flat, but how soothing was her low voice!

Besides—the Queen had often puzzled over it—if she loved Abigail it was justified related love and not mere fickleness. The Duchess might sneer at the upstart, but poor Masham was the Duchess' own cousin for all that. In fifty ways the relationship showed—in the droop of their lids, in their ugly wrists, above all in their rare pretty laugh. George even had noticed it. Her great Sarah and her humble Abigail were cousins right enough, though, as George had said, it was the relationship of a lioness to a poor mousing Tibbs. Still, she had first loved Abigail for the likeness' sake.

It was difficult to unlove a worshipped friend, but she could love in Abigail that which she had tried to find in the Duchess. And being a blood relation Abigail had a right to speak freely of the Duchess, yes, even to the Queen. And Abigail had spoken freely. Abigail had taught her that chains can be broken. Affection, association, time, gratitude—these were all chains to be broken. Mrs Freeman had never worn chains, though she had fettered Mrs Morley. Now, said Abigail, Mrs Morley need wear them no longer. Mrs Morley had only to remember that she was the Queen.

Abigail was right. She would remember. She stiffened as she bent over her writing. There was to be no appearance of welcome in the Duchess' reception. But she could not see the page, and plunged blindly at the standish with her quill. A cold paralysis was upon her, the suspension of her whole being by fear, pain, and re-awakening affection. She thrust her finger into her mouth and bit upon it, as, across the length of the room, the door reopened.

Then the lifelong friends were together once more, and the Queen's fear and pain shrank before the violent emotions which approached her. The room was alive with unseen passion, anger, greed. She heard a breathing behind her chair and in her nostrils was the faint, familiar smell of orris. But she would not turn. One—two—three—twenty ticks of the clock she counted. Then she felt her chair imperceptibly shaken as if a hand had clutched it. It was on that touch at second hand that Queen Anne took her finger out of her mouth and staring at the letter in front of her, still without turning, said meekly:

'I was going to write to you.'

The well-known voice answered her sharply:

'Upon what, Madam?'

The Queen drew a quick asthmatic breath and suddenly the quill bent in two beneath her hands as she said:

'I did not open your letter until now, and I was going to write to you.'

Once more she could feel the hand tightening upon the back of her chair. But the hectoring voice was lowered in humility:

'Was there anything in it, Madam, that you had in mind to answer?'

That was the question. Was there anything to answer? She had royal wounds to show. But her answer would be used against her. Better not answer. Aloud she said, stubbornly:

'Whatever you have to say, you may write it.'

There was a pause. Then the harsh voice faltered:

'I believe Your Majesty never did so hard a thing to anybody as to refuse to hear them speak.'

At that the Queen laid down her pen, and her thoughts began to drum angrily as flustered bees within her weary, sleepy head. Never so hard a thing to anybody as to refuse to hear them speak! No, the Duchess should not have said that. For King James had wanted to speak to his younger daughter Anne on that dreadful night when he heard beyond all disbelief that his elder daughter Mary and her Dutchman were sailing for England to dethrone him. And who had urged her, wicked daughter, to avoid that interview? Whose fault was it that never again had she heard her father speak? 'He can write to you if he chooses,' had been Mrs Freeman's view of it then. 'Tell him to put what he has to say in writing.' But he had never written again.

The Queen rose heavily, waited while obsequious hands pulled the chair away, then turned to the Duchess. But she would not look at her. In vain did the Duchess push back her golden hair with a distracted gesture. In vain did her full lower lip tremble. Her tears always came easily. They reddened her nose and increased the elusive likeness to poor Masham. But the Queen would not look.

'You know well enough,' said Queen Anne, her mouth drooping, 'that I do bid people put what they have to say in writing when I have a mind.' And continued to stare at the floor.

The Duchess could have struck her in her stupid face. How could one talk to the creature if she would not so much as look at you? Oh—how unfair are the weapons of royalty! How dared poor, faithful Mrs Morley refuse to look at her wonderful Mrs Freeman?

'Your Majesty!'

It was, in spite of her good resolutions, her parade voice, and she knew it as she saw the Queen shiver. Oh well, let her shrink! Let her ears have it if she would not lift her eyes!

'Your Majesty knows that I cannot hide my feelings. I cannot be obsequious. I must speak my mind. With all due respect let me say—Your Majesty is ungrateful! Where would Your Majesty be today without me and the Duke? You need not look at me like that. Who sacrificed all to your cause? Didn't Your Majesty's sister the late Queen bribe me to leave Your Majesty? And did I? Wasn't Mr Freeman—I would say "the Duke", but I love the old friendly titles, though Your Majesty has forgotten them, it seems—didn't the Duke risk his freedom? And when King William died, who put the crown on your head and kept it there? You may not believe it, Madam, but there are many people less quick than you to call your brother a bastard. There are many could tell you that there would be a King James the Third on the throne now if it were not for my husband and me. What would you have done without us? Who made you respected in your own Court? Who warned you of the treachery that Your Majesty's German cousin was plotting? Who fought your battles for you—Blenheim, Ramillies, Oudenarde, Malplaquet?'

The names rolled and reverberated like trumpet calls, but the Queen did not lift her eyes. She only sighed, a short asthmatic sigh, as the Duchess continued. And the Duchess had a brighter eye for speaking the names of her husband's victories, and in her cheek the red sat like a flag of battle.

'And what thanks has my husband got?' she demanded of her dumb sovereign. 'An empty title! For what is a title to him without Your Majesty's interest? There is he, risking his life on the muddy field of battle for you, and here am I, busying myself on Your Majesty's affairs—your robes, your purse, your servants—and what thanks do I get, does he get? While we slave for you, unthanked, this low creeping creature, this bed-chamber woman whom I raised from nothing, this Abigail whom I favoured, whom I put into Your Majesty's service, she slanders us to Your Majesty. Your Majesty need not trouble to deny it.

'When I come to see Your Majesty, Abigail Masham is with you, has been with you two hours, or you have crept into some low closet to talk with her. The creature forms a party against me and Your Majesty countenances it. The Duke would not believe it when he heard. He said to me, "Sure, my dear, the Queen knows what she owes you." I told him plainly that I would like to think you did. I am sure I deserve that you should. But what am I to think when Her Majesty is present at her bed-chamber woman's marriage with the Prince's page? Sam Masham and Abigail Hill! A pretty pair!

'To be sure, if Her Majesty is so fond of low company, it is no affair of mine. But are you aware, Madam, that the whole country talks of your passion for Abigail? That your brother in St Germains laughs at it and makes his plans? "Sure," he says, "my sister Anne will soon eat herself to death, she and her Masham, and then we enjoy our own again." Is it nothing to you, Ma'am, that they speak of you so?'

The Duchess paused, breathless, and on the whole pleased, for at least the Queen had lifted her head at last. How oddly she stared! Not even angry—not even sullen—as if she scarcely heard. What was the matter with the little card-playing automaton?

What was the matter? The Queen was saying goodbye to an illusion, farewell to a dream. Mechanically her mind had registered its protest to the Duchess' accusations, but she had not opened her mouth, because she could not put into words the clear pictures that rose in her mind at each stroke of speech. But she saw her pictures clearly. As she looked at the proud woman who railed at her, the Queen saw a woman thirty years younger. There that young Sarah Jennings knelt, her golden hair dishevelled, shaking with sobs as she poured out the story of her secret marriage. What refuge had she but her Mrs Morley? And, thought the Queen mildly, I did what I could. I gave her twelve hundred a year: I reconciled her to her relations: I gave her a place with me and loved her. What more could I do?

Then another picture arose in her slow mind. She saw her own sister's angry face lowering down upon her as she lay in child-bed, and the unfriendly voice of Mary II: 'I made the first step by coming to you. Now I look to you to do the rest, by sending Lady Marlborough away.' Cruel sister! She had been thrown into a fever by the visit, but she had not deserted the Marlboroughs. She had lost her own income, she had been thrust out of her royal lodgings, deprived of her guard, insulted in every way that Dutch malice could invent, but she had not sent away her Mrs Freeman. What more in love could she do?

The Duchess talked of her sorrows. But the Duchess had blooming children for whom to beg places. Where were Anne's children? Child-bed after child-bed passed before her dull eyes, bringing its recollection of useless pain endured. She remembered her one darling child, her eleven years old Knight of the Garter in his jewelled suit. What did the Duchess know of the heartbreak and grief of losing a child? If cards helped her to forget her sorrows for a little, must the Duchess mock at her?

Was Mrs Morley heartless? But she had given Mrs Freeman twenty years of passionate devotion. Thankless? But Mrs Freeman was the Duchess of Marlborough, with Blenheim, that king's estate, for her dwelling place, and ninety thousand pounds in her pocket. And yet she was greedy of her rights to the Queen's old gowns! She, the Queen, might not give a torn gown to a needy servant, or send a bottle of wine to a sick laundress, without a rating from her tyrant. Well, she had borne it, because as the Duke said, 'We all know the Duchess' temper. We all bear it.' But he was her husband and he loved her. Was it the Queen's place to bear it, though she loved her?

Yet the rages she had borne meekly, the insolent letters she had answered meekly, until that day six years ago when she found out that the Duchess had no love to give her back. Love? She was disliked, despised, all but hated after all these faithful years. Had the Duchess forgotten the gloves? Oh, the Duchess should never have put on Mrs Morley's gloves and taken them off again with an affected shudder, crying, 'What! Are they hers? As if I would wear anything that odious woman has touched!' She never knew that the door stood open, and that through the door Mrs Morley heard it all. But Masham knew. Masham had come to her, kind Masham had said, 'Don't cry, Your Majesty, she doesn't mean it. It's only her temper.'

Yes, the Duchess was inveterate against poor Masham, but Masham never railed against anyone. Masham was like herself, she only wanted peace. But the Duchess always wanted war. Very good, the Duchess should have it.

And staring with her cloudy eyes into the bright hard blue ones, the Queen, after a long pause, answered:

'You said that you required no answer and you shall have none.'

The Duchess' foot began to tap. 'Why is Your Majesty so angry? What have I done?'

'You said that you required no answer, and I will give you none.'

'If Your Majesty thinks my husband's demands go too far, he will withdraw them, though if he were Captain General of your army for life, it would save Your Majesty much trouble. This war will last for ever. Is Your Majesty afraid of trusting him?'

'You have said that you required no answer, and I will give you none.'

'Is it possible that Your Majesty has heard some garbled account of my letters to Hanover?'

The Queen stared but said nothing.

'What gossip has Your Majesty heard?'

The Queen began pulling on her gloves.

'Perhaps it is some small matter? I have noticed that Your Majesty is more concerned with little things than great. The Duke did not wear his full-bottomed wig, certainly, when he came to Your Majesty a week ago, but he rode in haste. Was Your Majesty offended?'

Finger by finger, the Queen fitted the fine leather to her fine hand.

'Is it because I sat the other day at the Sacheveral trial, in Your Majesty's presence?'

'If I had not liked you to sit, why would I have ordered it?'

'So I thought. I meant no harm. But the Duchess of Somerset did not see fit to sit. Her Grace would be glad to put me in the wrong—that Your Majesty cannot deny.'

No answer.

'Has Your Majesty no feeling left for me? Has Mrs Morley no word for Mrs Freeman?'

Said the Queen, busy with her second glove, 'You said you required no answer, and I will give you none.'

'Who has taught Your Majesty to answer me so?'

Suddenly the Queen smiled, as she worked upon the black finger of her glove. Who had taught her? Who had torn her from her husband's dead body, mocking at her grief? Who had denied her? When she longed for a shoulder to cry upon, who had refused to send for kind Masham, because 'it would make a disagreeable noise when there were bishops and ladies waiting without'? Who had mocked when she begged them to leave room in the vault so that her own body might one day lie beside her husband's? Who had sneered because, after three days' watching, she had taken a cup of broth? Who had tittered when she feared that the dear dead body might be shaken roughly on its long journey to Westminster? Who had told the world that Queen Anne's nature was hard, because she had learned to control her bitter tears?

And the Queen, when she thought of those losses and griefs, forgot for a moment the angry woman in front of her. And though she had learned control, the tears fell suddenly and rolled down her cheeks. But she should not have cried: her tears were swiftly taken for a weakness.

'I cannot help renewing my request for an explanation. Who has taught Your Majesty to answer me so?'

Who? Had the Duchess forgotten the drive to St Paul's to celebrate the victory of Oudenarde? For Anne it was a service of mourning for her dead soldiers. She, the widow, mourned with the widows of England. How could she wear the jewels the Duchess had set out for her? It was on Ludgate Hill that the Duchess first missed them. The Queen shivered as she remembered the rage of words that had hailed down on her then, as they slowly passed up the hill, through the yard and up the steps into the great church itself. And when Anne had tried to explain, it had been: 'Do not answer me! Hold your tongue!' Loyal subjects had heard it.

Well—her Grace's commands for the last time should be obeyed. She would not answer her. Turning aside, refusing to meet the compelling eyes, she repeated the phrase which was her shield, and repeated it with a bitterness which startled the Duchess:

'You said you required no answer, and I will give you none.'

'Then I am right! They have been lying to you.'

'There are many lies told,' said the Queen sullenly.

'Many lies!' retorted the Duchess. And there was such a note of rage in her voice that the Queen started. The beautiful face was contorted with passion, the cheeks were white and patchy.

'Ah, they tell you lies, Madam, but they tell me the truth. Perhaps Your Majesty would like to know what people *do* say? They say that though you fancied yourself a little sorry for your husband's death, you now taste the pleasures of freedom. Now you may carouse with your Masham! Now you may take orders from her, as once you took them from me! And another thing they say—they say that nobody is too infamous to be countenanced if they but apply to your great favourite. They say that she has taught you to keep low company, though, to be sure, Prince George taught you to do that. They say that you have played your wits away at cards, such wits indeed as you have not fuddled away already with eating and drinking. Perhaps Your Majesty knows the rhyme—

There's Mary the daughter and Willie the cheater, Georgie the drunkard and Annie the eater.

They say that your sister is more like King Lear's daughter than King James', and she got her deserts long ago—that they drink to the little gentleman in brown velvet. And that now that Prince George has drunk himself into his grave, you have only to eat yourself into yours. And then we shall be happy.'

'I will leave the room.'

The Queen's low voice cut through the scream of rage. The stout woman in her widow's dress was a queen at last. No longer did the lace head-dress wave in agitation, no longer did the gloved hand shake. King James the Second, King Charles the First, King James the First, Queen Mary of Scotland—Kings and Queens of three countries and three hundred years—Stuart, Tudor, Plantagenet—strengthened her sweet and thrilling voice as she repeated, with a mildness that frightened the Duchess more than wrath could do:

'I will leave the room.'

But the Duchess, having passed through the several stages of her celebrated passion, now burst into tears of anger, bafflement, despair. And even
while she felt for her handkerchief, uttering as she did so those sobbing
screams of rage which one hears in a nursery but seldom elsewhere, it ran
through her shrewd mind that even now all might not be lost.

The Queen had never liked to see her cry. A word of humility, after the
outburst of rage and jealous temper, could always soften Mrs Morley.
She would try it. Sobbing more gently and wiping her eyes, she raced for
the door and was there before the Queen's slow gouty tread had taken her
more than half across the room.

'I am sure I would not turn Your Majesty out of her own chamber,'
cried the Duchess of Marlborough, and retreated on to the landing,
closing the door behind her.

For long minutes she sat in the window seat staring at the door she had
closed upon herself. Had she closed it for ever? Had she won? What had
she said? Was it worse than anything she had said in any early outburst?
She had always been forgiven. Forgiven? She had always proved herself
right. She had only to wait for the letter signed 'Your poor unfortunate
faithful Morley.' It always came. It always begged forgiveness. But this
time perhaps it would become Mrs Freeman to beg forgiveness. Suppose
it were refused? She *must* be forgiven. Now she was out of the room, she
was dying to be back. She should not have let herself be driven out. She
had not begun to say a quarter of what was in her mind.

The page, carrying candles into the Queen's parlour, made her aware
of the time. She had come to the Queen a little before four, and now it
was dark. She had talked longer than she thought, and had accomplished—
face it!—nothing. If she went home now, would she ever see the Queen
again?

The page came out again and closed the door behind him. She watched
him out of sight down the corridor, then, slipping lightly off the window
seat, ran across the passage and scratched upon the door as he had done.
She was proud of her humility as she did so. She was the Duchess of
Marlborough, the mistress of the robes, with the gold keys at her girdle
which gave her the right to enter unannounced. Surely she showed her
repentance, her readiness to be friends, by scratching like a page or a bedchamber woman at the Queen's door? And what if Masham opened the
door?

But the Queen herself opened it. She had recovered her composure.
She looked at the Duchess mildly enough.

*

Said the great Sarah, meekly, winningly: 'I've been thinking that if Your Majesty comes to Windsor, you will not care to see me. I will take care to avoid being at the lodgings at the same time, lest people might talk if I do not wait upon you.'

The Queen looked at her, puzzled: then said pleasantly, 'You may come to me at the Castle. It will not make me uneasy.'

The Duchess knew what she meant. The Queen was not afraid of her when other folk were present. She ruffled up.

'To be sure, when Your Majesty has Masham at hand, nothing makes you uneasy.'

'I will not talk to you in private,' said Queen Anne.

'Will Your Majesty never speak to me again?'

'Not in private.'

The Queen turned from her to the candle on the table, and took up the snuffers.

'If I might for one moment,' said the Duchess, 'speak not to Your Majesty but to Mrs Morley?'

Slowly the Queen shook her head.

'If I spoke too freely just now,' hurriedly resumed the Duchess, 'it was because I know Mrs Morley's intentions are good. To let her run on in so many mistakes is just as if one should see a friend's house on fire and let them be burned in their beds.'

The Queen looked at her blankly.

'Mrs Morley——' said the Duchess. And she threw into her voice and glance that look of sunshine after rain which neither Mr Freeman nor Mrs Morley had ever been able to resist. The snuffers shook in the Queen's hand. She turned abruptly from the candle.

'Where's Masham?' said the Queen slowly. 'Where's Masham?'

'How can you be so inhuman?' cried the Duchess, between anger and entreaty. 'Your Majesty will suffer for it. I can tell you that.'

'That will be to myself,' replied Queen Anne, and as she spoke Abigail Masham came cheerfully into the room. Instantly the Queen turned to her. Neither of them saw the Duchess leave, but the room shook as the door banged behind her. Abigail, stunned by the noise, gave the Queen that quick questioning sideways look that was so like and so unlike the glance of her terrible cousin. Then her long lean arms were stretched out protectingly, and there at last was a lean bosom for a queen to cry upon.

'Oh, Masham, Masham,' cried poor Mrs Morley. 'I've lost a friend I never had.'

'What Your Majesty needs,' said Abigail comfortably, 'is a dish of tea.'

ELIZABETH GASKELL

The Squire's Story

In the year 1769 the little town of Barford was thrown into a state of great
excitement by the intelligence that a gentleman (and 'quite the gentleman',
said the landlord of the George Inn) had been looking at Mr Clavering's
old house. This house was neither in the town nor in the country. It stood
on the outskirts of Barford, on the roadside leading to Derby. The last
occupant had been a Mr Clavering—a Northumberland gentleman of
good family—who had come to live in Barford while he was but a younger
son; but when some elder branches of the family died, he had returned
to take possession of the family estate. The house of which I speak was
called the White House, from its being covered with a greyish kind of
stucco. It had a good garden to the back, and Mr Clavering had built
capital stables, with what were then considered the latest improvements.
The point of good stabling was expected to let the house, as it was in a
hunting county; otherwise it had few recommendations. There were many
bedrooms; some entered through others, even to the number of five, leading one beyond the other; several sitting-rooms of the small and poky
kind, wainscoted round with wood, and then painted a heavy slate colour;
one good dining-room, and a drawing-room over it, both looking into the
garden, with pleasant bow-windows.

Such was the accommodation offered by the White House. It did not
seem to be very tempting to strangers, though the good people of Barford
rather piqued themselves on it, as the largest house in the town; and as a
house in which 'townspeople' and 'county people' had often met at Mr
Clavering's friendly dinners. To appreciate this circumstance of pleasant
recollection, you should have lived some years in a little country town,
surrounded by gentlemen's seats. You would then understand how a bow
or a courtesy from a member of a county family elevates the individuals
who receive it almost as much, in their own eyes, as the pair of blue
garters fringed with silver did Mr Bickerstaff's ward. They trip lightly on
air for a whole day afterwards. Now Mr Clavering was gone, where could
town and county mingle?

I mention these things that you may have an idea of the desirability
of the letting of the White House in the Barfordites' imagination; and to

make the mixture thick and slab, you must add for yourselves the bustle, the mystery, and the importance which every little event either causes or assumes in a small town; and then, perhaps, it will be no wonder to you that twenty ragged little urchins accompanied the 'gentleman' aforesaid to the door of the White House; and that, although he was above an hour inspecting it, under the auspices of Mr Jones, the agent's clerk, thirty more had joined themselves on to the wondering crowd before his exit, and awaited such crumbs of intelligence as they could gather before they were threatened or whipped out of hearing distance. Presently, out came the 'gentleman' and the lawyer's clerk. The latter was speaking as he followed the former over the threshold. The gentleman was tall, well-dressed, handsome; but there was a sinister cold look in his quick-glancing, light blue eye, which a keen observer might not have liked. There were no keen observers among the boys, and ill-conditioned gaping girls. But they stood too near; inconveniently close; and the gentleman, lifting up his right hand, in which he carried a short riding-whip, dealt one or two sharp blows to the nearest, with a look of savage enjoyment on his face as they moved away whimpering and crying. An instant after, his expression of countenance had changed.

'Here!' said he, drawing out a handful of money, partly silver, partly copper, and throwing it into the midst of them. 'Scramble for it! fight it out, my lads! come this afternoon, at three, to the George, and I'll throw you out some more.' So the boys hurrahed for him as he walked off with the agent's clerk. He chuckled to himself, as over a pleasant thought. 'I'll have some fun with those lads,' he said; 'I'll teach 'em to come prowling and prying about me. I'll tell you what I'll do. I'll make the money so hot in the fire-shovel that it shall burn their fingers. You come and see the faces and the howling. I shall be very glad if you will dine with me at two; and by that time I may have made up my mind respecting the house.'

Mr Jones, the agent's clerk, agreed to come to the George at two, but, somehow, he had a distaste for his entertainer. Mr Jones would not like to have said, even to himself, that a man with a purse full of money, who kept many horses, and spoke familiarly of noblemen—above all, who thought of taking the White House—could be anything but a gentleman; but still the uneasy wonder as to who this Mr Robinson Higgins could be, filled the clerk's mind long after Mr Higgins, Mr Higgins's servants, and Mr Higgins's stud had taken possession of the White House.

The White House was re-stuccoed (this time of a pale yellow colour), and put into thorough repair by the accommodating and delighted land-lord; while his tenant seemed inclined to spend any amount of money on internal decorations, which were showy and effective in their character, enough to make the White House a nine days' wonder to the good people of Barford. The slate-coloured paints became pink, and were picked out with gold; the old-fashioned banisters were replaced by newly gilt ones; but, above all, the stables were a sight to be seen. Since the days of the Roman Emperor never was there such provision made for the care, the comfort, and the health of horses. But every one said it was no wonder, when they were led through Barford, covered up to their eyes, but curving their arched and delicate necks, and prancing with short high steps, in repressed eagerness. Only one groom came with them; yet they required the care of three men. Mr Higgins, however, preferred engaging two lads out of Barford; and Barford highly approved of his preference. Not only was it kind and thoughtful to give employment to the lounging lads themselves, but they were receiving such a training in Mr Higgins's stables as might fit them for Doncaster or Newmarket. The district of Derbyshire in which Barford was situated, was too close to Leicestershire not to support a hunt and a pack of hounds. The master of the hounds was a certain Sir Harry Manley, who was *aut* a huntsman *aut nullus*. He measured a man by the 'length of his fork', not by the expression of his countenance, or the shape of his head. But as Sir Harry was wont to observe, there was such a thing as too long a fork, so his approbation was withheld until he had seen a man on horseback; and if his seat there was square and easy, his hand light, and his courage good, Sir Harry hailed him as a brother.

Mr Higgins attended the first meet of the season, not as a subscriber but as an amateur. The Barford huntsmen piqued themselves on their bold riding; and their knowledge of the country came by nature; yet this new strange man, whom nobody knew, was in at the death, sitting on his horse, both well breathed and calm, without a hair turned on the sleek skin of the latter, supremely addressing the old huntsman as he hacked off the tail of the fox; and he, the old man, who was testy even under Sir Harry's slightest rebuke, and flew out on any other member of the hunt that dared to utter a word against his sixty years' experience as stable-boy, groom, poacher, and what not—he, old Isaac Wormeley, was meekly listening to the wisdom of this stranger, only now and then giving one of his quick, up-turning, cunning glances, not unlike the sharp o'er-canny looks of the poor deceased Reynard, round whom the hounds were howling, unadmonished by the short whip, which was now tucked into Wormeley's well-worn pocket. When Sir Harry rode into the copse—full of dead brushwood and wet tangled grass—and was followed by the members of the hunt, as one by one they cantered past, Mr Higgins took off his cap and bowed—half deferentially, half insolently—with a lurking smile in the corner of his eye at the discomfited looks of one or two of the laggards. 'A famous run, sir,' said Sir Harry. 'The first time you have hunted in our country; but I hope we shall see you often.'

'I hope to become a member of the hunt, sir,' said Mr Higgins.

'Most happy—proud, I am sure, to receive so daring a rider among us. You took the Cropper-gate, I fancy; while some of our friends here'—scowling at one or two cowards by way of finishing his speech. 'Allow me to introduce myself—master of the hounds.' He fumbled in his waistcoat pocket for the card on which his name was formally inscribed. 'Some of our friends here are kind enough to come home with me to dinner; might I ask for the honour?'

'My name is Higgins,' replied the stranger, bowing low. 'I am only lately come to occupy the White House at Barford, and I have not as yet presented my letters of introduction.'

'Hang it!' replied Sir Harry; 'a man with a seat like yours, and that good brush in your hand, might ride up to any door in the county (I'm a Leicestershire man!), and be a welcome guest. Mr Higgins, I shall be proud to become better acquainted with you over my dinner table.'

Mr Higgins knew pretty well how to improve the acquaintance thus begun. He could sing a good song, tell a good story, and was well up in practical jokes; with plenty of that keen worldly sense, which seems like an instinct in some men, and which in this case taught him on whom he might play off such jokes, with impunity from their resentment, and with a security of applause from the more boisterous, vehement, or prosperous. At the end of twelve months Mr Robinson Higgins was, out-and-out, the most popular member of the Barford hunt; had beaten all the others by a couple of lengths, as his first patron, Sir Harry, observed one evening, when they were just leaving the dinner-table of an old hunting squire in the neighbourhood.

'Because, you know,' said Squire Hearn, holding Sir Harry by the button—'I mean, you see, this young spark is looking sweet upon Catherine; and she's a good girl, and will have ten thousand pounds down, the day she's married, by her mother's will; and—excuse me, Sir Harry—but I should not like my girl to throw herself away.'

Though Sir Harry had a long ride before him, and but the early and short light of a new moon to take it in, his kind heart was so much touched by Squire Hearn's trembling, tearful anxiety, that he stopped and turned back into the dining-room to say, with more asseverations than I care to give:

'My good Squire, I may say, I know that man pretty well by this time; and a better fellow never existed. If I had twenty daughters he should have the pick of them.'

Squire Hearn never thought of asking the grounds for his old friend's opinion of Mr Higgins; it had been given with too much earnestness for any doubts to cross the old man's mind as to the possibility of its not being well founded. Mr Hearn was not a doubter, or a thinker, or suspicious by nature; it was simply his love for Catherine, his only daughter, that prompted his anxiety in this case; and, after what Sir Harry had said, the old man could totter with an easy mind, though not with very steady legs, into the drawing-room, where his bonny, blushing daughter Catherine and Mr Higgins stood close together on the hearth-rug—he whispering, she listening with downcast eyes. She looked so happy, so like her dead mother had looked when the Squire was a young man, that all his thought was how to please her most. His son and heir was about to be married, and bring his wife to live with the Squire; Barford and the White House were not distant an hour's ride; and, even as these thoughts passed through his mind, he asked Mr Higgins, if he could stay all night—the young moon was already set—the roads would be dark—and Catherine looked up with a pretty anxiety, which, however, had not much doubt in it, for the answer.

With every encouragement of this kind from the old Squire, it took everybody rather by surprise when, one morning, it was discovered that Miss Catherine Hearn was missing; and when, according to the usual fashion in such cases, a note was found, saying that she had eloped with 'the man of her heart', and gone to Gretna Green, no one could imagine why she could not quietly have stopped at home and been married in the parish church. She had always been a romantic, sentimental girl; very pretty and very affectionate, and very much spoiled, and very much wanting in common sense. Her indulgent father was deeply hurt at this want of confidence in his never-varying affection; but when his son came, hot with indignation from the Baronet's (his future father-in-law's house, where every form of law and of ceremony was to accompany his own impending marriage), Squire Hearn pleaded the cause of the young couple with imploring cogency, and protested that it was a piece of spirit in his daughter, which he admired and was proud of. However, it ended with Mr Nathaniel Hearn's declaring that he and his wife would have nothing to do with his sister and her husband. 'Wait till you've seen him, Nat!' said the old Squire, trembling with his distressful anticipations of family discord. 'He's an excuse for any girl. Only ask Sir Harry's opinion of him.' 'Confound Sir Harry! So that a man sits his horse well, Sir Harry cares nothing about anything else. Who is this man—this fellow? Where does he come from? What are his means? Who are his family?'

'He comes from the south—Surrey or Somersetshire, I forget which; and he pays his way well and liberally. There's not a tradesman in Barford but says he cares no more for money than for water; he spends like a prince, Nat. I don't know who his family are, but he seals with a coat of arms, which may tell you if you want to know—and he goes regularly to collect his rents from his estates in the south. Oh, Nat! if you would but be friendly, I should be as well pleased with Kitty's marriage as any father in the county.'

Mr Nathaniel Hearn gloomed, and muttered an oath or two to himself. The poor old father was reaping the consequences of his weak indulgence to his two children. Mr and Mrs Nathaniel Hearn kept apart from Catherine and her husband; and Squire Hearn durst never ask them to Levison Hall, though it was his own house. Indeed, he stole away as if he were a culprit whenever he went to visit the White House; and if he passed a night there, he was fain to equivocate when he returned home the next day; an equivocation which was well interpreted by the surly, proud Nathaniel. But the younger Mr and Mrs Hearn were the only people who did not visit at the White House. Mr and Mrs Higgins were decidedly more popular than their brother and sister-in-law. She made a very pretty, sweet-tempered hostess, and her education had not been such as to make her intolerant of any want of refinement in the associates who gathered round her husband. She had gentle smiles for townspeople as well as county people; and unconsciously played an admirable second in her husband's project of making himself universally popular.

But there is someone to make ill-natured remarks, and draw ill-natured conclusions from very simple premises, in every place; and in Barford this bird of ill-omen was a Miss Pratt. She did not hunt—so Mr Higgins's admirable riding did not call out her admiration. She did not drink— so the well-selected wines, so lavishly dispensed among his guests, could never mollify Miss Pratt. She could not bear comic songs, or buffo stories —so, in that way, her approbation was impregnable. And these three secrets of popularity constituted Mr Higgins's great charm. Miss Pratt sat and watched. Her face looked immovably grave at the end of any of Mr Higgins's best stories; but there was a keen, needle-like glance of her unwinking little eyes, which Mr Higgins felt rather than saw, and which made him shiver, even on a hot day, when it fell upon him. Miss Pratt was a dissenter, and, to propitiate this female Mordecai, Mr Higgins asked the dissenting minister whose services she attended, to dinner; kept himself and his company in good order; gave a handsome donation to the poor of the chapel. All in vain—Miss Pratt stirred not a muscle more of her face towards graciousness; and Mr Higgins was conscious that, in spite of all his open efforts to captivate Mr Davis, there was a secret influence on the other side, throwing in doubts and suspicions, and evil interpretations of all he said or did. Miss Pratt, the little, plain old maid, living on eighty pounds a year, was the thorn in the popular Mr Higgins's side, although she had never spoken one uncivil word to him; indeed, on the contrary, had treated him with a stiff and elaborate civility.

The thorn—the grief to Mrs Higgins was this. They had no children! Oh! how she would stand and envy the careless, busy motion of half a dozen children; and then, when observed, move on with a deep, deep sigh of yearning regret. But it was as well.

It was noticed that Mr Higgins was remarkably careful of his health. He ate, drank, took exercise, rested, by some secret rules of his own; occasionally bursting into an excess, it is true, but only on rare occasions— such as when he returned from visiting his estates in the south, and collecting his rents. That unusual exertion and fatigue—for there were no stagecoaches within forty miles of Barford, and he, like most country gentlemen of that day, would have preferred riding if there had been— seemed to require some strange excess to compensate for it; and rumours went through the town that he shut himself up, and drank enormously for some days after his return. But no one was admitted to these orgies.

One day—they remembered it well afterwards—the hounds met not far from the town; and the fox was found in a part of the wild heath, which was beginning to be enclosed by a few of the more wealthy towns- people, who were desirous of building themselves houses rather more in the country than those they had hitherto lived in. Among these, the prin- cipal was a Mr Dudgeon, the attorney of Barford, and the agent for all the county families about. The firm of Dudgeon had managed the leases, the marriage-settlements, and the wills, of the neighbourhood for generations. Mr Dudgeon's father had the responsibility of collecting the landowners' rents just as the present Mr Dudgeon had at the time of which I speak: and as his son and his son's son have done since. Their business was an hereditary estate to them; and with something of the old feudal feeling was mixed a kind of proud humility at their position towards the squires whose family secrets they had mastered, and the mysteries of whose fortunes and estates were better known to the Messrs Dudgeon than to themselves.

Mr John Dudgeon had built himself a house on Wildbury Heath; a mere cottage as he called it: but though only two storeys high, it spread out far and wide, and workpeople from Derby had been sent for on purpose to make the inside as complete as possible. The gardens too were exquisite in arrangement, if not very extensive; and not a flower was grown in them but of the rarest species. It must have been somewhat of a mortification to the owner of this dainty place when, on the day of which I speak, the fox, after a long race, during which he had described a circle of many miles, took refuge in the garden; but Mr Dudgeon put a good face on the matter when a gentleman hunter, with the careless insolence of the squires of those days and that place, rode across the velvet lawn, and tapping at the window of the dining-room with his whip-handle, asked permission—no! that is not it—rather, informed Mr Dudgeon of their intention—to enter his garden in a body, and have the fox unearthed. Mr Dudgeon compelled himself to smile assent, with the grace of a masculine Griselda; and then he hastily gave orders to have all that the house afforded of provision set out for luncheon, guessing rightly enough that a six hours' run would give even homely fare an acceptable welcome. He bore without wincing the entrance of the dirty boots into his exquisitely clean rooms; he only felt grateful for the care with which Mr Higgins strode about, laboriously and noiselessly moving on the tip of his toes, as he reconnoitred the rooms with a curious eye.

'I'm going to build a house myself, Dudgeon; and, upon my word, I don't think I could take a better model than yours.'

'Oh! my poor cottage would be too small to afford any hints for such a house as you would wish to build, Mr Higgins,' replied Mr Dudgeon, gently rubbing his hands nevertheless at the compliment.

'Not at all! not at all! Let me see. You have dining-room, drawing-room,'—he hesitated, and Mr Dudgeon filled up the blank as he expected.

'Four sitting-rooms and the bedrooms. But allow me to show you over the house. I confess I took some pains in arranging it, and, though far smaller than what you would require, it may, nevertheless, afford you some hints.'

So they left the eating gentlemen with their mouths and their plates quite full, and the scent of the fox overpowering that of the hasty rashers of ham; and they carefully inspected all the ground-floor rooms. Then Mr Dudgeon said:

'If you are not tired, Mr Higgins—it is rather my hobby, so you must pull me up if you are—we will go upstairs, and I will show you my sanctum.'

Mr Dudgeon's sanctum was the centre room, over the porch, which formed a balcony, and which was carefully filled with choice flowers in pots. Inside, there were all kinds of elegant contrivances for hiding the real strength of all the boxes and chests required by the particular nature of Mr Dudgeon's business: for although his office was in Barford, he kept (as he informed Mr Higgins) what was the most valuable here, as being safer than an office which was locked up and left every night. But, as Mr Higgins reminded him with a sly poke in the side, when next they met, his own house was not over-secure. A fortnight after the gentlemen of the Barford hunt lunched there, Mr Dudgeon's strong-box—in his sanctum upstairs, with the mysterious spring-bolt to the window invented by himself, and the secret of which was only known to the inventor and a few of his most intimate friends, to whom he had proudly shown it—this strong-box, containing the collected Christmas rents of half a dozen landlords (there was then no bank nearer than Derby), was rifled; and the secretly rich Mr Dudgeon had to stop his agent in his purchases of paintings by Flemish artists, because the money was required to make good the missing rents.

The Dogberries and Verges of those days were quite incapable of obtaining any clue to the robber or robbers; and though one or two vagrants were taken up and brought before Mr Dunover and Mr Higgins, the magistrates who usually attended in the court-room at Barford, there was no evidence brought against them, and after a couple of nights' durance in the lock-ups they were set at liberty. But it became a standing joke with Mr Higgins to ask Mr Dudgeon, from time to time, whether he could recommend him a place of safety for his valuables; or, if he had made any more inventions lately for securing houses from robbers.

About two years after this time—about seven years after Mr Higgins had been married—one Tuesday evening, Mr Davis was sitting reading the news in the coffee-room of the George Inn. He belonged to a club of gentlemen who met there occasionally to play at whist, to read what few newspapers and magazines were published in those days, to chat about the market at Derby, and prices all over the country. This Tuesday night it was a black frost; and few people were in the room. Mr Davis was anxious to finish an article in the *Gentleman's Magazine*; indeed, he was making extracts from it, intending to answer it, and yet unable with his small income to purchase a copy. So he stayed late; it was past nine, and at ten o'clock the room was closed. But while he wrote, Mr Higgins came in. He was pale and haggard with cold. Mr Davis, who had had for some time sole possession of the fire, moved politely on one side, and handed to the newcomer the sole London newspaper which the room afforded. Mr Higgins accepted it, and made some remark on the intense coldness of the weather; but Mr Davis was too full of his article, and intended reply, to fall into conversation readily. Mr Higgins hitched his chair nearer to the fire, and put his feet on the fender, giving an audible shudder. He put the newspaper on one end of the table near him, and sat gazing into the red embers of the fire, crouching down over them as if his very marrow were chilled. At length he said:

'There is no account of the murder at Bath in that paper?' Mr Davis, who had finished taking his notes, and was preparing to go, stopped short, and asked:

'Has there been a murder at Bath? No! I have not seen anything of it—who was murdered?'

'Oh! it was a shocking, terrible murder!' said Mr Higgins, not raising his look from the fire, but gazing on with his eyes dilated till the whites were seen all round them. 'A terrible, terrible murder! I wonder what will become of the murderer? I can fancy the red glowing centre of that fire—look and see how infinitely distant it seems, and how the distance magnifies it into something awful and unquenchable.'

'My dear sir, you are feverish; how you shake and shiver!' said Mr Davis, thinking privately that his companion had symptoms of fever, and that he was wandering in his mind.

'Oh, no!' said Mr Higgins. 'I am not feverish. It is the night which is so cold.' And for a time he talked with Mr Davis about the article in the *Gentleman's Magazine*, for he was rather a reader himself, and could take more interest in Mr Davis's pursuits than most of the people at Barford. At length it drew near to ten, and Mr Davis rose up to go home to his lodgings.

'No, Davis, don't go. I want you here. We will have a bottle of port together, and that will put Saunders into good humour. I want to tell you about this murder,' he continued, dropping his voice, and speaking hoarse and low. 'She was an old woman, and he killed her, sitting reading her Bible by her own fireside!' He looked at Mr Davis with a strange searching gaze, as if trying to find some sympathy in the horror which the idea presented to him.

'Who do you mean, my dear sir? What is this murder you are so full of? No one has been murdered here.'

'No, you fool! I tell you it was in Bath!' said Mr Higgins, with sudden passion; and then calming himself to most velvet-smoothness of manner, he laid his hand on Mr Davis's knee, there, as they sat by the fire, and gently detaining him, began the narration of the crime he was so full of; but his voice and manner were constrained to a stony quietude: he never looked in Mr Davis's face; once or twice, as Mr Davis remembered afterwards, his grip tightened like a compressing vice.

'She lived in a small house in a quiet old-fashioned street, she and her maid. People said she was a good old woman; but for all that, she hoarded and hoarded, and never gave to the poor. Mr Davis, it is wicked not to give to the poor—wicked—wicked, is it not? I always give to the poor, for once I read in the Bible that "Charity covereth a multitude of sins". The wicked old woman never gave, but hoarded her money, and saved, and saved. Someone heard of it; I say she threw a temptation in his way, and God will punish her for it. And this man—or it might be a woman, who knows?—and this person—heard also that she went to church in the mornings, and her maid in the afternoons; and so—while the maid was at church, and the street and the house quite still, and the darkness of a winter afternoon coming on—she was nodding over the Bible—and that, mark you! is a sin, and one that God will avenge sooner or later; and a step came in the dusk up the stair, and that person I told you of stood in the room. At first he—no! At first, it is supposed—for, you understand, all this is mere guesswork—it is supposed that he asked her civilly enough to give him her money, or to tell him where it was; but the old miser defied him, and would not ask for mercy and give up her keys, even when he threatened her, but looked him in the face as if he had been a baby— Oh, God! Mr Davis, I once dreamt when I was a little innocent boy that I should commit a crime like this, and I wakened up crying; and my mother comforted me—that is the reason I tremble so now—that and the cold, for it is very very cold!'

'But did he murder the old lady?' asked Mr Davis. 'I beg your pardon, sir, but I am interested by your story.'

'Yes! he cut her throat; and there she lies yet in her quiet little parlour, with her face upturned and all ghastly white, in the middle of a pool of blood. Mr Davis, this wine is no better than water; I must have some brandy!'

Mr Davis was horror-struck by the story, which seemed to have fascinated him as much as it had done his companion.

'Have they got any clue to the murderer?' said he. Mr Higgins drank down half a tumbler of raw brandy before he answered.

'No! no clue whatever. They will never be able to discover him; and I should not wonder, Mr Davis—I should not wonder if he repented after all, and did bitter penance for his crime; and if so—will there be mercy for him at the last day?'

'God knows!' said Mr Davis, with solemnity. 'It is an awful story,' continued he, rousing himself; 'I hardly like to leave this warm light room and go out into the darkness after hearing it. But it must be done,' buttoning on his greatcoat—'I can only say I hope and trust they will find out the murderer and hang him. If you'll take my advice, Mr Higgins, you'll have your bed warmed, and drink a treacle-posset just the last thing; and, if you'll allow me, I'll send you my answer to Philologus before it goes up to old Urban.'

The next morning, Mr Davis went to call on Miss Pratt, who was not very well; and, by way of being agreeable and entertaining, he related to her all he had heard the night before about the murder at Bath; and really he made a very pretty connected story out of it, and interested Miss Pratt very much in the fate of the old lady—partly because of a similarity in their situations; for she also privately hoarded money, and had but one servant, and stopped at home alone on Sunday afternoons to allow her servant to go to church.

'And when did all this happen?' she asked.

'I don't know if Mr Higgins named the day; and yet I think it must have been on this very last Sunday.'

'And today is Wednesday. Ill news travels fast.'

'Yes, Mr Higgins thought it might have been in the London newspaper.'

'That it could never be. Where did Mr Higgins learn all about it?'

'I don't know; I did not ask. I think he only came home yesterday: he had been south to collect his rents, somebody said.'

Miss Pratt grunted. She used to vent her dislike and suspicions of Mr Higgins in a grunt whenever his name was mentioned.

'Well, I shan't see you for some days. Godfrey Merton has asked me to go and stay with him and his sister; and I think it will do me good. Besides,' added she, 'these winter evenings—and these murderers at large in the country—I don't quite like living with only Peggy to call to in case of need.'

Miss Pratt went to stay with her cousin, Mr Merton. He was an active magistrate, and enjoyed his reputation as such. One day he came in, having just received his letters.

'Bad account of the morals of your little town here, Jessy!' said he, touching one of his letters. 'You've either a murderer among you, or some friend of a murderer. Here's a poor old lady at Bath had her throat cut last Sunday week; and I've a letter from the Home Office, asking to lend them "my very efficient aid", as they are pleased to call it, towards finding out the culprit. It seems he must have been thirsty, and of a comfortable jolly turn; for before going to his horrid work he tapped a barrel of ginger wine the old lady had set by to work; and he wrapped the spigot round with a piece of a letter taken out of his pocket, as may be supposed; and this piece of a letter was found afterwards; there are only these letters on the outside, "*ns, Esq., -arford, -egworth*", which some one has ingeniously made out to mean Barford, near Kegworth. On the other side there is some allusion to a racehorse, I conjecture, though the name is singular enough: "Church-and-King-and-down-with-the-Rump." '

Miss Pratt caught at this name immediately; it had hurt her feelings as a dissenter only a few months ago, and she remembered it well.

'Mr Nat Hearn has—or had (as I am speaking in the witness-box, as it were, I must take care of my tenses), a horse with that ridiculous name.'

'Mr Nat Hearn,' repeated Mr Merton, making a note of the intelligence; then he recurred to his letter from the Home Office again.

'There is also a piece of a small key, broken in the futile attempt to open a desk—well, well. Nothing more of consequence. The letter is what we must rely upon.'

'Mr Davis said that Mr Higgins told him—' Miss Pratt began.

'Higgins!' exclaimed Mr Merton, '*ns*. Is it Higgins, the blustering fellow that ran away with Nat Hearn's sister?'

'Yes!' said Miss Pratt. 'But though he has never been a favourite of mine—'

'*ns*,' repeated Mr Merton. 'It is too horrible to think of; a member of the hunt—kind old Squire Hearn's son-in-law! Who else have you in Barford with names that end in *ns*?'

'There's Jackson, and Higginson, and Blenkinsop, and Davis, and Jones. Cousin! One thing strikes me—how did Mr Higgins know all about it to tell Mr Davis on Tuesday what had happened on Sunday afternoon?'

There is no need to add much more. Those curious in lives of the highwayman may find the name of Higgins as conspicuous among those annals as that of Claude Duval. Kate Hearn's husband collected his rents on the highway, like many another 'gentleman' of the day; but, having been unlucky in one or two of his adventures, and hearing exaggerated accounts of the hoarded wealth of the old lady at Bath, he was led on from robbery to murder, and was hung for his crime at Derby, in 1775.

He had not been an unkind husband; and his poor wife took lodgings in Derby to be near him in his last moments—his awful last moments. Her old father went with her everywhere but into her husband's cell; and wrung her heart by constantly accusing himself of having promoted her marriage with a man of whom he knew so little. He abdicated his squireship in favour of his son Nathaniel. Nat was prosperous, and the helpless silly father could be of no use to him; but to his widowed daughter the foolish fond old man was all in all; her knight, her protector, her companion— her most faithful loving companion. Only he ever declined assuming the office of her counsellor—shaking his head sadly, and saying—

'Ah! Kate, Kate! if I had had more wisdom to have advised thee better, thou need'st not have been an exile here in Brussels, shrinking from the sight of every English person as if they knew thy story.'

I saw the White House not a month ago; it was to let, perhaps for the twentieth time since Mr Higgins occupied it; but still the tradition goes in Barford that once upon a time a highwayman lived there, and amassed untold treasures; and that the ill-gotten wealth yet remains walled up in some unknown concealed chamber; but in what part of the house no one knows.

Will any of you become tenants, and try to find out this mysterious closet? I can furnish the exact address to any applicant who wishes for it.

LILLIAN DE LA TORRE

The Missing Shakespeare Manuscript

ɷ

'Twas Dr Sam: Johnson in the end, who restored the missing Shakespeare manuscript at the Stratford Jubilee in the year 1769; though in the
beginning he would not so much as look at it. In that rainy September, he
preferred to hug the fire at the Red Lion Inn.

There he stood, bulky and immovable, holding forth his large, well-
shaped hands to the glow of the coals and turning a deaf ear to my
perswasions. But if he was stubborn, I was pertinacious.

'Do, Dr Johnson,' I urged, 'give me your company to Mr Ararat's
though you come but to scoff.'

'I shall not remain to pray, I promise you,' rejoined the great Cham of
literature intransigeantly.

'So much is unnecessary,' I replied, 'but indeed I have promised we
would meet there with Dr Percy and his young friend Malone, the Irish
lawyer.'

'This is very proper for Thomas Percy and his scavenging friends,'
remarked Dr Johnson, lifting his coat-tails before the blaze, 'for they are
very methodists in the antiquarian *enthusiasm*. But truly this is ill for a
scholar, to run with the vulgar after a parcel of old waster paper.'

'Sir, sir,' I protested, 'the antiquarian zeal of Mr Ararat has preserved
to us a previously unknown tragedy of Shakespeare, "Caractacus; or, the
British Hero".'

'Which little Davy Garrick is to represent in the great amphitheatre
tomorrow night. Let him do so. Let us see him do so. Let us not meddle
with the musty reliques of the writing desk.'

'Musty!' I cried. 'Let me tell you this is no musty old dog's-eared folio
that has lost its wrappings for pyes or worse, like the ballad-writings
Percy cherishes, but a manuscript as fair and unblemished, so Dr Warton
assures me, as the day it came from the bard's own hand. By singular
good luck Mr Ararat is of antiquarian mind, and the manuscript was
preserved from a noisome fate in the out-house.'

'That it was preserved for Garrick to play and Dodsley to publish, this
is luck indeed; but now that the playhouse copies are taken off, it may end
in the out-house for all of me,' replied my learned friend. 'No, sir; let a

good play be well printed and well played; but to idolize mere paper and ink is rank superstition and idolatry.'

'Why, sir, you need not adore it, nor look at it if you will not; but pray let us not disappoint Dr Percy and his young friend.'

Dr Johnson's good nature was not proof against this appeal to friendship; he consented to walk along with me to Mr Ararat's.

I made haste to don my hat and be off before anything could supervene. As we set off on foot from the yard of the Red Lion, my revered friend peered at me with puckered eyes.

'Pray, Mr Boswell,' he enquired in tones of forced forbearance, 'what is the writing inserted in your hat?'

I doffed the article in question and gazed admiringly at the neatly inscribed legend which adorned it.

'CORSICA BOSWELL,' read off my learned friend in tones of disgust. 'Corsica Boswell! Pray, what commodity are you touting, Mr Boswell, that you advertise the world of your name in this manner?'

'A very precious commodity,' I retorted with spirit, 'liberty for downtrodden Corsica. Do but attend the great masquerade tonight, you shall see how I speak for Corsica.'

'Well, sir, you may speak for whom you will, and advertise Stratford of your name as you please. For me, let me remain incognito. I should be loath to parade about Stratford as DICTIONARY JOHNSON.'

'Say rather,' I replied, 'as SHAKESPEARE JOHNSON, for your late edition of the Bard must endear you to the town of his birth.'

'I come to Stratford,' remarked Dr Johnson with finality, 'to observe men and manners, and not to tout for my wares.'

'Be it so,' I replied, 'here is material most proper for your observation.'

As I spoke, we were crossing the public square, which teemed with bewildered Stratfordians and jostling strangers. The centre of a milling crowd, a trumpeter was splitting the air with his blasts and loudly proclaiming:

'Ladies and gentlemen! The famous Sampson is just going to begin— just going to mount four horses at once with his feet upon two saddles— also the most wonderful surprizing feats of horsemanship by the most *notorious* Mrs Sampson.'

A stringy man and an Amazon of a woman seconded his efforts by giving away inky bills casting further light on their own notorious feats. As we strolled on, we met a man elbowing his way through the press beating a drum and shouting incessantly:

'The notified Porcupine Man, and all sorts of outlandish birds and other beasts to be seen without loss of time on the great meadow near the amphitheatre at so small a price as one shilling a piece. Alive, alive, alive, ho.'

Behind him came a man leading a large bin, and a jostling crowd following. Dr Johnson smiled.

'This foolish fellow will scarce make his fortune at the Jubilee,' he remarked. 'Who will pay a shilling to see strange animals in a house, when a man may see them for nothing going along the streets, alive, alive, ho?'

As we walked along, Dr Johnson marvelled much at the elegant art of the decorations displayed about the town. The town hall was adorned with five transparencies on silk—in the centre Shakespeare, flanked by Lear and Caliban, Falstaff and Pistol. The humble cottage where Shakespeare was born, gave me those feelings which men of enthusiasm have on seeing remarkable places; and I had a solemn and serene satisfaction in contemplating the church in which his body lies.

Dr Johnson, however, took a more lively interest in the untutored artistry of the townsfolk of Stratford, who had everywhere adorned their houses, according to their understanding and fantasy, in honour of their Bard. We read many a rude legend displayed to the glorification of Shakespeare and Warwickshire. We beheld many a crude portrait intended for the great playwright, and only a few less libels on the lineaments of David Garrick, as we strolled down to Mr Ararat's.

'This is Garrick's misfortune, that as steward of the Jubilee, he is man of the hour,' remarked Dr Johnson, 'for the admiration of Warwickshire has done him no less wrong than the lampoons of London.'

'In Shakespeare he has a notable fellow-sufferer,' I replied.

Johnson 'You say true, Bozzy. Alack, Bozzy, do my eyes inform me true as to the nature of the small building, set apart, which someone has seen fit to adorn with the honoured features of the Bard?'

Boswell 'Your eyes inform you truly. We are approaching the stationer's shop of Mr Ararat, whose zeal for Shakespeare extends even to adorning the exterior of his out-house with the counterfeit presentment of the Bard.'

Johnson 'Better his face without than his works within.'

Boswell 'Sir, the antiquarian zeal of Mr Ararat, 'tis said, extends even so far, for he provides for the convenience of his household a pile of old accounts of wonderful and hoary antiquity. The Stratfordians are long dead and gone who bought the paper for which the reckoning still awaits a last usefulness.'

Johnson 'Let Mr Ararat keep Thomas Percy out of here. Last year he published the Earl of Northumberland's reckonings for bread and cheese from the year 1512; next year, unless he's watched, I'll be bound, he'll rush to the press with a parcel of stationer's accounts he's *borrowed* from Ararat's out-house.'

Boswell 'Sir, you wrong Thomas Percy. He's a notable antiquarian and his works are much sought after.'

Johnson 'He's a snapper-up of unconsidered trifles, and that young Irishman who's followed him hither is no better. Sir, be it a Shakespeare manuscript or a publican's reckoning, just so it be old, I'd watch it narrowly while Percy is about.'

Speaking thus, we turned the corner, when the full complexity of Mr Ararat's decorative scheme struck us at once. Limned by an unskilful hand, the characters of Shakespeare's plays crowded the ancient façade, dominated under the gabled roof by the lineaments of the Bard, for which the portrait on the necessary-house was clearly a preliminary study. Hamlet leaned a melancholy elbow on the steep gable of the window, Macbeth and Macduff fought with claymores over the front door, a giant warrior guarded the corner post, all endued with a weird kind of life in the grey glare of the sky, for a storm was threatening.

'Ha,' said Dr Johnson, 'who is this painted chieftain? Can it be Cymbeline?'

'No, sir, this is Caractacus, hero of the new play just recovered.'

Johnson 'Why has he painted himself like an Onondaga?'

Boswell 'Sir, he is an ancient Briton. He has painted himself with woad.'

Johnson 'Will little Davy Garrick paint himself blue?'

Boswell 'I cannot say, sir, though 'tis known he means to present the character in ancient British dress.'

Johnson 'This is more of your *antiquarianism*. Let Davy Garrick but present a *man*, he may despise the fribbles of the tiring-room.'

As we thus stood chatting before the stationer's shop, a strange creature insinuated himself before us. From his shoulder depended a tray full of oddments.

'Toothpick cases, needle cases, punch ladles, tobacco stoppers, inkstands, nutmeg graters, and all sorts of boxes, made out of the famous mulberry tree,' he chanted.

'Pray, sir, shall we venture?'

'Nay, Bozzy, the words of the bard are the true metal, his mulberry tree is but dross. You seem determined to make a papistical idolator of me.'

'Yet perhaps this box—' I indicated a wooden affair large enough for a writing-desk—'this box is sufficiently useful in itself—'

With a resentful scowl the man snatched it rudely from my hand.

''Tis not for sale,' he mumbled, and ran down the street with his boxes hopping.

'Are all the people mad?' quoted Dr Johnson from the 'Comedy of

Errours'; and the shop-bell tinkled to herald our entrance into the stationer's shop of Mr Ararat.

Behind the counter in the dim little shop stood a solid-built man in a green baize apron. He had a sanguine face and thin, gingery hair. This was Mr Ararat, stationer of Stratford, Shakespearian enthusiast, and owner of the precious manuscript of 'Caractacus; or, the British Hero'. He spelled out the sign on my hat and gave me a low bow.

'Welcome, Mr Boswell, to you and your friend.'

We greeted Mr Ararat with suitable distinction. Being made known to Dr Johnson, he greeted him with surprised effusion.

'This is indeed an unlooked-for honour, Dr Johnson,' cried Mr Ararat.

'Percy is late,' I observed to Dr Johnson.

'Dr Percy was here, and has but stepped out for a moment,' Mr Ararat informed us.

We whiled away the time of waiting by examining the honest stationer's stock, and Dr Johnson purchased some of his laid paper, much to my surprise to good advantage. As the parcel was wrapping Thomas Percy put his long nose in at the door, and followed it by his neat person attired in clerical black. He laid his parcel on the counter and took Dr Johnson by both his hands.

'We must count ourselves fortunate,' he cried, 'to have attracted Dr Johnson hither. I had feared we could never lure you from Brighthelmstone, where the witty and fair conspired to keep you.'

'Why, sir, the witty and fair, if by those terms you mean to describe Mrs Thrale, took a whim that the sea air gave her a megrim, and back she must post to Streatham; and I took a whim not to wait upon her whims, so off I came for Stratford.'

'We are the gainers,' cried Percy.

Dr Johnson's eye fell on the counter, where lay his package of paper and the exactly similar parcel Percy had laid down. He picked up the latter.

'Honest Mr Ararat does well by us Londoners,' he remarked, 'to sell us fine paper so cheap.'

'Yes, sir,' replied Percy, possessing himself of his parcel with more haste than was strictly mannerly, 'you see I know how to prize new folios as well as old, ha ha.'

He gripped his parcel, and during the whole of our exciting trans-actions in the house of Mr Ararat it never left his hands again.

At that moment the shop-bell tinkled to admit a stranger. I saw a fresh-faced Irishman with large spiritual eyes the colour of brook water, a straight nose long at the tip, and a delicate smiling mouth. He was shabbily dressed in threadbare black. The newcomer nodded to Percy, and made a low bow to my venerable friend.

'Your servant, Dr Johnson,' he exclaimed in a soft mellifluous voice. 'Permit me to recall myself—Edmond Malone, at your service I had the honour to be made known to you some years since by my countryman Edmund Southwell.'

'I remember it well,' replied Dr Johnson cordially, "Twas at the Grecian, in the Strand. I had a kindness for Southwell.'

'He will be happy to hear it,' replied Malone.

'Twas thus that I, James Boswell, the Scottish advocate, not quite twenty-nine, met Edmond Malone, the Irish lawyer, then in the twenty-eighth year of his age, who was destined to become—but I digress.

Our party being complete, we repaired into the inner room and were accommodated with comfortable chairs. Seated by the chimney-piece was a boy of about sixteen, a replica of old Mr Ararat, with a rough red mop of hair and peaked red eyebrows. He looked at us without any expression on his round face.

"Tis Anthony,' said his father with pride, 'Anthony's a good boy.'

'What do you read so diligently, my lad?' enquired Dr Johnson kindly, peering at the book the boy held. 'Johnson's *Shakespeare*! I am honoured!'

'Nay, sir, 'tis we who are honoured,' said Malone fervently. 'To inspect the Shakespeare manuscript in the company of him who knows the most in England of the literature of our country and the plays of the Bard, to read the literature of yesterday in the presence of Dictionary Johnson, who knows the age and lineage of every English word from the oldest to the word minted but yesterday, this is to savour the fine flower of scholarship.'

The red-haired boy turned his eyes towards Dr Johnson.

'Pray, sir,' replied Dr Johnson, 'don't cant. In restoring a lost play this worthy boy has deserved as well as I of his fellow-Englishmen.'

'Anthony's a good boy,' said his father with pride, 'he knows the plays of Shakespeare by heart, "Caractacus" included.'

I looked at Anthony, and doubted it.

'Shall you make him a stationer, like his fathers before him?' enquired Dr Percy politely.

'No, sir,' replied Ararat, 'he's prenticed to old Mr Quiney the scrivener over the way. Here, Anthony, fetch my strong-box, we'll show the gentleman what they came to see.'

Anthony nodded, and went quickly out of the room.

'This is a great good fortune,' said Dr Percy eagerly, 'to see the very writing of Shakespeare himself. We are your debtors that it has been preserved.'

"Tis nothing,' but old Ararat began to swell like a turkey-cock. He launched into the story: 'The first Anthony Ararat was a stationer in

Stratford, like me, and Will Shakespeare was his neighbour. Anthony saved his life in the Avon, and in recompense he had of Will the manuscript of this very play, "Caractacus; or, the British Hero", to be his and his children's forever. Old Anthony knew how to value it, for he folded it in silk, and laid with it a writing of how he came by it, and laid it away with his accounts and private papers.'

'Then how came it to be lost?' enquired Dr Percy.

''Twas my grandmother, sir, who took the besom to all the old papers together, and bundled one with another into the shed, and there they lay over the years with the lumber and the stationer's trash. I played in there when I was a boy, and so did Anthony after me. I remember, there was paper in there my father said his grandfather had made when he was prenticed in the paper-mills. But I never turned over the old accountings, nor paid them any heed. But to make a long story short, gentlemen, come Jubilee time I thought to turn an honest penny letting lodgings, so I bade Anthony turn out the lumber in the shed and make a place where the horses could stand. Anthony turned out a quantity of waste paper and lumber, and my mother's marriage lines that went missing in the '28, and the manuscript of "Caractacus", wrapped in silk as the first Anthony had laid it by. He had the wit to bring it to me, and I took it over to old Mr Quiney the engrosser, and between us we soon made out what we had. Warton of Trinity rode over from Oxford, and Mr Garrick came down from London and begged to play it . . .'

The words died in his throat. I followed his gaze towards the inner door. There stood young Anthony, pale as death. Tears were streaming down his wet face. Angrily he dashed the drops from his shoulder. In his hand he held a brass-bound coffer, about the size of the mulberry-wood box the pedlar had snatched from us. Wordlessly, though his throat constricted, he held out the strong-box towards his father. It was empty. We saw the red silk lining, and the contorted metal where the lock had been forced.

The manuscript of 'Caractacus' had vanished quite away.

Old Ararat was beside himself. Thomas Percy was racked between indignation and pure grief. Only Dr Johnson maintained a philosophical calm.

'Pray, Mr Ararat, compose yourself. Remember the playhouse copies are safely taken off. You have lost no more than a parcel of waste paper.'

'But, sir,' cried Malone, 'the very hand of the Bard!'

'And a very crabbed hand too,' rejoined Johnson, 'old Quiney over the way will engross you a better for a crown.'

'But, pray, Dr Johnson,' I enquired, 'is not its value enormous?'

'Its value is nil. 'Tis so well known, and so unique, that the thief can never sell it; he can only feed his fancy, that it is now his. Let him gloat

'Caractacus' is ours. Tomorrow we shall see Garrick play the British hero; the day after tomorrow it will be given to the world in an elegant edition. The thief has gained, Mr Ararat has lost, nothing but old paper.'

But Percy and the Ararats thought otherwise. We deployed like an army through the domain of the good stationer, and left no corner unsearched. We had up the red satin lining of the coffer; we turned over the stationer's stock-in-trade; we searched the house from top to bottom; all to no purpose. In the end we went away without finding anything, leaving young Anthony stupefied by the chimney-piece and old Ararat red with rage and searching, blaming the whole thing on the Jubilee.

We were a dreary party as we walked back to the Red Lion in the rain. Percy and Malone stalked on in heart-broken silence. Having given his parcel into Percy's keeping, Dr Johnson swayed along muttering to himself and touching the palings as we passed. Alone retaining my wonted spirits, I broached in vain half a dozen cheerful topicks, and at last fell silent like the rest.

Arrived in the courtyard of the Red Lion, Dr Johnson took his parcel from Percy's hand and vanished without a word. I lingered long enough to take a dram for the prevention of the ague. Percy and Malone were sorry company, quaffing in silence by my side, and soon by mutual consent we parted to shift our wet raiment.

In the chamber I shared with Dr Johnson (dubbed, according to the fancy of Mr Peyton the landlord, after one of Shakespeare's plays, *Much Ado about Nothing*) I found my venerable friend, shifted to dry clothing, muffled in a counterpane and staring at the fire.

I ventured to enquire where in his opinion the sacred document had got to.

'Why, Bozzy,' replied he, 'some Shakespeare-maniac has got it, you may depend upon it, or as it might be, some old-paper maniac. Some scavenging antiquarian has laid hold of it and gloats over it in secret.'

'I cry your pardon,' said Dr Percy, suddenly appearing at our door. He was white and uneasy still. In his hand he carried a parcel.

'Pray, Dr Johnson, do you not have my parcel that I brought from Mr Ararat's?'

'I, Dr Percy? I have my own parcel.' Dr Johnson indicated it where it lay still wrapped on the table.

Percy seized it, and scrutinized the wrappings narrowly.

'You are deceived, Dr Johnson. This parcel is mine. Here is yours, which I retained in errour for my own. I fear I have disarranged it in opening. Pray forgive me. I see you have opened mine more neatly.'

''Tis as I had it of you,' replied Dr Johnson.

'You have not opened it!' cried Percy. 'Well, Dr Johnson, now we each have our own again, and no harm's done, eh? We lovers of good paper have done a shrewd day's bargaining, have we not, ha ha ha!'

'I will wager mine was the better bargain,' said Dr Johnson good-humouredly. 'Come, open up, let us see.'

'No, no, Dr Johnson, I must be off,' and Percy whipped through the door before either of us could say a word.

'Now,' remarked Dr Johnson, ''tis seen that Peyton was well advised to name our chamber "Much Ado about Nothing".'

The rain continued in a dreary stream, so that boards had to be laid over the kennel to transport the ladies dry-shod into the amphitheatre; but for all that, the great masquerade that night was surely the finest entertainment of the kind ever witnessed in Britain. I was sorry that Dr Johnson elected to miss it. There were many rich, elegant, and curious dresses, many beautiful women, and some characters well supported. Three ladies personated Macbeth's three witches with devastating effect, while a person dressed as the devil gave inexpressible offence.

I own, however, that 'twas my own attire that excited the most remark. Appearing in the character of an armed Corsican chief, I wore a short, dark-coloured coat of coarse cloth, scarlet waistcoat and breeches, and black spatterdashes, and a cap of black cloth, bearing on its front, embroidered in gold letters, VIVA LA LIBERTA, and on its side a blue feather and cockade. I also wore a cartridge-pouch, into which was stuck a stiletto, and on my left side a pistol. A musket was slung across my shoulder, and my black hair, unpowdered, hung plaited down my neck, ending in a knot of blue ribands. In my right hand I carried a long vine staff, with a bird curiously carved at the long curving upper end, emblematical of the sweet bard of Avon. In this character of a Corsican chief I delivered a poetical address on the united subjects of Corsica and the Stratford Jubilee.

I cannot forbear to rehearse the affecting peroration:

> 'But let me plead for LIBERTY distrest,
> And warm for her each sympathetick Breast:
> Amongst the splendid Honours which you bear,
> To save a Sister Island! be your Care:
> With generous Ardour make US also FREE;
> And give to CORSICA, a NOBLE JUBILEE.'

As I came to an applauded close, I heard a resonant voice at my elbow.

'Pray, Bozzy,' demanded Dr Johnson, peering at me with disfavour, 'what is the device on your coat? The head of a blackamoor upon a charger, garnished with watercress?'

'That, sir,' I replied stiffly, 'is the crest of Corsica, a Moor's head surrounded by branches of laurel. But what brings you from your bed, whither you were bound when I left you?'

'Sir,' replied Dr Johnson, 'somebody in Stratford is in possession of the missing manuscript of Mr Ararat. Here I have them all gathered under one roof, and all out of character, or into another character, which is just as revealing. I am here to observe. Let us retire into this corner and watch how they go on. To him who will see with his eyes, all secrets are open.'

'Toothpick cases, needle cases, punch ladles, tobacco stoppers, inkstands, nutmeg graters, and all sorts of boxes, made out of the famous mulberry tree,' chanted a musical voice behind us. We turned to behold the very figure of the man with the tray. His brilliant eyes twinkled behind his mask.

'Goods from the mulberry tree,' he chanted, 'made out of old chairs and stools and stained according, toothpick cases, needle cases, punch ladles—'

A blast from a trumpet cut him off. Beside him stood a second mask, garbed 'like Rumour painted full of tongues,' impersonating Fame with trumpet and scroll.

'Pray, sir,' said Dr Johnson, entering into the spirit of the occasion, 'let us glimpse your scroll, whether our names be not inscribed thereon.'

The mask withheld the scroll, and spoke in a husky voice:

'Nay, sir, my scroll is blank.'

'Why, sir, then you are the prince of cynics. What, not one name? Not *Corsica* Boswell? Not Garrick? Not Shakespeare? Sir, were I to betray this to the Corporation, you should stand in the pillory.'

'Therefore I shall not reveal myself—even to Dr Johnson—' replied the mask in his husky voice. He would have slipped away, when one of those spasmodic movements which cause my venerable friend so much distress hurled to the ground both trumpet and scroll. In a contest of courtesy, Fame retrieved the trumpet and my venerable friend the scroll.

'You say true,' remarked the last-named sadly, re-rolling the scroll, 'on the roster of Fame, my name is not inscribed.'

He restored the scroll with a bow, and Fame made off with the mulberry-wood vendor.

'I interest myself much in the strange personages of this assemblage,' remarked my philosophical friend. 'Alack, there's a greater guy than you, *Corsica* Boswell, for he's come out without his breeches.'

I recognized with surprise the fiery mop and blank face of young Ararat, whom I had last seen that morning weeping for the lost manuscript. He was robed in white linen, and carried scrip and claymore. He wore no mask, but his face was daubed with blue.

''Tis Anthony,' said I, 'he personates Caractacus, the British hero. Sure he trusts in vain if he thinks to conceal his identity behind a little blue paint.'

'To the man with eyes, the heaviest mask is no concealment,' replied Dr Johnson; 'sure you smoaked our friends with the scroll and the mulberry wood in spite of their valences.'

'Not I, trust me. Fame's husky voice was no less strange to me than the wizened figure of the pedlar.'

'The husky voice, the bent figure, were assumed for disguise,' replied Dr Johnson, 'but Percy's long nose was plain for all to see, and Malone's mellifluous tones were no less apparent. They thought to quiz me; but I shall quiz them tomorrow.'

I was watching young Ararat, with his father the centre of a sycophantic group of masks who made *lions* of them. Young Anthony was as impassive as ever, but his face was as red as his father's. Lady Macbeth plucked at his elbow; the three Graces fawned upon him; in the press about him I saw the trumpet of Fame and the tray of the mulberry pedlar.

'A springald Caractacus,' remarked Dr Johnson, following my gaze, 'how long, think you, could he live in equal combat if his life depended on that dull-edged claymore?'

'Yet see,' I commented severely, 'how the ladies flatter him, whose only claim on their kindness amounts to this, that through no merit of his own he found a dusty bundle of papers in his father's shed.'

'While those who can compose, ay and declaim, verses upon *liberty*,' supplied Dr Johnson slyly, 'stand neglected save by a musty old scholar.'

'Nay, sir,' I protested; but Dr Johnson cut me off:

'Why, sir, we are all impostors here. Fame with an empty scroll, mulberry wood cut from old chairs and stools! Sir, I have canvassed the abilities of the company, and I find that but one sailor out of six can dance a hornpipe, and but one more box his compass. Not one conjuror can inform me whether he could tell my fortune better by chiromancy or catoptromancy. None of four farmers knows how a score of runts sells now; and the harlequin is as stiff as a poker. So your Caractacus is an impostor among impostors, and we must not ask too much of him.'

I looked at the press around the finder and the owner of the missing manuscript, buzzing like bees with talk and laughter. There was a sudden silence, broken by a bellow from old Ararat. The buzzing began again on a higher note, and the whole swarm bore down on our corner, old Ararat in the lead. He brandished in his hand an open paper.

Wordlessly he extended the paper to my friend. Peering over his shoulder, I read with him:

Sir,

   The manuscript of Caractacus is safe, and I have a mind to profit from it in spite of your teeth. Lay £100 in the font at the church, and you shall hear further.

   Look to it; for if the value of the manuscript is nil, and profits me nothing, as God is my judge I will destroy it. I do not steal in sport.

   I am,
      Sir,
         Your obliged humble servant,

                                                                    Ignotus

'The scoundrel!' cried old Ararat. 'Where am I to find £100?'

'This is more of your antiquarianism,' I remarked, 'like a knight of old, the miscreant holds his captive to *ransom*.'

Dr Johnson turned the letter in his hands, and held it against the lights of the great chandelier. 'Twas writ in a fair hand on ordinary laid paper, and sealed with yellow wax; but instead of using a seal, the unknown writer had set his thumb in the soft wax.

'Why,' says he, 'the thief has signed himself with *hand* and *seal* indeed. Now were there but some way to match this seal to the thumb that made it, we should lay the robber by the heels and have back the manuscript that Shakespeare wrote.'

'Alack, sir,' I replied, 'there is no way.'

'Nevertheless, let us try,' said Dr Johnson sturdily. 'Pray, Mr Malone, set your thumb in this seal.'

'I?' said the mulberry-wood pedlar, drawing back.

'I will,' said I, and set my thumb in the waxen matrix. It fitted perfectly. The eyes of the maskers turned to me, and I felt my ears burning. Dr Johnson held out the seal to old Ararat, who with a stormy mutter of impatience tried to crowd his huge thumb into the impression. 'Twas far too broad.

Dr Johnson tried in turn the thumb of each masker. The ladies' thumbs were too slender, Malone's too long; but there were many in the group that fitted. Dr Johnson shook his head.

'This is the fallacy of the undistributed middle term,' said he. 'Some other means must be found than gross measurement, to fit a thumb to the print it makes. Pray, how came you by this letter?'

''Twas tossed at my feet by someone in the press,' replied old Ararat. 'Come, Dr Johnson, advise me, how am I to come by £100 to buy back my lost manuscript?'

'A subscription!' cried Fame. 'The price is moderate for so precious a prize. I myself will undertake to raise the sum for you.'

So it was concerted. Dr Johnson enjoined secrecy upon the maskers, and Fame with his visor off, revealed as Dr Percy indeed, bustled off to open the subscription books.

We lay late the next day in the 'Much Ado about Nothing' chamber. Dr Johnson was given over to indolence, and declined to say what he had learned at the masquerade, or whether he thought that the mysterious communication held out any hope that the missing manuscript might be recovered.

The rain continuing, the pageant was dispensed with. We whiled away the hours comfortably at the Red Lion, while Percy and Malone spent a damp day with their subscription books. Representing the collection merely as 'for the Ararats', they found the sum of £100 not easy to be amassed. Toward evening, however, they returned to the Red Lion with £87 in silver and copper, and Garrick's promise to make up the sum when the play's takings should be counted.

Dr Johnson spurned at the idea of buying back mere paper and faded ink. In his roaring voice he *tossed* and *gored* Dr Percy for his magpie love of old documents, adverting especially to Percy's recent publication of *The Household Book of the Earls of Northumberland.*

'Pray, sir,' he demanded with scorn, 'of what conceivable utility to mankind can the "Household Book" be supposed to be? The world now knows that a dead-and-gone Percy had beef to the value of twelve pence on a Michaelmas in 1512. Trust me, 'twill set no beef on the table of any living Percy.'

The young Irish lawyer came to the unfortunate clergyman's defence, and fared no better. Johnson was in high good spirits as we dined off a veal pye and a piece of good beef (which the living Percy relished well).

We then repaired to the amphitheatre, where Percy had concerted to meet the Ararats with Caractacus's ransom.

Old Ararat would have none of Dr Johnson's advice, to ignore Ignotus's letter. He was hot to conclude the business, and would hear of no other plan, than to deposit the £100 in the font as soon as the play should be over and the takings counted.

'Then, sir,' said Dr Johnson in disgust, 'at least let us entrap Ignotus, and make him Gnotus. Mr Boswell and I will watch by the font and take him as he comes for his ill-gotten gains.'

'We must stand watch and watch,' cried Percy. 'Malone and I will relieve you.'

'Nay, let me,' cried old Ararat.

'So be it,' assented Dr Johnson; and we repaired to our respective boxes to see the play.

We shared a box with Percy and Malone. Dr Johnson grunted to himself when David Garrick made his first entrance on the battlements, wearing white linen kilts and bedaubed with blue paint. In spite of this antiquarianism, I found myself moved deeply by the noble eloquence, the aweful elevation of soul, with which Garrick spoke the words of this play so strangely preserved for our generation. I was most affected by the solemn soliloquy which concluded the first act:

> O sovereign death,
> Thou hast for thy domain this world immense:
> Churchyards and charnel-houses are thy haunts,
> And hospitals thy sumptuous palaces;
> And when thou would'st be merry, thou dost chuse
> The gaudy chamber of a dying King.
> O! Then thou dost ope wide thy boney jaw
> And with rude laughter and fantastick tricks,
> Thou clapp'st thy rattling fingers to thy sides:
> And when this solemn mockery is o'er,
> With icy hand thou tak'st him by the feet,
> And upward so, till thou dost reach the heart,
> And wrap him in the cloak of lasting night.

As the act ended, from the stage box the Ararats, father and son, rose to share the plaudits of the huzzaing crowd.

'Davy Garrick,' remarked Dr Johnson in my ear, 'has surpassed himself; and King is inimitable as the Fool.'

The second act opened with another scene of King's.

> 'Alack,' cries the lovelorn Concairn,
> 'Alack, I will write verses of my love,
> They shall be hung on every tree . . .'

King turned a cartwheel, ending with a resounding smack on the rump.

'Say rather,' he cried, 'they shall be used in every jakes, for by'r lakin, such fardels does thy prentice hand compose, they are as caviare to the mob. I can but compliment thee thus, they do go to the *bottom* of the matter.'

The pit roared.

'Ha, what?' exclaimed Dr Johnson. 'Bozzy, Bozzy, where's my hat?'

'Your hat, sir? Why, the play is not half over.'

Dr Johnson fumbled around in the dark.

'No matter. Do you stay and see it through. Where's this hat of mine?'

'Here, sir.' I handed it to him.

'Whither do you go, sir?' enquired Malone eagerly.

'To do what must be done. Fool that I was, not to see—but 'tis not yet too late.' Dr Johnson lumbered off as the pit began to cry for silence.

We were on pins and needles in our box, but we sat through till Davy Garrick had blessed the land of the Britons and died a noble death, and we joined in the plaudits that rewarded the great actor and the great playwright and the finders of the manuscript. The Ararats were the cynosure of all eyes. It was long till we brought them away from their admirers and down to the church. Percy carried the £100 in a knitted purse. The rain had ceased, and a pale round moon contended with the clouds.

The solemn silence oppressed me as we pushed back the creaking door and entered, and my heart leaped to my mouth when a shadowy figure moved in the silent church. 'Twas Dr Johnson. He had wrapped himself in his greatcoat, and armed himself with a dark lanthorn. I could smell it, but it showed no gleam.

Without ceremony old Ararat dropped the heavy purse in the empty font and carried young Anthony off for home, promising to return and relieve our watch. I envied Percy and Malone as they, too, departed, with the Red Lion's mulled ale in their minds. They promised to return in an hour's time. Dr Johnson quenched the lanthorn, and we were left alone in the dark.

I own I liked it little, alone in the dark with the bones of dead men under our feet, and a desperate thief who knows how near? There was no sound. Dead Shakespeare lay under our feet, his effigy stared into the dark above our heads.

We sat in the shadow, back from the font. I fixed my eyes on its pale gleam, whereon the cloudy moon dropped a fitful light through the open door.

I will swear I saw nothing, no shadow on the font, no stealing figure by the open door; I heard nothing, I neither nodded nor closed my eyes. Dr Johnson fought sleep by my side. The hour was gone, and he was beginning to snore, when the light of a link came towards us, and Percy and Malone came in with the Ararats. Johnson awoke with a snort.

'For this relief much thanks,' he muttered. 'What, all four of you?'

'Ay,' returned Percy, extinguishing the link, 'for the Ararats are as eager as we to stand the next watch.'

'Let it be so,' replied Dr Johnson, approaching the font, 'we will but verify it, that the money is here, and passes from our keeping into yours.'

He bent over the font, and his voice changed.

'Pray, gentlemen, step over here.'

We did so as he made a light and opened his dark lanthorn.

The money was gone. In its place lay a pile of yellowed papers, thick-writ in a fair court-hand.

Beholding with indescribable feelings this relique of the great English Bard, I fell on my knees and thanked heaven that I had lived to see this day.

'Get up, Bozzy,' said Dr Johnson, 'and cease this flummery.'

'Oh, sir,' I exclaimed, 'the very handwriting of the great Bard of Stratford!'

''Tis not the handwriting of the great Bard of Stratford,' retorted Dr Johnson.

Old Ararat's jaw fell. The boy Anthony opened his mouth and closed it again. By the light of the lanthorn Dr Percy peered at the topmost page.

'Yet the paper is old,' he asserted.

'The paper may be old,' replied Dr Johnson, 'yet the words are new.'

'Nay, Dr Johnson,' cried old Mr Ararat, 'this is merely to affect singularity. Eminent men from London have certified that my manuscript is genuine, including David Garrick and Dr Warton.'

'Garrick and Warton are deceived,' returned Dr Johnson sternly. '"Caractacus; or, the British Hero" is a modern forgery, and no ancient play.'

'Pray, sir, how do you make that good?' enquired Malone.

'I knew it,' replied Dr Johnson, 'when I heard King use a word Shakespeare never heard—"mob"—a word shortened from "mobile" long after Shakespeare died. Nor would Shakespeare have understood the verb "to compliment".'

'Then,' said I, 'the thief has had his trouble for his pains, for he has stolen but waste paper indeed.'

'Not so,' replied Dr Johnson, 'the thief has come nigh to achieving his object, for the thief and the forger are one.'

'Name him,' cried Dr Percy. All eyes turned to old Ararat. His face showed the beginnings of a dumb misery, but no guilt. Anthony's face might have been carved out of a pumpkin.

'If,' said Dr Johnson slowly, 'if there were in Stratford a young man, apprenticed to a scrivener and adept with his pen; a young man who has the plays of Shakespeare by heart; and if that young man found as it might be a packet of old paper unused among the dead stationers' gear; is it unreasonable to suppose that that young man was tempted to try out his skill at writing like Shakespeare? And when his skill proved more than adequate, and the play "Caractacus" was composed and indited, and the Jubilee had raised interest in Shakespeare to fever pitch—what must have been the temptation to put forward the manuscript as genuine?'

'Yet why should he steal his own manuscript?'

'For fear of what has happened,' replied Dr Johnson, 'for fear that Dictionary Johnson, the editor of Shakespeare, with his special knowledge might scrutinize the manuscript and detect the imposture.'

Old Ararat's face was purple.

'Pray, sir,' said Dr Johnson, 'moderate your anger. The boy is a clever boy, and full of promise. Let him be honest from this time forward.'

Old Ararat looked at his son, and his jaw worked.

'But, Dr Johnson,' cried Percy, 'the hundred pounds!'

Anthony Ararat fell on his knees and raised his hand to Heaven.

'I swear before God,' he cried vibrantly, 'that I never touched the hundred pounds.'

It was the first word I had heard out of Anthony. By the fitful light of the lanthorn I stared in amazement at the expressionless face. The boy spoke like a player.

'Believe me, father,' cried Anthony earnestly, still on his knees by the font, 'I know nothing of the hundred pounds; nor do I know how the manuscript came to be exchanged for the money, for indeed I never meant to restore it until Dr Johnson was once more far from Stratford.'

'He speaks truth,' said Dr Johnson, 'for here is the hundred pounds, and it was I who laid the manuscript in the font.'

He drew the purse from his capacious pocket and handed it to Dr Percy.

'How came you by the manuscript?' asked Percy, accepting of the purse.

'It was not far to seek. The forger was the thief. It was likely that the finder was the forger. If Malone's panegyric on my learning frightened him into sequestering the manuscript to prevent it from falling under my eye, then it must have been hid between the time young Anthony left the shop and the time he returned with the empy coffer. He was gone long enough for Mr Ararat to spin us his long-winded tale. In that space of time he hid the manuscript—surely no further afield than his father's out-buildings. When he came in to us his face and shoulders were wet with rain.'

'Tears, surely?'

'Why, his eyes were full of tears. The boy is a comedian. But the drops on his shoulders never fell from his eyes; they were raindrops.'

'But, Dr Johnson,' put in Edmond Malone, 'we searched the out-buildings thoroughly, and the manuscript was not to be found.'

'The manuscript,' replied Dr Johnson, 'lay in plain sight before your eyes, and you passed it by without seeing it.'

'How could we?' cried Malone, 'we turned over the old papers in the shed.'

'Did you turn over the other old papers?'

'There were no other old papers.'

'There were,' said Dr Percy suddenly, 'for when I visited the—the necessary-house, I turned over a pile of old accounts of the greatest interest, put to this infamous use by the carelessness of the householder. I—ah—' his voice trailed off.

'The forged sheets of "Caractacus" were hastily thrust among them,' said Dr Johnson. 'I guessed so much when I heard the allusion to the jakes as the destination of bad poetry. What more likely hiding-place for a day or two, till Dr Johnson be far from Stratford once more? In short, I left the play and hurried thither, and found the pages undisturbed where young Ararat had thrust them into the heart of the pile.'

'Yet if you only meant to sequester the writings, boy,' said Dr Percy sternly, 'how came you to offer to barter them for money?'

Anthony rose to his feet.

'Sir,' he said respectfully, 'I never meant to touch the money. But Dr Johnson saw clearly, and said so, that 'twas no theft for profit; and I feared that such thoughts might lead him to me. I saw a way by which a thief might profit, and I wrote the letter and dropped it at my father's feet that the deed might seem after all the work of a real thief. Consider my apprehension, sir,' he turned to Dr Johnson, 'when you fitted my thumb into the impression it had made.'

Dr Johnson shook his head.

'Too many thumbs fitted it,' he said. 'Another way must be found to fit a thumb to its print. 'Twas so, too, with the paper. 'Twas clearly from your father's shop; but Percy and I and half Stratford were furnished with the same paper. Again the undistributed middle term.'

'Pray, sir, how came you to spare me in your thoughts?' enquired old Ararat.

'I acquitted you,' replied Dr Johnson, 'because after Malone's eulogy you never left my side; nor did your thumb fit the print in the wax.'

'Pray, Dr Johnson,' added Malone, 'coming down here from Mr Ararat's necessary-house with the manuscript in your pocket, why did you play out the farce? Why not reveal all at once?'

'To amuse Mr Boswell,' replied my friend with a broad smile. 'I thought an hour's watch by the bones of Shakespeare, and a dramatic discovery at its end, would give him a rich range of those sensations native to a man of sensibility, and enrich those notes he is constantly taking of my proceedings.'

In the laugh that followed at my expense, the Ararats sullenly took themselves off, and we four repaired to the Red Lion.

'Sir,' said young Malone, taking leave of us at the door of 'Much Ado about Nothing', 'this is a lesson in the detection of imposture which I will never forget.'

'Sir,' said Dr Johnson, 'you are most obliging. Be sure, sir, that I shall stand by you in your every endeavour to make known the truth. Pray, Dr Percy, accept of the forged manuscript as a memento of the pitfalls of *antiquarianism*.'

Dr Percy accepted with a smile, and we parted on most cordial terms.

'I blush to confess it,' I remarked as we prepared to retire, 'but I made sure that Dr Percy was carrying stolen documents about with him in yonder folio-sized packet he was so particular with.'

'So he was,' remarked Dr Johnson. 'Therefore I exchanged packets with him. I knew with certainty then that Thomas Percy had not stolen the Shakespeare manuscript, for all his antiquarian light fingers.'

'How so?' I enquired.

'Because I knew what he *had* stolen.'

'What?'

'A household reckoning of the first Anthony Ararat, showing that the good stationer's family consumed an unconscionable quantity of small beer during the year 1614. The magpie clergyman had filched it from old Ararat's necessary-house!'

BRINSLEY MOORE

'My Dear Clarissa'

ഇരു

There are certain interesting incidents connected with George Scattergood's visit to Bath and his miraculous recovery which, so far as I know, have never been made public. We get faint hints of them in various contemporary allusions, which have hitherto puzzled investigators. But the real truth could never have come to light but for the discovery, after the fire at Shipley Manor, of Dick Mortimer's love letters.

Now, this Mr Scattergood, you must understand, was an elderly bachelor of considerable wealth. Always of a shy and retiring disposition, he had lived a secluded life at Little Podlington, chiefly concerned with his books and a rather famous collection of old china. Latterly, with gout and growing infirmities, he became more and more a recluse, till at the age of fifty-six he is described as 'a confirmed invalid, seldom venturing outside his own room'.[1]

The chief interest in a rather colourless life centres around George Scattergood's money. When you find a rich uncle in failing health and a crowd of expectant nephews and nieces you know what to look for. Mrs Somerville, in her gossipy reminiscences, hints with a quiet humour at the persistent siege of Uncle George: the little presents, the affectionate letters, the frequent visits of enquiry, and the schemes and jealousies of the various members of the family circle. It was out of this general scheming and planning that there arose the unexpected developments revealed in Dick Mortimer's correspondence.

Now Mortimer's letters, as I said, are love letters—written to a Miss Clarissa Haywood, of Shipley Manor. This, of course, is to our advantage, for it ensures a frank and intimate account of what was really going on. On the other hand, being love letters, they are exceedingly verbose, and I shall therefore take the liberty of cutting down the correspondence by the omission of most of the love-making, which, after all, only concerned the fair recipient—merely giving you from each epistle such extracts as properly belong to our story.

The first of these letters, however, is so informative and helpful that I think you had better see it in full.

[1] *Life in a Gloustershire Village*, by Mary Somerville, p. 137.

'*Kensington,*
May 3rd, 1772.

'My Dearest Clarissa,—Alas! at the last moment all our pleasant plans are overturned, and I cannot come to Shipley as arranged.

'Mother went down to see Uncle George last week. The gout is bad, and he seemed very weak and feeble. She thinks he is breaking up, and he talked of making a new will.

'Somebody has recommended a course of the waters at Bath, and the old boy wants *me* to go there with him! Of course, it won't do him any good—he is too far gone for that; but mother thinks his fancy ought to be humoured. As she says, 'tis a great score to be selected out of all the nephews and nieces for this mark of preference; all the more surprising as I have been rather out of favour since the Oxford rumpus.

'My first impulse was to refuse, being so bent on spending my holiday with you. But, on second thoughts, it seemed madness to refuse. You see, to get a whole month alone with him apart from all the rest will be a splendid opportunity to work myself high in his good books, and I fancy I have enough wits about me to make the most of it. I think, at any rate, I could gradually interest him in my painting and persuade him about Italy. Perhaps, if I make myself extra devoted and affectionate, he might even be moved to increase my allowance sufficiently to enable us to be married at once!

'Now I hope, dear, you will see that I am acting wisely. If I followed my own inclinations, I should certainly spend the next few weeks with you. But there are times when personal inclination must yield to policy, and I think you will agree that, considering uncle's precarious state of health, it is wise to show him every consideration. Do write and say that you understand and approve.

'We start for Bath on Thursday, and shall make the journey in two stages, so as not to tire the invalid. I will write to you upon our arrival, and by my usual constant letters will keep you posted up as to how things progress. In the meantime accept, my dearest Clarissa, the assurance of the undying admiration and constancy of your devoted lover,

DICK MORTIMER.

'P.S.—Won't Cousin Frank feel his nose put out of joint when he hears about it!'

It will explain some allusions in this letter, and at the same time complete your mental picture of the writer, if I insert here a brief extract from a very competent observer. A certain Mrs Maynard, writing in January, 1772, to her sister, says:

'. . . Of old friends there is little to report. I found Alice Mortimer much concerned about her boy. He has been sent down from Oxford, overwhelmed with debts, which his uncle has had to pay. Friends have found him several good posts in the City; but in each case he throws up the work after a brief trial as uncongenial and degrading. His latest idea is to be an artist, and he dabbles about with paints and brushes—to little purpose, as it seems, save as an excuse for idleness and a gay life. He is bent now on joining a party of young gentlemen who are going out in the autumn on a foreign tour, and says that a year's work in the galleries of Italy is absolutely essential for success in his art; though where the money is to come from, goodness knows.

'The case is an illustration of the mischief of vague expectations. Richard Mortimer is quite a nice young man: not so clever as he thinks, but quite clever enough, if he were really thrown upon his own resources, to set to work and do well. It is the prospect of sharing in the wealth of that rich uncle that encourages him to be so extravagant in the present and so careless about the future. And now, to make matters worse, he has just got engaged to a charming girl who has not a penny in the world. No wonder his mother is worried.'[2]

Dick Mortimer's second letter, dated 9 May, reports a safe arrival at Bath. The invalid has borne the journey fairly well, but has spent two days in bed to recover from the fatigue. This sounds rather a bad beginning; but, as Dick points out, it has left the young man free 'to get about and see things'. He is charmed with Bath and its many attractions. And as he has found 'some old friends and made several new ones', there is every prospect of his 'having quite a pleasant time'. If only 'dearest Clarissa' were there, all would be perfect. But 'amidst all the noted belles of Bath, there is not one to compare with her own dear self, whose image is for ever imprinted on the heart' . . . etc., etc.

Two days later there is a further report.

'Uncle George is distinctly better. He has bathed twice and confesses that he feels another man. He spent quite a long time at the Pump Room yesterday talking to a lady he has met here, a Mrs Montmorency, and actually planned for us to drive out with her next week. In the afternoon he went out for two hours in a chair; and in the evening, to my surprise, instead of retiring to bed, invited me to a game of cribbage.

'This unexpected activity is rather a trial, as it upsets all my private plans and keeps me tied; but—of course, within limits—I shall try to meet all his requirements as a matter of policy. And I think my attentions

[2] *Belford Memoirs*, p. 57.

have already gained his approval: only this morning he said, quite graciously for him, that he was glad I am able to be with him.'

On 20 May, the news is still more encouraging. Uncle George now manages to get without assistance from one room to another, and is constantly surprising his attendants by the little things he can do for himself. He has even been for a short stroll on the sunny side of the parade with no more support than his stick and the occasional use of Dick Mortimer's arm.

But even more surprising than this physical improvement is the mental alertness displayed. 'The daily visit to the Pump Room is now quite a pleasure to him. He will sit for half the morning, gossiping with a bevy of people with whom he has struck up an acquaintance, and finds much satisfaction in discussing symptoms and remedies with the various people drinking the waters. And he feels so much better and finds Bath so agreeable that he has decided not to leave at the end of the month as arranged, but has booked our rooms for another four weeks at least.'

There was evidently something—possibly the last sentence—in this letter which displeased 'dearest Clarissa'. It would seem, too, that she has heard some news from Bath which suggests that Master Richard is hardly playing the part of a disconsolate lover. Anyhow, Mortimer has now to pen a letter of expostulation and protest.

'My Dearest Clarissa,—How could you write so petulantly and cruelly? If I followed my own inclination I should fly to your side. Don't you understand that it is duty and not inclination that ties me here? It is uncle, and not I, who decrees that we prolong our stay. And to leave him now would be to undo all that I have achieved by my solicitous attention and self-denial.

'As to a counter-attraction here, I vow that in all Bath there is no lady fit to hold a candle to my lovely Clarissa; and for Mrs Montmorency— why, my dear, she is old enough to be my mother and has two nearly grown-up children! Please do put such foolish notions out of your pretty head. With all the ardour of a devoted lover, I vow and protest . . .

Which Master Richard proceeds to do at such length that this, to an outsider, is quite the most tiresome letter of the series. It is only near the end, when for a moment he turns from sentiment to practical details, that we get anything really to our purpose.

'To show you that my view is right and my efforts effective, I may tell you that every day I get proof of a steady advance in the old man's favour. He has been looking over my sketches and was pleased to say "they were not bad", which is high praise from him. And when I ventured to hint at the Italian trip, he seemed quite favourable and said, "Well, we must see what can be arranged." I flatter myself I have played my cards pretty well to achieve such a result already.'

It would appear that these explanations satisfied the lady and that harmony was restored; for the next letter is almost boisterously jubilant.

'. . . Things are going splendidly. Did I tell you that uncle had bought a new wig and ordered an entire outfit from the tailor in the latest fashions? The things came on Monday, and what does the old boy do but dress up in all his new finery—plum-coloured coat, flowered vest, silk stockings, and silver-buckled shoes—and march off with me in the evening to the Rooms to display his splendour! And he enjoyed himself so much that the visit has been repeated almost every evening since.

'The astonishing thing is that he has become so popular—especially with the ladies. There is quite a flutter when he hobbles into the Card Room, and they rush forward to assist him to the most comfortable chair, and almost quarrel for the honour of being his partner in a hand at whist. He seems to find all this attention most agreeable and quite looks forward to his evening game. And, of course, the arrangement suits me very well; for when I have settled him down at cards, I can get away for a dance or two, just to keep myself in practice. I only hope my endeavours may be successful enough to enable us to fix the happy date when I may claim you for my own. To this end, all my efforts must be directed to keep my venerable uncle in his present amiable mood.'

The same jubilant note marks the next letter, dated 12 June.

'You would be astonished to see how Uncle George has come out of his shell. And such a popular person, especially with the old ladies! it is really quite extraordinary.

'On our daily perambulation of the parade we are stopped every few yards by bevies of ladies, anxious to greet "dear Mr Scattergood" and to congratulate him on the improvement in his health. It is most amusing to see his comical attempts to respond to these polite attentions, and to hear him exchanging compliments and chitchat with the fair sex. I never expected that an old hermit like him would develop into a ladies' man.'

Now I must confess that, as I read these last two letters, I felt some suspicion that, for all his frankness, Dick Mortimer might not be here quite a reliable witness. I know that in many cases gout has been wonderfully relieved, and I can quite accept the account of Uncle George's physical recovery; but I doubt whether the most enthusiastic partisan would claim for the waters of Bath the power of transforming a shy recluse into the elderly Don Juan here depicted. And since Dick is writing to a lady who is already suspicious, it seemed possible that the narrative might be somewhat coloured by a desire to divert her attention and to discount in advance any further rumours that might reach her from Bath. In a word, it looked to me far more likely that these 'bevies of ladies' on the parade and elsewhere might really be concerned with the sprightly young nephew rather than with the gouty old uncle.

This may seem a small point. But inasmuch as Dick Mortimer is our chief witness and indeed our only source of information for the essentials of the story, it becomes important to establish his accuracy. I am glad therefore to be able to give you from a contemporary record reliable confirmation of the picture he draws.

Anne Somers, whose acute observation misses little that takes place in Bath, writes as follows:

'*June 15th, 1772.*
'Really I don't know what the world is coming to. To see the way in which old maids and dowagers, mammas with marriageable daughters and widows old enough to know better, crowd round that old Mr Scattergood and vie with one another in attempts to captivate his affections is positively disgusting. Here is a doddering, broken-down creature, who hobbles about on two sticks and almost creaks with gout every time he rises from his chair, with neither intellect, social gifts, or conversation to recommend him, and hardly a notion in his head except old china. And yet, because he is a rich man, these shameless ladies, throwing modesty and decorum to the winds, quite openly lay siege and make no secret of the nets they spread for his capture. It makes me positively ashamed for my sex.'[3]

This extract has a threefold helpfulness. It confirms Dick Mortimer's account. It furnishes an explanation. And it prepares us for the note of alarm in Master Richard's next letter, when he suddenly realized what the explanation was.

'*Bath,*
June 27th, 1772.
'My Dear Clarissa,—I have received the shock of my life! You know I told you how attentive the ladies have been to our convalescent, and my satisfaction that his time was so pleasantly filled. But the other day Mrs Montmorency took me aside and asked me bluntly if I realized what it all meant? She said these old ladies were really after uncle and his money, and if I didn't look out one of them would certainly catch him and marry him before we know where we are!

[3] *Diary of Anne Somers*, vol. 1, p. 204.

'Well, it almost took my breath away; for never has my wildest imagination connected a dried up old stick like Uncle George with so romantic an idea as marriage. And it would, of course, be absolutely fatal to all our hopes.

'My first thought was to get him away from Bath as speedily as possible. But the moment I began to talk of Harrogate and to suggest he should complete his cure there, he said that Bath suited him excellently, and that he intended to stay on—adding that he was now quite well enough to look after himself, and that if I were tired of the place, I could go away at once!

'You may imagine how worrying all this is, and I am really at my wits' end. I am wondering whether your feminine instinct can suggest any means by which I can choke off these designing creatures. I am prepared to take any steps, however drastic, to ward off the danger; for the case is really desperate.

'Do, my dearest Clarissa, give the matter your best consideration and let me know what you think.

'With all love and devotion from your distracted

DICK.'

I know very little about Miss Clarissa Haywood, and cannot say how far she was competent to advise on such a perplexing problem. If Dick Mortimer on the spot could find no solution, it would hardly be fair to expect much from a young girl at a distance. Indeed, if you will but suppose for the moment that you yourself had been the recipient of the foregoing epistle and try to sketch out the sort of reply you would make, you will appreciate the difficulty. For if an obstinate old gentleman is blindly bent on rushing into this kind of danger, it is not easy to see what anybody can do to avert the risk.

I think, therefore, it was really quite clever of Miss Haywood to be able to suggest anything at all, even though, as you will see from the next letter, Dick thinks so little of her plan.

'*Bath,*
July 5th, 1772.

'Dear Clarissa,—I am surprised that it has taken you a whole week to think over my worries, and that even now you have nothing better to suggest than the unpractical idea in your letter.

'Of course, it is quite plain that uncle's money has been the attraction, and that if convinced he was a pauper the attentions of the tiresome old ladies would immediately cease. But everybody knows that he is rolling in luxury with a landed estate and a large income, to say nothing of his collection of porcelain which must be of fabulous value. No one would ever believe that all this could have disappeared in a moment; the suggestion would only make me ridiculous.

'But I need not discuss this further, for long before your letter arrived —indeed, the very night after I last wrote to you—as I lay and tossed upon my bed, unable to sleep for worry, there suddenly flashed upon me as an inspiration the one and only solution of the problem . . .

'Accordingly, next morning I sought out Miss Chatterley. Miss Chatterley is pre-eminently *the* gossip of Bath—a lady who must always have something interesting to tell and quite incapable of holding her tongue. So, under a pledge of secrecy, *I told her that Uncle George was already engaged to be married!*

'She seemed rather surprised and asked such a lot of questions that it taxed all my powers of imagination to satisfy her. And when she still appeared incredulous and pressed me for the lady's name, I was rather at a loss for a safe invention.

'Then, most fortunately, I thought of Miss Elizabeth Stride. I have heard mother speak of her as a friend of Uncle George's in the old days before he became a hermit; but he quarrelled with her because she would not sell her Meissen vase which he wanted to complete his set. So I borrowed her name for the occasion; and to round off the story for Miss Chatterley's benefit, I said the attachment was of old standing, though suspended by uncle's illness; but now that he had made so wonderful a recovery, the marriage would probably be announced for an early date.

'It rather takes it out of one to invent at this rate. But as I always say, if you must fib at all, you had better do it thoroughly. And my choice of Miss Chatterley for these confidences was fully justified. The results were so quickly manifested that I think she must have had quite a busy afternoon.

'When we went to the Rooms as usual that evening, there was no rushing forward to greet us. With averted eyes the old ladies ignored our entry; Mrs Carmichael even retained uncle's favourite chair, and he was forced into a draughty corner with an old frump whom nobody knows. And there was so much difficulty in finding anyone to play with him that Mrs Montmorency and I had to stay to make up his set; an arrangement so dull that he soon decided to go home, declaring that he would never again enter the Rooms to be treated in such an unsociable fashion.

'Still more marked was the coldness of his reception on the parade next morning. These scheming ladies, who used to be so gushing, passed us with the barest acknowledgments; some pointedly looked aside as they met us; not one stopped to return our greeting. Everywhere our progress was so shorn of its glory that uncle soon got tired of it and now seldom cares to go out at all.

'All this, of course, is a great relief to me. I can now take much more time to myself and get about for a little amusement on my own, without the constant dread that something might happen in my absence. And after the strain of the last few weeks I really need some relaxation.

'I find, too, that now uncle is freed from the constant adulation of these designing females, he seems much more appreciative of such little attentions as I am able to pay him, and our intercourse is much more intimate than of late. So that my little scheme has now removed a most appalling danger and has improved the situation all round.

'And now, my dear Clarissa, please rest assured . . .

'P.S.—I wish you could see how much I am sought after by certain gossips of this place, and how they pump for particulars. My line is to be very mysterious and to refuse to add any further details. But wasn't it lucky I gave the name of a lady so far away? Otherwise, so insatiable is the curiosity of these chatterers that they would certainly search her out. I believe the smartest thing I ever did was to fix on the name of Miss Elizabeth Stride, of Cumberland.'

Up to this point, as you will see, Dick Mortimer's letters run in a fairly regular sequence, so that the course of events is easy to follow. Here there comes rather an awkward break. Whether there were letters which have not been preserved, or whether Master Richard, now free 'to get about for a little amusement on his own', found less time to write, I cannot say. But for the next three weeks I can find nothing except one short note reporting that 'all goes well, with no more trouble from the harpies. Uncle is regular at the baths and daily improves in health. But he complains now that Bath is rather a dull place and will I think be quite ready to leave when the doctor so advises.'

To fill the gap, I put in here a brief note by Anne Somers.

'*July 16th.*

'Quite the most startling event of the month is the news of old Mr Scattergood's secret engagement. It fell like a bombshell amongst the little coterie of ladies who were so assiduous in their attentions, and it is most comical to see their dismay and indignation. Where a few days ago it was "dear Mr Scattergood", it is now all black looks and violent vituperation. They speak of him as "that base deceiver", and I am told that our bulky Miss Venn muttered as the little man hobbled past her at the Pump Room, "Perfidious monster!" I should never have thought the doddering old gentleman capable of carrying out such a neat little comedy; but he kept his secret well and must be chuckling to himself at the result of its disclosure. I hope it will teach those shameless creatures a salutary lesson.

'Rumour has it that this is quite a long-standing attachment, and there is much curiosity about the lady. This Miss Stride lives so far away that nobody seems to know anything about her, except that, like old Scattergood himself, she is a collector of pottery and porcelain; I suppose it was this common interest that brought them together. But the excitement aroused by this unexpected romance is so widespread that I have no doubt our diligent gossips will speedily unearth further particulars.'[4]

This entry not only shows the wide acceptance of Mortimer's daring canard, it will also prepare the discerning reader for the catastrophe revealed in the next letter.

'Bath,
July 25th, 1772.

'Dear Clarissa,—Here is a pretty kettle of fish! You thought nobody would ever believe my little invention about Miss Stride. As a matter of fact, the thing has spread in a manner I never anticipated and is discussed by our busy gossips all over the place, with the result that it has come to uncle's ears.

'When I got in from the ball last night, to my surprise he was sitting up for me. He was purple with rage, and at once accused me of spreading falsehoods about himself and of taking unwarrantable liberty with the name of an innocent and estimable lady.

'I was so taken aback that I had no time to get any reasonable explanation ready, and I am afraid I made rather a mess of it. But I don't think he really listened. He said my conduct was either that of an incorrigible liar, or of a half-witted fool, and told me to clear out at once. So I am to leave today by the afternoon coach.

'It is unfortunate that there is nobody here to smooth the old bear over a bit, and to persuade him that the total withdrawal of my allowance is an excessive punishment. Mrs Montmorency, who had some influence over him, left last week for Tunbridge Wells, and my other friends, for various reasons, are unsuitable for the purpose. And I don't think it is any good to bother mother about it.

'So I am going to Tunbridge Wells to take counsel with Mrs Montmorency. She has always been so friendly and interested, and understands all about my difficulties. I know she will be kind and sympathetic, and perhaps she may be able to suggest something. I will write again when we have decided what is best to be done.

'I am, as you may guess, terribly upset by this unlooked-for disaster, just when all seemed to be going so well. But whatever may be the outcome, you may rest assured that no possible effort to overcome the present difficulties of the situation will be lacking on the part of your ever devoted admirer.

RICHARD MORTIMER.'

---

[4] *Diary of Anne Somers*, vol. 1, p. 224.

Strictly speaking, this is the last of Dick Mortimer's love letters. For though there is one more note of a later date, it is of a character, as you will see, that can hardly be reckoned in that category. And before it could be written, events had moved rapidly and to a startling conclusion.

Our evidence here becomes rather disjointed, being derived from different sources. But I think, with a little imagination, you will have no difficulty in filling the gaps.

You may take first Anne Somers' account of how things were going at Bath after our young gentleman's hurried departure.

'*August 4th*.

'It is difficult to know the exact truth about the Scattergood affair. Last week the whole thing was denied, and the nephew was reported to have been dismissed in disgrace for spreading an unfounded story of his uncle's engagement—a pretty setback for our busy chatterers. But this morning I hear that the announcement was merely premature, not incorrect, and that Mr Scattergood is returning home almost immediately to prepare for the reception of his bride. 'Tis said, that to save the old gentleman the fatigue of a long journey, the lady will travel down to Little Podlington, and that the marriage will be solemnized in the early autumn. But Bath is such a place for gossip that I dare say all this is but a new canard.'[5]

Next I should like you to see an extract from the memoirs of the Hon. Mrs Ponsonby.

'Taking my first walk on the Parade this morning, I had a great surprise. Who should I meet but old George Scattergood. And, my dear, I hardly knew him! Such a change as you would never think possible. He was walking quite briskly and seemed years younger. The Bath treatment has certainly worked wonders in his case. And such a chatterer! He was loud in his praises of Bath's attractions, and seemed delighted with its stir and liveliness—all so different from the silent George Scattergood we used to know. And when I tell you that he finished up by telling me that he was going to be married, you will draw your own conclusions. My own opinion is that the excitement of Bath has upset his wits; anyhow, that idea of getting married must be a fiction.'

But whatever the state of George Scattergood's wits the idea of getting married was not a fiction, as you shall see from the letter which Uncle George himself wrote about this time to an old bachelor friend.[6]

[5] *Diary of Anne Somers*, vol. 1, p. 236.
[6] *Life and Correspondence of the late Henrietta Ponsonby*, p. 77.

'*Bath,*
August 9th, 1772.

'My Dear Ned,—I don't understand why you should find the news incredible. After all, I am only fifty-six; at least three years younger than you, and still at an age when many men marry.

'Of course, as you say, in my case the step is unexpected, as for many years past the state of my health had kept any such idea out of my thoughts. But Bath has done great things for me. Not only am I now practically free of my old enemy, the gout, and able to get about with renewed vigour, but my sojourn here has also revealed to me how much I have lost by my solitary life in the past, particularly by my isolation from the polite attentions of female society.

'The prospect of returning to my dull routine at Podlington was extremely distasteful, and I was already contemplating a radical change in my household arrangements. But up to this point, I can honestly declare, such a thing as matrimony never entered my head: my intention was the engagement of a lady-housekeeper who should bring a touch of feminine brightness and cheerfulness into my home, and enable me to entertain and enjoy the society of my neighbours.

'It was then that an extraordinary rumour was spread abroad that I was about to marry Miss Stride. My first thought was one of indignation at the invention and anger at the liberty taken with the lady's name, and I visited due punishment on the perpetrator of the unseemly joke. But upon further consideration, the widespread acceptance of this rumour gradually suggested to me that the idea could not be so absurd as I had supposed, and that it might really indicate the best solution of my domestic problem.

'Why, after all, should I be at the mercy of hired menials, of doubtful services, and constant changes: offering at best an uncertain present and in the future the prospect of a solitary old age untended by loving care and devotion?

'Elizabeth Stride and I were old friends, and, but for her obstinacy about the Meissen vase, would still have been so. She is cheerful, active, and domesticated, and in every way qualified to minister to the needs and comfort of a middle-aged man. In opinions and tastes we have much in common, and the combination of her small but select cabinet with my own treasures will give me the most complete collection of Meissen porcelain in the kingdom. So I penned a careful letter, explaining fully my views, and making a formal proposal of marriage, which, after a brief interval for consideration and enquiry, she was graciously pleased to accept.

'I hope I have now fully answered your rather vulgar enquiry, "How the dickens I have got myself into such a hole?" It is not a hole at all, but the happiest and most sensible step of my life. And I must always think of Bath with gratitude for two great blessings: the wonderful restoration of my health and the possession of a charming bride. For certainly but for the oft-censured gossip of this place, I should never have thought of proposing to Elizabeth Stride.

'Yours very sincerely,
GEORGE SCATTERGOOD.'

I think, having read this enlightening epistle, you are fairly posted up in events to date and will be able to understand, though I don't expect you to approve, Dick Mortimer's final letter to Clarissa.

'*Tunbridge Wells,
Sept. 10th, 1772.*

'Dear Clarissa,—I suppose you have heard the news about Uncle George. This, of course, is the final blow, and fatal to all our hopes. Bereft as I am now of both income and expectations, the continuance of our engagement would be futile. As an honourable man, my only course is to release you at once and set you free from your pledge.

'If in the near future you should hear that I have made other arrangements for my own settlement in life, I would ask you to remember that I did not take this step until Fate had rendered impossible the realization of our romantic youthful dreams.

'Yours very truly,
RICHARD MORTIMER.'

I find in the parish register of Little Podlington an entry certifying that George Scattergood was duly married to Miss Elizabeth Stride on 18 September 1772. I trust that the acquisition of a charming bride and the Meissen vase brought Uncle George all the satisfaction he anticipated.

With reference to the 'other arrangements' for his own future foreshadowed in Dick Mortimer's last letter, I cannot tell you exactly when and where he married his rich and sympathetic widow. But, to round off the record, I give you a little picture of the result as penned by Mrs Maynard some two years later.

'. . . Richard Mortimer is much sobered by his marriage and seems quite cured of his wild foolishness. I always said that he needed the control of a firm hand, and by all accounts he now has it. In public his wife treats him with a sort of motherly affection and speaks of him as her "handsome boy"; but privately I hear she rules him with a rod of iron and never allows him to forget that she was once Mrs Montmorency and holds the purse strings. And when Archie Bevan was congratulating him on the lavish appointments of their new house in Kensington he smiled grimly, and said, "Yes, and I am the prisoner in the gilded cage." [7]

For myself, I am inclined to agree with Mrs Maynard that Dick did need the control of a firm hand; and, remembering some of those letters to Clarissa, I am not altogether sorry if things were a bit uncomfortable for him. But as I arrange these last pieces of the story and try to picture him as the prisoner in the gilded cage, I find myself wondering if he recalled those gay times at Bath, and if he still thought that the smartest thing he ever did was to concoct that story about Miss Stride, of Cumberland.

[7] *Belford Memoirs*, p. 142.

JOHN BUCHAN

The Company of the Marjolaine[1]

∽∞∾

'Qu'est-c' qui passe ici si tard,
Compagnons de la Marjolaine?'

CHANSONS DE FRANCE

I

. . . I came down from the mountains and into the pleasing valley of the
Adige in as pelting a heat as ever mortal suffered under. The way underfoot
was parched and white; I had newly come out of a wilderness of white
limestone crags, and a sun of Italy blazed blindingly in an azure Italian
sky. You are to suppose, my dear aunt, that I had had enough and something more of my craze for foot-marching. A fortnight ago I had gone to
Belluno in a post-chaise, dismissed my fellow to carry my baggage by way
of Verona, and with no more than a valise on my back plunged into the
fastnesses of those mountains. I had a fancy to see the little sculptured
hills which made backgrounds for Gianbellin, and there were rumours of
great mountains built wholly of marble which shone like the battlements
of the Celestial City. So at any rate reported young Mr Wyndham, who
had travelled with me from Milan to Venice. I lay the first night at Piave,
where Titian had the fortune to be born, and the landlord at the inn
displayed a set of villainous daubs which he swore were the early works
of that master. Thence up a toilsome valley I journeyed to the Ampezzan
country, where indeed I saw my white mountains, but, alas! no longer
Celestial. For it rained like Westmoreland for five endless days, while
I kicked my heels in an inn and turned a canto of Ariosto into halting

---

[1] This extract from the unpublished papers of the Manorwater family has seemed to the
Editor worth printing for its historical interest. The famous Lady Molly Carteron became
Countess of Manorwater by her second marriage. She was a wit and a friend of wits, and
her nephew, the Honourable Charles Hervey-Townshend (afterwards our ambassador at
The Hague), addressed to her a series of amusing letters while making, after the fashion of
his contemporaries, the Grand Tour of Europe. Three letters, written at various places in
the Eastern Alps and dispatched from Venice, contain the following short narrative. [Author's
note.]

English couplets. By and by it cleared, and I headed westward towards Bozen, among the tangle of wild rocks where the Dwarf King had once his rose garden. The first night I had no inn, but slept in the vile cabin of a forester, who spoke a tongue half Latin, half Dutch, which I could not master. The next day was a blaze of heat, the mountain paths lay thick with dust, and I had no wine from sunrise to sunset. Can you wonder that, when the following noon I saw Santa Chiara sleeping in its green circlet of meadows, my thought was only of a deep draught and a cool chamber? I protest that I am a great lover of natural beauty, of rock and cascade, and all the properties of the poet; but the enthusiasm of M. Rousseau himself would sink from the stars to earth if he had marched since breakfast in a cloud of dust with a throat like the nether millstone.

Yet I had not entered the place before Romance revived. The little town—a mere wayside halting-place on the great mountain road to the North—had the air of mystery which foretells adventure. Why is it that a dwelling or a countenance catches the fancy with the promise of some strange destiny? I have houses in my mind which I know will some day and somehow be intertwined oddly with my life; and I have faces in memory of which I know nothing save that I shall undoubtedly cast eyes again upon them. My first glimpses of Santa Chiara gave me this earnest of romance. It was walled and fortified, the streets were narrow pits of shade, old tenements with bent fronts swayed to meet each other. Melons lay drying on flat roofs, and yet now and then would come a high-pitched northern gable. Latin and Teuton met and mingled in the place, and, as Mr Gibbon has taught us, the offspring of this admixture is something fantastic and unpredictable. I forgot my grievous thirst and my tired feet in admiration and a certain vague expectation of wonders. Here, ran my thought, it is fated, maybe, that Romance and I shall at last compass a meeting. Perchance some princess is in need of my arm, or some affair of high policy is afoot in this jumble of old masonry. You will laugh at my folly, but I had an excuse for it. A fortnight in strange mountains disposes a man to look for something at his next encounter with his kind, and the sight of Santa Chiara would have fired the imagination of a judge in Chancery.

I strode happily into the courtyard of the Tre Croci, and presently had my expectation confirmed. For I found my fellow, Gianbattista—a faithful rogue I got in Rome on a Cardinal's recommendation—hot in dispute with a lady's-maid. The woman was old, harsh-featured—no Italian clearly, though she spoke fluently in the tongue. She rated my man like a pick-pocket, and the dispute was over a room.

'The signor will bear me out,' said Gianbattista. 'Was not I sent to Verona with his baggage, and thence to this place of ill manners? Was I not bidden engage for him a suite of apartments? Did I not duly choose these fronting on the gallery, and dispose therein the signor's baggage? And lo! an hour ago I found it all turned into the yard and this woman installed in its place. It is monstrous, unbearable! Is this an inn for travellers, or haply the private mansion of these Magnificences?'

'My servant speaks truly,' I said, firmly yet with courtesy, having no mind to spoil adventure by urging rights. 'He had orders to take these rooms for me, and I know not what higher power can countermand me.'

The woman had been staring at me scornfully, for no doubt in my dusty habit I was a figure of small count; but at the sound of my voice she started, and cried out, 'You are English, signor?'

I bowed an admission.

'Then my mistress shall speak with you,' she said, and dived into the inn like an elderly rabbit.

Gianbattista was for sending for the landlord and making a riot in that hostelry; but I stayed him, and bidding him fetch me a flask of white wine, three lemons, and a glass of *eau de vie*, I sat down peaceably at one of the little tables in the courtyard and prepared for the quenching of my thirst. Presently, as I sat drinking that excellent compound which was my own invention, my shoulder was touched, and I turned to find the maid and her mistress. Alas for my hopes of a glorious being, young and lissom and bright with the warm riches of the south! I saw a short, stout little lady, well on the wrong side of thirty. She had plump red cheeks, and fair hair dressed indifferently in the Roman fashion. Two candid blue eyes redeemed her plainness, and a certain grave and gentle dignity. She was notably a gentlewoman, so I got up, doffed my hat, and awaited her commands.

She spoke in Italian. 'Your pardon, signor, but I fear my good Cristine has done you unwittingly a wrong.'

Cristine snorted at this premature plea of guilty, while I hastened to assure the fair apologist that any rooms I might have taken were freely at her service.

I spoke unconsciously in English, and she replied in a halting parody of that tongue. 'I understand him,' she said, 'but I do not speak him happily. I will discourse, if the signor pleases, in our first speech.'

She and her father, it appeared, had come over the Brenner, and arrived that morning at the Tre Croci, where they purposed to lie for some days. He was an old man, very feeble, and much depending upon her constant care. Wherefore it was necessary that the rooms of all the party should adjoin, and there was no suite of the size in the inn save that which I had taken. Would I therefore consent to forgo my right, and place her under an eternal debt?

I agreed most readily, being at all times careless where I sleep, so the bed be clean, or where I eat, so the meal be good. I bade my servant see the landlord and have my belongings carried to other rooms. Madame thanked me sweetly, and would have gone, when a thought detained her.

'It is but courteous,' she said, 'that you should know the names of those whom you have befriended. My father is called the Count d'Albani, and I am his only daughter. We travel to Florence, where we have a villa in the environs.'

'My name,' said I, 'is Hervey-Townshend, an Englishman travelling abroad for his entertainment.'

'Hervey?' she repeated. 'Are you one of the family of Miladi Hervey?'

'My worthy aunt,' I replied, with a tender recollection of that preposterous woman.

Madame turned to Cristine, and spoke rapidly in a whisper.

'My father, sir,' she said, addressing me, 'is an old frail man, little used to the company of strangers; but in former days he has had kindness from members of your house, and it would be a satisfaction to him, I think, to have the privilege of your acquaintance.'

She spoke with the air of a vizier who promises a traveller a sight of the Grand Turk. I murmured my gratitude, and hastened after Gianbattista. In an hour I had bathed, rid myself of my beard, and arrayed myself in decent clothing. Then I strolled out to inspect the little city, admired an altar-piece, chaffered with a Jew for a cameo, purchased some small necessaries, and returned early in the afternoon with a noble appetite for dinner.

The Tre Croci had been in happier days a bishop's lodging, and possessed a dining-hall ceiled with black oak and adorned with frescos. It was used as a general *salle à manger* for all dwellers in the inn, and there accordingly I sat down to my long-deferred meal. At first there were no other diners, and I had two maids, as well as Gianbattista, to attend on my wants. Presently Madame d'Albani entered, escorted by Cristine and by a tall gaunt serving-man, who seemed no part of the hostelry. The landlord followed, bowing civilly, and the two women seated themselves at the little table at the further end. 'Il Signor Conte dines in his room,' said Madame to the host, who withdrew to see to that gentleman's needs.

I found my eyes straying often to the little party in the cool twilight of that refectory. The man-servant was so old and battered, and yet of such a dignity, that he lent a touch of intrigue to the thing. He stood stiffly behind Madame's chair, handing dishes with an air of silent reverence— the lackey of a great noble, if ever I had seen the type. Madame never glanced towards me, but conversed sparingly with Cristine, while she pecked delicately at her food. Her name ran in my head with a tantalizing flavour of the familiar. Albani! D'Albani! It was a name not uncommon in the Roman States, but I had never heard it linked to a noble family. And yet I had—somehow, somewhere; and in the vain effort at recollection I had almost forgotten my hunger. There was nothing bourgeois in the little lady. The austere servants, the high manner of condescension, spake of a stock used to deference, though, maybe, pitifully decayed in its fortunes. There was a mystery in these quiet folk which tickled my curiosity. Romance after all was not destined to fail me at Santa Chiara.

My doings of the afternoon were of interest to myself alone. Suffice it to say that when I returned at nightfall I found Gianbattista the trustee of a letter. It was from Madame, written in a fine thin hand on a delicate paper, and it invited me to wait upon the signor, her father, that evening at eight o'clock. What caught my eye was a coronet stamped in a corner. A coronet, I say, but in truth it was a crown, the same as surmounts the Arms Royal of England on the signboard of a Court tradesman. I marvelled at the ways of foreign heraldry. Either this family of d'Albani had higher pretensions than I had given it credit for, or it employed an unlearned and imaginative stationer. I scribbled a line of acceptance and went to dress.

The hour of eight found me knocking at the Count's door. The grim serving-man admitted me to the pleasant chamber which should have been mine own. A dozen wax candles burned in sconces, and on the table, among fruits and the remains of supper, stood a handsome candelabra of silver. A small fire of logs had been lit on the hearth, and before it in an armchair sat a strange figure of a man. He seemed not so much old as aged. I should have put him at sixty, but the marks he bore were clearly less those of Time than of Life. There sprawled before me the relics of noble looks. The fleshy nose, the pendulous cheek, the drooping mouth, had once been cast in the lines of manly beauty. Heavy eyebrows above and heavy bags beneath spoiled the effect of a choleric blue eye, which age had not dimmed. The man was gross and yet haggard; it was not the padding of good living which clothed his bones, but a heaviness as of some dropsical malady. I could picture him in health a gaunt loose-limbed being, high-featured and swift and eager. He was dressed wholly in black velvet, with fresh ruffles and wristbands, and he wore heeled shoes with antique silver buckles. It was a figure of an older age which rose slowly to greet me, in one hand a snuff-box and a purple handker-chief, and in the other a book with finger marking place. He made me a great bow as Madame uttered my name, and held out a hand with a kindly smile.

'Mr Hervey-Townshend,' he said, 'we will speak English, if you please. I am fain to hear it again, for 'tis a tongue I love. I make you welcome, sir, for your own sake and for the sake of your kin. How is her honourable ladyship, your aunt? A week ago she sent me a letter.'

I answered that she did famously, and wondered what cause of correspondence my worthy aunt could have with wandering nobles of Italy.

He motioned me to a chair between Madame and himself, while a servant set a candle on a shelf behind him. Then he proceeded to catechize me in excellent English, with now and then a phrase of French, as to the doings in my own land. Admirably informed this Italian gentleman proved himself. I defy you to find in Almack's more intelligent gossip. He enquired as to the chances of my Lord North and the mind of my Lord Rockingham. He had my Lord Shelburne's foibles at his fingers' ends. The habits of the Prince, the aims of their ladyships of Dorset and Buckingham, the extravagance of this noble Duke and that right honourable gentleman were not hid from him. I answered discreetly yet frankly, for there was no ill-breeding in his curiosity. Rather it seemed like the enquiries of some fine lady, now buried deep in the country, as to the doings of a forsaken Mayfair. There was humour in it and something of pathos.

'My aunt must be a voluminous correspondent, sir,' I said.

He laughed. 'I have many friends in England who write to me, but I have seen none of them for long, and I doubt I may never see them again. Also in my youth I have been in England.' And he sighed as at a sorrowful recollection.

Then he showed the book in his hand. 'See,' he said, 'here is one of your English writings, the greatest book I have ever happened on.' It was a volume of Mr Fielding.

For a little he talked of books and poets. He admired Mr Fielding profoundly, Dr Smollett somewhat less, Mr Richardson not at all. But he was clear that England had a monopoly of good writers, saving only my friend M. Rousseau, whom he valued, yet with reservations. Of the Italians he had no opinion. I instanced against him the plays of Signor Alfieri. He groaned, shook his head, and grew moody.

'Know you Scotland?' he asked suddenly.

I replied that I had visited Scotch cousins, but had no great estimation for the country. 'It is too poor and jagged,' I said, 'for the taste of one who loves colour and sunshine and suave outlines.'

He sighed. 'It is indeed a bleak land, but a kindly. When the sun shines at all he shines on the truest hearts in the world. I love its bleakness too. There is a spirit in the misty hills, and the harsh sea-wind which inspires men to great deeds. Poverty and courage go often together, and my Scots, if they are poor, are as untamable as their mountains.'

'You know the land, sir?' I asked.

'I have seen it, and I have known many Scots. You will find them in

Paris and Avignon and Rome, with never a plack in their pockets. I have a feeling for exiles, sir, and I have pitied these poor people. They gave their all for the cause they followed.'

Clearly the Count shared my aunt's views of history—those views which have made such sport for us often at Carteron. Stalwart Whig as I am, there was something in the tone of the old gentleman which made me feel a certain majesty in the lost cause.

'I am Whig in blood and Whig in principle,' I said, 'but I have never denied that those Scots who followed the Chevalier were too good to waste on so trumpery a leader.'

I had no sooner spoken the words than I felt that somehow I had been guilty of a *bêtise*.

'It may be so,' said the Count. 'I did not bid you here, sir, to argue on politics, on which I am assured we should differ. But I will ask you one question. The King of England is a stout upholder of the right of kings. How does he face the defection of his American possessions?'

'The nation takes it well enough, and as for His Majesty's feelings, there is small inclination to enquire into them. I conceive of the whole war as a blunder out of which we have come as we deserved. The day is gone by for the assertion of monarchic rights against the will of a people.'

'Maybe. But take note that the King of England is suffering today as—how do you call him?—the Chevalier suffered forty years ago. "The wheel has come full circle," as your Shakespeare says. Time has wrought his revenge.'

He was staring into a fire, which burned small and smokily.

'You think the day for kings is ended. I read it differently. The world will ever have need of kings. If a nation cast out one it will have to find another. And mark you, those later kings, created by the people, will bear a harsher hand than the old race who ruled as of right. Some day the world will regret having destroyed the kindly and legitimate line of monarchs and put in their place tyrants who govern by the sword or by flattering an idle mob.'

This belated dogma would at other times have set me laughing, but the strange figure before me gave no impulse to merriment. I glanced at Madame, and saw her face grave and perplexed, and I thought I read a warning gleam in her eye. There was a mystery about the party which irritated me, but good breeding forbade me to seek a clue.

'You will permit me to retire, sir,' I said. 'I have but this morning come down from a long march among the mountains east of this valley. Sleeping in wayside huts and tramping those sultry paths make a man think pleasantly of bed.'

The Count seemed to brighten at my words. 'You are a marcher, sir, and love the mountains? Once I would gladly have joined you, for in my youth I was a great walker in hilly places. Tell me, now, how many miles will you cover in a day?'

I told him thirty at a stretch.

'Ah,' he said, 'I have done fifty, without food, over the roughest and mossiest mountains. I lived on what I shot, and for drink I had spring water. Nay, I am forgetting. There was another beverage, which I wager you have never tasted. Heard you ever, sir, of that *eau de vie* which the Scots call *usquebagh*? It will comfort a traveller as no thin Italian wine will comfort him. By my soul, you shall taste it. Charlotte, my dear, bid Oliphant fetch glasses and hot water and lemons. I will give Mr Hervey-Townshend a sample of the brew. You English are all *têtes-de-fer*, sir, and are worthy of it.'

The old man's face had lighted up, and for the moment his air had the jollity of youth. I would have accepted the entertainment had I not again caught Madame's eye. It said, unmistakably and with serious pleading, 'Decline'. I therefore made my excuses, urged fatigue, drowsiness, and a delicate stomach, bade my host good-night, and in deep mystification left the room.

Enlightenment came upon me as the door closed. There on the threshold stood the manservant whom they called Oliphant, erect as a sentry on guard. The sight reminded me of what I had once seen at Basle when by chance a Rhenish Grand Duke had shared the inn with me. Of a sudden a dozen clues linked together—the crowned notepaper, Scotland, my aunt Hervey's politics, the tale of old wanderings.

'Tell me,' I said in a whisper, 'who is the Count d'Albani, your master?' and I whistled softly a bar of 'Charlie is my darling'.

'Ay,' said the man, without relaxing a muscle of his grim face. 'It is the King of England—my king and yours.'

II

In the small hours of the next morning I was awoke by a most unearthly sound. It was as if all the cats on all the roofs of Santa Chiara were sharpening their claws and wailing their battle-cries. Presently out of the noise came a kind of music—very slow, solemn, and melancholy. The notes ran up in great flights of ecstasy, and sunk anon to the tragic deeps. In spite of my sleepiness I was held spellbound, and the musician had concluded with certain barbaric grunts before I had the curiosity to rise. It came from somewhere in the gallery of the inn, and as I stuck my head out of my door I had a glimpse of Oliphant, nightcap on head and a great bagpipe below his arm, stalking down the corridor.

The incident, for all the gravity of the music, seemed to give a touch of farce to my interview of the past evening. I had gone to bed with my mind full of sad stories of the deaths of kings. Magnificence in tatters has always affected my pity more deeply than tatters with no such antecedent, and a monarch out at elbows stood for me as the last irony of our mortal life. Here was a king whose misfortunes could find no parallel. He had been in his youth the hero of a high adventure, and his middle age had been spent in fleeting among the courts of Europe, and waiting as pensioner on the whims of his foolish but regnant brethren. I had heard tales of a growing sottishness, a decline in spirit, a squalid taste in pleasures. Small blame, I had always thought, to so ill-fated a princeling. And now I had chanced upon the gentleman in his dotage, travelling with a barren effort at mystery, attended by a sad-faced daughter and two ancient domestics. It was a lesson in the vanity of human wishes which the shallowest moralist would have noted. Nay, I felt more than the moral. Something human and kindly in the old fellow had caught my fancy. The decadence was too tragic to prose about, the decadent too human to moralize on. I had left the chamber of the—shall I say *de jure* King of England?—a sentimental adherent of the cause. But this business of the bagpipes touched the comic. To harry an old valet out of bed and set him droning on pipes in the small hours smacked of a theatrical taste, or at least of an undignified fancy. Kings in exile, if they wish to keep the tragic air, should not indulge in such fantastic serenades.

My mind changed again when after breakfast I fell in with Madame on the stair. She drew aside to let me pass, and then made as if she would speak to me. I gave her good-morning, and, my mind being full of her story, addressed her as 'Excellency'.

'I see, sir,' she said, 'that you know the truth. I have to ask your forbearance for the concealment I practised yesterday. It was a poor requital for your generosity, but it is one of the shifts of our sad fortune. An uncrowned king must go in disguise or risk the laughter of every stable-boy. Besides, we are too poor to travel in state, even if we desired it.'

Honestly, I knew not what to say. I was not asked to sympathize, having already revealed my politics, and yet the case cried out for sympathy. You remember, my dear aunt, the good Lady Culham, who was our Dorsetshire neighbour, and tried hard to mend my ways at Carteron? This poor Duchess—for so she called herself—was just such another. A woman made for comfort, housewifery, and motherhood, and by no means for racing about Europe in charge of a disreputable parent. I could picture her settled equably on a garden seat with a lapdog and needlework, blinking happily over green lawns and mildly rating an errant gardener. I could fancy her sitting in a summer parlour, very orderly and dainty, writing

lengthy epistles to a tribe of nieces. I could see her marshalling a household in the family pew, or riding serenely in the family coach behind fat
bay horses. But here, on an inn staircase, with a false name and a sad air
of mystery, she was woefully out of place. I noted little wrinkles forming
in the corners of her eyes, and the ravages of care beginning in the plump
rosiness of her face. Be sure there was nothing appealing in her mien. She
spoke with the air of a great lady, to whom the world is matter only for
an afterthought. It was the facts that appealed and grew poignant from
her courage.

'There is another claim upon your good-nature,' she said. 'Doubtless
you were awoke last night by Oliphant's playing upon the pipes. I rebuked
the landlord for his insolence in protesting, but to you, a gentleman and
a friend, an explanation is due. My father sleeps ill, and your conversation
seems to have cast him into a train of sad memories. It has been his habit
on such occasions to have the pipes played to him, since they remind him
of friends and happier days. It is a small privilege for an old man, and he
does not claim it often.'

I declared that the music had only pleased, and that I would welcome
its repetition. Whereupon she left me with a little bow and an invitation
to join them that day at dinner, while I departed into the town on my own
errands. I returned before midday, and was seated at an arbour in the
garden, busy with letters, when there hove in sight the gaunt figure of
Oliphant. He hovered around me, if such a figure can be said to hover,
with the obvious intention of addressing me. The fellow had caught my
fancy, and I was willing to see more of him. His face might have been
hacked out of grey granite, his clothes hung loosely on his spare bones,
and his stockinged shanks would have done no discredit to Don Quixote.
There was no dignity in his air, only a steady and enduring sadness. Here,
thought I, is the one of the establishment who most commonly meets the
shock of the world's buffets. I called him by name and asked him his
desires.

It appeared that he took me for a Jacobite, for he began a rigmarole
about loyalty and hard fortune. I hastened to correct him, and he took the
correction with the same patient despair with which he took all things.
'Twas but another of the blows of Fate.

'At any rate,' he said in a broad Scotch accent, 'ye come of kin that has
helpit my maister afore this. I've many times heard tell o' Herveys and
Townshends in England, and a' folk said they were on the richt side.
Ye're maybe no a freend, but ye're a freend's freend, or I wadna be
speirin' at ye.'

I was amused at the prologue, and waited on the tale. It soon came.
Oliphant, it appeared, was the purse-bearer of the household, and woeful

straits that poor purse-bearer must have been often put to. I questioned him as to his master's revenues, but could get no clear answer. There were payments due next month in Florence which would solve the difficulties for the winter, but in the meantime expenditure had beaten income. Travelling had cost much, and the Count must have his small comforts. The result, in plain words, was that Oliphant had not the wherewithal to frank the company to Florence; indeed, I doubted if he could have paid the reckoning in Santa Chiara. A loan was therefore sought from a friend's friend, meaning myself.

I was very really embarrassed. Not that I would not have given willingly, for I had ample resources at the moment and was mightily concerned about the sad household. But I knew that the little Duchess would take Oliphant's ears from his head if she guessed that he had dared to borrow from me, and that, if I lent, her back would for ever be turned against me. And yet, what would follow on my refusal? In a day or two there would be a pitiful scene with mine host, and as like as not some of their baggage detained as security for payment. I did not love the task of conspiring behind the lady's back, but if it could be contrived 'twas indubitably the kindest course. I glared sternly at Oliphant, who met me with his pathetic, dog-like eyes.

'You know that your mistress would never consent to the request you have made of me?'

'I ken,' he said humbly. 'But payin' is *my* job, and I simply havena the siller. It's no' the first time it has happened, and it's a sair trial for them both to be flung out o' doors by a foreign hostler because they canna meet his charges. But, sir, if ye can lend to me, ye may be certain that her leddyship will never hear a word o't. Puir thing, she takes nae thocht o' where the siller comes frae, ony mair than the lilies o' the field.'

I became a conspirator. 'You swear, Oliphant, by all you hold sacred, to breathe nothing of this to your mistress, and if she should suspect, to lie like a Privy Councillor?'

A flicker of a smile crossed his face. 'I'll lee like a Scots packman, and the Father o' lees could do nae mair. You need have no fear for your siller, sir. I've aye repaid when I borrowed, though you may have to wait a bittock.' And the strange fellow strolled off.

At dinner no Duchess appeared till long after the appointed hour, nor was there any sign of Oliphant. When she came at last with Cristine, her eyes looked as if she had been crying, and she greeted me with remote courtesy. My first thought was that Oliphant had revealed the matter of the loan, but presently I found that the lady's trouble was far different. Her father, it seemed, was ill again with his old complaint. What that was I did not ask, nor did the Duchess reveal it.

We spoke in French, for I had discovered that this was her favourite speech. There was no Oliphant to wait on us, and the inn servants were always about, so it was well to have a tongue they did not comprehend. The lady was distracted and sad. When I enquired feelingly as to the general condition of her father's health she parried the question, and when I offered my services she disregarded my words. It was in truth a doleful meal, while the faded Cristine sat like a sphinx staring into vacancy. I spoke of England and of her friends, of Paris and Versailles, of Avignon where she had spent some years, and of the amenities of Florence, which she considered her home. But it was like talking to a nunnery door. I got nothing but 'It is indeed true, sir,' or 'Do you say so, sir?' till my energy began to sink. Madame perceived my discomfort, and, as she rose, murmured an apology. 'Pray forgive my distraction, but I am poor company when my father is ill. I have a foolish mind, easily frightened. Nay, nay!' she went on when I again offered help, 'the illness is trifling. It will pass off by tomorrow, or at the latest the next day. Only I had looked forward to some ease at Santa Chiara, and the promise is belied.'

As it chanced that evening, returning to the inn, I passed by the north side where the windows of the Count's room looked over a little flower garden abutting on the courtyard. The dusk was falling, and a lamp had been lit which gave a glimpse into the interior. The sick man was standing by the window, his figure flung into relief by the lamplight. If he was sick, his sickness was of a curious type. His face was ruddy, his eye wild, and, his wig being off, his scanty hair stood up oddly round his head. He seemed to be singing, but I could not catch the sound through the shut casement. Another figure in the room, probably Oliphant, laid a hand on the Count's shoulder, drew him from the window, and closed the shutter.

It needed only the recollection of stories which were the property of all Europe to reach a conclusion on the gentleman's illness. The legitimate King of England was very drunk.

As I went to my room that night I passed the Count's door. There stood Oliphant as sentry, more grim and haggard than ever, and I thought that his eye met mine with a certain intelligence. From inside the room came a great racket. There was the sound of glasses falling, then a string of oaths, English, French, and for all I knew, Irish, rapped out in a loud drunken voice. A pause, and then came the sound of maudlin singing. It pursued me along the gallery, an old childish song, delivered as if 'twere a pot-house catch—

> Qu'est-c' qui passe ici si tard,
> Compagnons de la Marjolaine—

One of the late-going company of the Marjolaine hastened to bed. This king in exile, with his melancholy daughter, was becoming too much for him.

III

It was just before noon next day that the travellers arrived. I was sitting in the shady loggia of the inn, reading a volume of De Thou, when there drove up to the door two coaches. Out of the first descended very slowly and stiffly four gentlemen; out of the second four servants and a quantity of baggage. As it chanced there was no one about, the courtyard slept its sunny noontide sleep, and the only movement was a lizard on the wall and a buzz of flies by the fountain. Seeing no sign of the landlord, one of the travellers approached me with a grave inclination.

'This is the inn called the Tre Croci, sir?' he asked.

I said it was, and shouted on my own account for the host. Presently that personage arrived with a red face and a short wind, having ascended rapidly from his own cellar. He was awed by the dignity of the travellers, and made none of his usual protests of incapacity. The servants filed off solemnly with the baggage, and the four gentlemen set themselves down beside me in the loggia and ordered each a modest flask of wine.

At first I took them for our countrymen, but as I watched them the conviction vanished. All four were tall and lean beyond the average of mankind. They wore suits of black, with antique starched frills to their shirts; their hair was their own and unpowdered. Massive buckles of an ancient pattern adorned their square-toed shoes, and the canes they carried were like the yards of a small vessel. They were four merchants, I had guessed, of Scotland, maybe, or of Newcastle, but their voices were not Scotch, and their air had no touch of commerce. Take the heavy-browed preoccupation of a Secretary of State, add the dignity of a bishop, the sunburn of a fox-hunter, and something of the disciplined erectness of a soldier, and you may perceive the manner of these four gentlemen. By the side of them my assurance vanished. Compared with their Olympian serenity my person seemed fussy and servile. Even so, I mused, must Mr Franklin have looked when baited in Parliament by the Tory pack. The reflection gave me the cue. Presently I caught from their conversation the word 'Washington', and the truth flashed upon me. I was in the presence of four of Mr Franklin's countrymen. Having never seen an American in the flesh, I rejoiced at the chance of enlarging my acquaintance.

They brought me into the circle by a polite question as to the length of road to Verona. Soon introductions followed. My name intrigued them, and they were eager to learn of my kinship to Uncle Charles. The eldest of the four, it appeared, was Mr Galloway out of Maryland. Then came two brothers, Sylvester by name, of Pennsylvania, and last Mr Fish, a lawyer of New York. All four had campaigned in the late war, and all four were members of the Convention, or whatever they call their rough-and-ready Parliament. They were modest in their behaviour, much disinclined to speak of their past, as great men might be whose reputation was world-wide. Somehow the names stuck in my memory. I was certain that I had heard them linked with some stalwart fight or some moving civil deed or some defiant manifesto. The making of history was in their steadfast eye and the grave lines of the mouth. Our friendship flourished mightily in a brief hour, and brought me the invitation, willingly accepted, to sit with them at dinner.

There was no sign of the Duchess or Cristine or Oliphant. Whatever had happened, that household today required all hands on deck, and I was left alone with the Americans. In my day I have supped with the Macaronies, I have held up my head at the Cocoa Tree, I have avoided the floor at hunt dinners, I have drunk glass to glass with Tom Carteron. But never before have I seen such noble consumers of good liquor as those four gentlemen from beyond the Atlantic. They drank the strong red Cyprus as if it had been spring water. 'The dust of your Italian roads takes some cleansing, Mr Townshend,' was their only excuse, but in truth none was needed. The wine seemed only to thaw their iron decorum. Without any surcease of dignity they grew communicative, and passed from lands to peoples and from peoples to constitutions. Before we knew it we were embarked upon high politics.

Naturally we did not differ on the war. Like me, they held it to have been a grievous necessity. They had no bitterness against England, only regret for her blunders. Of His Majesty they spoke with respect, of His Majesty's advisers with dignified condemnation. They thought highly of our troops in America; less highly of our generals.

'Look you, sir,' said Mr Galloway, 'in a war such as we have witnessed the Almighty is the only strategist. You fight against the forces of Nature, and a newcomer little knows that the success or failure of every operation he can conceive depends not upon generalship, but upon the conformation of a vast country. Our generals, with this in mind and with fewer men, could make all your schemes miscarry. Had the English soldiery not been of such stubborn stuff, we should have been victors from the first. Our leader was not General Washington, but General America, and his brigadiers were forests, swamps, lakes, rivers, and high mountains.'

'And now,' I said, 'having won, you have the greatest of human experiments before you. Your business is to show that the Saxon stock is adaptable to a republic.'

It seemed to me that they exchanged glances.

'We are not pedants,' said Mr Fish, 'and have no desire to dispute about the form of a constitution. A people may be as free under a king as under a senate. Liberty is not the lackey of any type of government.'

These were strange words from a member of a race whom I had thought wedded to the republicanism of Helvidius Priscus.

'As a loyal subject of a monarchy,' I said, 'I must agree with you. But your hands are tied, for I cannot picture the establishment of a House of Washington, and—if not, where are you to turn for your sovereign?'

Again a smile seemed to pass among the four.

'We are experimenters, as you say, sir, and must go slowly. In the meantime, we have an authority which keeps peace and property safe. We are at leisure to cast our eyes round and meditate on the future.'

'Then, gentlemen,' said I, 'you take an excellent way of meditation in visiting this museum of old sovereignties. Here you have the relics of any government you please—a dozen republics, tyrannies, theocracies, merchant confederations, kingdoms, and more than one empire. You have your choice. I am tolerably familiar with the land, and if I can assist you I am at your service.'

They thanked me gravely. 'We have letters,' said Mr Galloway; 'one in especial is to a gentleman whom we hope to meet in this place. Have you heard in your travels of the Count of Albany?'

'He has arrived,' said I, 'two days ago. Even now he is in the chamber above us at dinner.'

The news interested them hugely.

'You have seen him?' they cried. 'What is he like?'

'An elderly gentleman in poor health, a man who has travelled much, and, I judge, has suffered something from fortune. He has a fondness for the English, so you will be welcome, sirs; but he was indisposed yesterday, and may still be unable to receive you. His daughter travels with him and tends his old age.'

'And you—you have spoken with him?'

'The night before last I was in his company. We talked of many things, including the late war. He is somewhat of your opinion on matters of government.'

The four looked at each other, and then Mr Galloway rose.

'I ask your permission, Mr Townshend, to consult for a moment with my friends. The matter is of some importance, and I would beg you to await us.' So saying, he led the others out of doors, and I heard them withdraw to a corner of the loggia. Now, thought I, there is something afoot, and my long-sought romance approaches fruition. The company of the Marjolaine, whom the Count had sung of, have arrived at last.

Presently they returned and seated themselves at the table.

'You can be of great assistance to us, Mr Townshend, and we would fain take you into our confidence. Are you aware who is this Count of Albany?'

I nodded. 'It is a thin disguise to one familiar with history.'

'Have you reached any estimate of his character or capabilities? You speak to friends, and, let me tell you, it is a matter which deeply concerns the Count's interests.'

'I think him a kindly and pathetic old gentleman. He naturally bears the mark of forty years' sojourn in the wilderness.'

Mr Galloway took snuff.

'We have business with him, but it is business which stands in need of an agent. There is no one in the Count's suite with whom we could discuss affairs?'

'There is his daughter.'

'Ah, but she would scarcely suit the case. Is there no man—a friend, and yet not a member of the family, who can treat with us?'

I replied that I thought that I was the only being in Santa Chiara who answered the description.

'If you will accept the task, Mr Townshend, you are amply qualified. We will be frank with you and reveal our business. We are on no less an errand than to offer the Count of Albany a crown.'

I suppose I must have had some suspicion of their purpose, and yet the revelation of it fell on me like a thunderclap. I could only stare owlishly at my four grave gentlemen.

Mr Galloway went on unperturbed. 'I have told you that in America we are not yet republicans. There are those among us who favour a republic, but they are by no means a majority. We have got rid of a king who misgoverned us, but we have no wish to get rid of kingship. We want a king of our own choosing, and we would get with him all the ancient sanctions of monarchy. The Count of Albany is of the most illustrious royal stock in Europe—he is, if legitimacy goes for anything, the rightful King of Britain. Now, if the republican party among us is to be worsted, we must come before the nation with a powerful candidate for its favour. You perceive my drift? What more potent appeal to American pride than to say: "We have got rid of King George; we choose of our own free will the older line and King Charles"?'

I said foolishly that I thought monarchy had had its day, and that 'twas idle to revive it.

'That is a sentiment well enough under a monarchical government; but we, with a clean page to write upon, do not share it. You know your ancient historians. Has not the repository of the chief power always been the rock on which republicanism has shipwrecked? If that power is given to the chief citizen, the way is prepared for the tyrant. If it abides peacefully in a royal house, it abides with cyphers who dignify, without obstructing, a popular constitution. Do not mistake me, Mr Townshend. This is no whim of a sentimental girl, but the reasoned conclusion of the men who achieved our liberty. There is every reason to believe that General Washington shares our views, and Mr Hamilton, whose name you may know, is the inspirer of our mission.'

'But the Count is an old man,' I urged; for I knew not where to begin in my exposition of the hopelessness of their errand.

'By so much the better. We do not wish a young king who may be fractious. An old man tempered by misfortune is what our purpose demands.'

'He has also his failings. A man cannot lead his life for forty years and retain all the virtues.'

At that one of the Sylvesters spoke sharply. 'I have heard such gossip, but I do not credit it. I have not forgotten Preston and Derby.'

I made my last objection. 'He has no posterity—legitimate posterity— to carry on his line.'

The four gentlemen smiled. 'That happens to be his chiefest recommendation,' said Mr Galloway. 'It enables us to take the House of Stuart on trial. We need a breathing-space and leisure to look around; but unless we establish the principle of monarchy at once the republicans will forestall us. Let us get our king at all costs, and during the remaining years of his life we shall have time to settle the succession problem. We have no wish to saddle ourselves for good with a race who might prove burdensome. If King Charles fails he has no son, and we can look elsewhere for a better monarch. You perceive the reason of my view?'

I did, and I also perceived the colossal absurdity of the whole business. But I could not convince them of it, for they met my objections with excellent arguments. Nothing save a sight of the Count would, I feared, disillusion them.

'You wish me to make this proposal on your behalf?' I asked.

'We shall make the proposal ourselves, but we desire you to prepare the way for us. He is an elderly man, and should first be informed of our purpose.'

'There is one person whom I beg leave to consult—the Duchess, his daughter. It may be that the present is an ill moment for approaching the Count, and the affair requires her sanction.'

They agreed, and with a very perplexed mind I went forth to seek the lady. The irony of the thing was too cruel, and my heart ached for her. In the gallery I found Oliphant packing some very shabby trunks, and when I questioned him he told me that the family were to leave Santa

Chiara on the morrow. Perchance the Duchess had awakened to the true state of their exchequer, or perchance she thought it well to get her father on the road again as a cure for his ailment.

I discovered Cristine, and begged for an interview with her mistress on an urgent matter. She led me to the Duchess's room, and there the evidence of poverty greeted me openly. All the little luxuries of the menage had gone to the Count. The poor lady's room was no better than a servant's garret, and the lady herself sat stitching a rent in a travelling cloak. She rose to greet me with alarm in her eyes.

As briefly as I could I set out the facts of my amazing mission. At first she seemed scarcely to hear me. 'What do they want with him?' she asked. 'He can give them nothing. He is no friend to the Americans or to any people who have deposed their sovereign.' Then, as she grasped my meaning, her face flushed.

'It is a heartless trick, Mr Townshend. I would fain think you no party to it.'

'Believe me, dear madame, it is no trick. The men below are in sober earnest. You have but to see their faces to know that theirs is no wild adventure. I believe sincerely that they have the power to implement their promise.'

'But it is madness. He is old and worn and sick. His day is long past for winning a crown.'

'All this I have said, but it does not move them.' And I told her rapidly Mr Galloway's argument.

She fell into a muse. 'At the eleventh hour! Nay, too late, too late. Had he been twenty years younger, what a stroke of fortune! Fate bears too hard on us, too hard!'

Then she turned to me fiercely. 'You have no doubt heard, sir, the gossip about my father, which is on the lips of every fool in Europe. Let us have done with this pitiful make-believe. My father is a sot. Nay, I do not blame him. I blame his enemies and his miserable destiny. But there is the fact. Were he not old, he would still be unfit to grasp a crown and rule over a turbulent people. He flees from one city to another, but he cannot flee from himself. That is his illness on which you condoled with me yesterday.'

The lady's control was at breaking-point. Another moment and I expected a torrent of tears. But they did not come. With a great effort she regained her composure.

'Well, the gentlemen must have an answer. You will tell them that the Count, my father—nay, give him his true title if you care—is vastly obliged to them for the honour they have done him, but would decline on account of his age and infirmities. You know how to phrase a decent refusal.'

'Pardon me,' said I, 'but I might give them that answer till doomsday and never content them. They have not travelled many thousand miles to be put off by hearsay evidence. Nothing will satisfy them but an interview with your father himself.'

'It is impossible,' she said sharply.

'Then we must expect the renewed attentions of our American friends. They will wait till they see him.'

She rose and paced the room.

'They must go,' she repeated many times. 'If they see him sober he will accept with joy, and we shall be the laughing-stock of the world. I tell you it cannot be. I alone know how immense is the impossibility. He cannot afford to lose the last rags of his dignity, the last dregs of his ease. They must not see him. I will speak with them myself.'

'They will be honoured, madame, but I do not think they will be convinced. They are what we call in my land "men of business". They will not be content till they get the Count's reply from his own lips.'

A new Duchess seemed to have arisen, a woman of quick action and sharp words.

'So be it. They shall see him. Oh, I am sick to death of fine sentiments and high loyalty and all the vapouring stuff I have lived among for years. All I ask for myself and my father is a little peace, and, by Heaven! I shall secure it. If nothing will kill your gentlemen's folly but truth, why, truth they shall have. They shall see my father, and this very minute. Bring them up, Mr Townshend, and usher them into the presence of the rightful King of England. You will find him alone.' She stopped her walk and looked out of the window.

I went back in a hurry to the Americans. 'I am bidden to bring you to the Count's chamber. He is alone and will see you. These are the commands of madame his daughter.'

'Good!' said Mr Galloway, and all four, grave gentlemen as they were, seemed to brace themselves to a special dignity as befitted ambassadors to a king. I led them upstairs, tapped at the Count's door, and, getting no answer, opened it and admitted them.

And this was what we saw. The furniture was in disorder, and on a couch lay an old man sleeping a heavy drunken sleep. His mouth was open and his breath came stertorously. The face was purple, and large purple veins stood out on the mottled forehead. His scanty white hair was draggled over his cheek. On the floor was a broken glass, wet stains still lay on the boards, and the place reeked of spirits.

The four looked for a second—I do not think longer—at him whom they would have made their king. They did not look at each other. With one accord they moved out, and Mr Fish, who was last, closed the door very gently behind him.

In the hall below Mr Galloway turned to me. 'Our mission is ended, Mr Townshend. I have to thank you for your courtesy.' Then to the others, 'If we order the coaches now, we may get well on the way to Verona ere sundown.'

    .   .   .

An hour later two coaches rolled out of the courtyard of the Tre Croci. As they passed, a window was half-opened on the upper floor, and a head looked out. A line of a song came down, a song sung in a strange quavering voice. It was the catch I had heard the night before:

> Qu'est-c' qui passe ici si tard,
> Compagnons de la Marjolaine—e?

It was true. The company came late indeed—too late by forty years. . . .

EMMA DONOGHUE

Words for Things

The day before the governess came was even longer. Over a dish of cooling tea, Margaret watched her mother. Not the eyes, but the stiff powdery sweep of hair. She answered two questions—on the progress of her cross-stitch, and a French proverb—but missed the final one, on the origin of the word 'October'. Swallowing the tea noiselessly, Margaret allowed her eyes to unlatch the window, creep across the lawn. She thought she could smell another thatch singeing.

The next morning woke her breathless; one rib burned under the weight of whalebone. The dark was lifting reluctantly, an inch of wall at a time. Practised at distracting herself, Margaret reached down with one hand. She scrabbled under the mattress edge for the buckled volume. But it was gone, as if absorbed into the feathers. Confiscated on her mother's orders, no doubt. Clamping her eyes shut, Margaret focused on the rib, bending her anger into a manageable line. She lay flat until the room was full of faint light that snagged on the shapes of two small girls in the next bed.

Her belly rumbled. Margaret was hungry for words. None on the walls, except an edifying motto in cross-stitch hung over the ewer. Curling patterns on the curtains could sometimes be suggested into letters and then acrobatic words, but by now the light was too honest. She shut her eyes again, and called up a grey, wavering page with an ornate printer's mark at the top. 'The History of the Primdingle Family,' she spelled out, 'Part the Fifth.' Once she worked her way into the flow, she no longer needed to imagine the letters into existence one by one; the lines formed themselves, neat and crisp and believable. Her eyes flickered under their lids, scattering punctuation.

The black trunk sat in the hall, its brass worn at the edges. Dot caught her winter petticoat on it as she scuttled by. The governess was in the parlour, sipping cold tea. Mistress Mary, her employer called her; it was to be understood that the Irish preferred this traditional form of address, and besides, it avoided the outlandish surname. Her Ladyship showed no interest in wages, nor in the little school Mistress Mary ran in London, nor in her recent treatise on female education. Her Ladyship's questions sounded like statements. She outlined the children's day, hour by hour.

They had been let run wild too long, and now it was a race to make the eldest presentable for Dublin Castle in a bare two years. The girl was somewhat perverse, her Ladyship mentioned over the silver teapot, and seemed to be growing larger by the day.

Mistress Mary watched a minute grain of powder from her Ladyship's widow's peak drift down and alight on the surface of the tea. She had been here one hour and felt light with fatigue already. The three children were the kind of hoydens she liked least, the fourteen-year-old in particular having an unrestrained guffaw certain to set on edge the nerves of any potential suitors. The governess asked herself again why she had exiled herself among the wild Irish rather than scour pots for a living. But your mother was a native of this country, was she not, Mistress Mary? You are half one of us, then. Oh, your Ladyship, I would not presume. But that bony voice did remind the governess of her mother's limper tones. Bending her head over the tea, Mistress Mary heard in her gut the usual battle between gall and compassion.

Behind an oak, Margaret was shivering as she nipped her muslin skirts between her knees. If she stood narrow as a sapling she would not be seen. The outraged words of two languages carried across the field, equally indistinguishable. Dot would carry the news later: who said what, which of the usual threats and three-generation curses were made, which fists shaken in which faces. It had to be time for dinner, Margaret thought. She would go when the smoke rose white as feathers from the second thatch.

Stand up straight, her mother told her. You have been telling your sisters wicked make-believe once more. How can I persuade you of the difference between what is real and what is not?

I do not know madam.

You will run mad before the age of sixteen and then I will be spared the trouble of finding a husband for you.

Yes madam.

Have you forgotten who you are, girl?

Margaret King of the family of Lord and Lady Kingsborough of Kingsborough Estate.

Of which county?

Of the county of Cork in the kingdom of Ireland in the year of our Lord one thousand seven hundred and eighty six.

Now go and wash your face so your new governess will not think you a peasant.

In November the evictions were more plentiful, and Margaret wearied of them. The apple trees stooped under their cloaks of rain. Mistress Mary had been here three weeks. She and the girls were kept busy all day from half past six to half past five with a list of nonsensical duties. So-called accomplishments being in her view those things which were never fully accomplished. Mistress Mary kept biting her soft lips and thinking of Lisbon. Between them the children churned out acres of lace, lists of the tributaries of the Nile, piles of netted purses, and an assortment of complaints from violin, flute, and harpsichord. The two small girls could sing five songs in French without understanding any of the words, and frequently did. The harpsichord was often silent, on days when Margaret, blankfaced and docile, slipped away with a message for the cook and was not seen again.

Sometimes, losing herself along windy corridors, her air of calm efficiency beginning to slip, Mistress Mary caught sight of a long booted ankle disappearing round a corner. Dot denied all knowledge. On the first occasion, the matter was mentioned to her Ladyship. Bruises slowed Margaret's walk for a week, though no one referred to them. After that, Mistress Mary kept silent about her pupil's comings and goings. I will conquer her with kindness, she promised herself. Tenderness will lead where birch cannot drive.

Need the girl sleep in her stays, your Ladyship? She heaves so alarmingly at night when I look in on her. As your Ladyship wishes.

Considering her governess's animated face bent over a letter, Margaret decided that here was one doll worth playing with. She knew how to do it. They always began stiff and proper but soon they went soft over you, and then every smile pulled their strings. Mostly they were lonely and despised themselves for not being mothers. The Mademoiselle in her last boarding-school was the easiest. Watch for the first signs—vague laughter, fingers against your cheek, a fuzziness about the hairline—and seize on her weakness. Ask her to help you tie the last bow of ribbon on your stomacher. Take her arm in yours while walking, beg to sit next to her at supper, even bring her apples if the case requires it. Mademoiselle, the poor toad, had reached the stage of hiding pears in Margaret's desk.

By December, Mistress Mary was astonished at her power over this sweet little girl. The dreadful laugh had muted to a wheeze of merriment between wide lips. Margaret sat on a footstool below her governess's needle, and chanted French songs of which she understood half the words. At night, she had taken to pleading for Mistress Mary to come into her bedchamber; not Dot, nor any of the servingmaids, nor either of her sisters, only Mistress Mary might lift off the muslin cap and brush Margaret's long coarse hair. My little friend, the woman called her, as a final favour, though when Margaret stood up she had a good inch over her governess. They shared a smile, then. Mistress Mary would have liked to be sure that they were smiling at the same thing.

Another thing the governess could not understand is why Margaret loved to read and hated to write.

Words were a treasure to be hoarded and never shared.

On Margaret's twelfth birthday, an event marked only by one of her Ladyship's sudden visits to the schoolroom, she had been discovered sitting under a desk with a very blunt quill, writing 'The History of the Quintumbly Family, The Third Chapter'. Under interrogation in the parlour, Margaret could offer no reason for such an outlay of precious time. She licked the corner of her lips. Nor could she explain her knowledge of such an unsuitable family, who, it seemed, kept tame weasels and sailed down the Nile. Her Ladyship was digusted at last to find that the Quintumblys were merely fanciful, a pretend family. The rest of the journal pleased her even less, being a daily account of Margaret's less filial thoughts and sentiments. Having read a sample of them aloud in a tone of wonder, her Ladyship picked the limp book up by one corner and held it out to her daughter. Margaret was halfway to the parlour fire before her mother's voice tightened around her: did she mean to smoke them out entirely? Dot was called for to carry the manuscript to the gardener's bonfire, where it would do no harm.

From that day on the girl would write no words of her own, only lists of French verbs and English wars. In the strongbox behind her eyes, she stacked volumes of stories about her pretend families. She sealed her lashes and reread the adventures only in bed before it was light, in case any-one might catch her lips moving. By day she stole other people's words: romances, newspapers, treatises, anything left beside an armchair or in an unlocked cabinet in the library. Margaret swallowed up the words and would give none back.

Laid low by one of her periodic fevers in January, the girl was starving for a story but could not concentrate enough to form the letters. What she did remember to do was to clamour for Mistress Mary to hold her hand. The governess flushed, and consented, after a little show of reluctance. Humouring the patient, she agreed to petty things, like brushing her own dark hair with Margaret's ivory brush. By suppertime, the girl was too worn out to think of anything endearing to say, but Mistress Mary leaned over and hushed her most tenderly.

In her sleep, the girl made hoarse cries, and threw the gentlest hand off her forehead. There was a girl in Dublin used to sleep in her stays, Dot remarked, that died from compunction of the organs. The governess heaved a breath and knocked on the parlour door. Most dangerous in her state of health, your Ladyship, eminent medical gentlemen agree. The stays came off. Margaret tossed still, picking at invisible ribbons across her chest.

The girl woke one February twilight to find that her lie had come true. She could not bear to let the governess out of her sight. Her puzzled eyes followed every movement, and her voice was cracked and fractious. She insisted that she could not remember how to sew. There was nothing comfortable about this love.

Scrawny children plucked at Margaret's skirts as she walked between the burnt cottages, wheezing. She shared her pocketful of French grapes among them. Their ginger freckles stood out bold on transparent skin.

Hungry for the familiarity of words, the girl stole into the library. She knelt on the moving steps and pressed her face against the glass cases, following their bevelled edges with her lips. One of the cases would be left unlocked, if she had prayed hard enough the night before. Titles in winking gold leaf reached for her fingers. At first she looked for story-books and engravings, but by March she had got a taste for books of words about words: dictionaries and lexicons and medical encyclopedias. One strange fish of a word leapt into the mouth of another and that one into another, meanings hooked on each other, confusing and enticing her, until, after the hour or so she could steal from each day, Margaret was netted round with secret knowledge.

There was a farmer went for the bailiff with a pitchfork last month, Dot said. When they hanged him in Cork town his bit stuck up. Margaret knew about bits and the getting of babies and the nine months; Mistress Mary, flushing slightly, said every girl should know the words for things.

Why was I not born a boy, Margaret asked her governess while walking in the orchard, or why was I born at all? Mistress Mary was bewildered by the question. Margaret explained: girls are good for nothing in particular. In all the stories, boys can run and leap and save wounded animals. My mother says I am a mannish little trull. Already I am taller than ladies like you. So why may I not be a boy?

There is nothing wrong, began Mistress Mary cautiously, with being manly, in the best sense. Manly virtues, you know, and masculine fortitude. You must not be afraid. No matter what anyone says. Even if they say things which, no doubt unintentionally, may seem unkind.

Margaret kicked at a rotten apple.

You must stand tall, like a tree, explained the governess, gathering confidence. No matter how tall you grow you will be my little girl, and your head will always fit on my shoulder. Tall as a young tree, she went on hurriedly, and you should move like one too. Why do you not romp and bound when I say you may; why do you cling to my side like a little doll?

My mother forbids it.

She cannot see through the garden wall. I give you permission.

She may ask why we spend so long in the orchard.

We are studying the names of the insects.

And the governess tagged her on the shoulder and, picking up her skirts, hared off down a damp grassy path. Margaret was still considering the matter when she found her legs leaping away with her.

In the April evenings, Mistress Mary entertained the household with recitations and English country dances. Her Ladyship looked on, her hair whiter by the day. They argued over the number of buttons on the girls' boots.

When she grew up, Margaret had decided, she would make the bailiffs give all the rent back. The redheaded children would grow fat, and clap their hands when they saw her coming.

By May the air was white with blossom. Margaret could not swallow when she looked at her governess. Their hairs were mingled on the brush; Margaret teased them with her cold fingers. She could not seem to learn the rules of bodies. What Mistress Mary called innocent caresses were allowable, and these were: cheeks and foreheads, lips on the backs of hands, arms entwined in the orchard, heads briefly nested in grey satin skirts. But when Margaret slipped into the governess's chamber one morning and found her in shift and stockings, she earned a scathing glance. Gross familiarity, Mistress Mary called it, and immodest forwardness. When she had fastened her last button and called the girl back into the room, she explained more calmly that one must never forget the respect which one human creature owed another. Margaret could not see what respect had to do with dressing in separate rooms. She hung her head and thought of breakfast. Had any of the servants ever tried to teach her dirty indecent tricks? No mistress, Dot was always busy, and besides, she knew no tricks.

It was June by the time she found that there were words for girls like her. Words tucked away in the library, locked only until you looked for them. Romp and hoyden she knew already. Tomboy was when she ran down the front staircase with her bootlaces undone. But there were sharper words as well, words that cut when she lifted them into her mouth to taste and whisper them. Tommie was when women kissed and pressed each other to their hearts, it said so in a dirty poem on the top shelf of the cabinet. Tribad was the same only worse. The word had to mean, she reasoned, along the lines of triangle and trimester, that she was three times as bad as other girls.

Margaret knelt up on the moving steps as the page fell open to that word again; her legs shook and her belly-rumbles echoed under the whalebone. Tribad meant if she let the badness take her, she would grow and grow. Already she was taller than anyone in the house except her mother. The book said she would grow down there until she became a hermaphrodite shown for pennies at the fair, or ran away in her brother's breeches (but she had no brother) and married a Dutch widow. The change was coming already. When the girl lay in bed on hot mornings the bit between her legs stirred and leapt like a minnow.

One noon she limped into the bedchamber, phantom blows from her mother's rod still landing on her calves. Dot was sweeping the cold floor, her broom trailing now as she gazed into the frontispiece of a book of travels.

Give it, said Margaret.

Dot regarded her, then stared at the book again, at its pages flattened by the grey morning light. She looked back at the girl as if trying to remember her name. Seizing the besom. Margaret threaded her fingers between the twigs, and set to bludgeoning the maid's thick body with the handle. The coarse petticoats dulled the impact; it sounded like a rug being beaten. She pursued Dot to the window with a constant hiss of phrases, from idle ignoramus to tell my mother to dirty good-for-nothing inch of life. Dot broke into a wail at last, expressive less of pain than of a willingness to get it over and done with. She stood in the corner, hunched over to protect her curves. Tears plummeted to the floorboards.

Beg pardon, Margaret instructed. Her ribs heaved and sank under the creaking corset.

Beg pardon miss, Dot repeated, her tone neutral.

The broom was lowered but the eyes held.

Margaret had made it to the door before, with a lurch, she found herself sorry. She turned to see Dot industriously brushing her tears into the floor. She was so sorry it swamped her, left her feeble. Was there any comfort to give? A lump in one of the unmade beds reminded her, and she scrabbled under the coverlet. The doll she pulled out was missing one eye, but her pink damask skippers were good as new. The girl walked up behind Dot and tapped her on the shoulder with the doll's powdered head. Take her, she said graciously, and leave off crying.

Dot turns a face that was almost dry. What am I to do with that, miss?

Margaret was disconcerted. Play with her, she could have said, but when? Dress her up, but in what? A shaky, benevolent smile. Perhaps you could beat her when you are angry.

Weary, Dot considered the two faces. Get away out of that, miss, she remarked at last, and walked from the room, trailing her broom behind her.

It was in the corner of the bedchamber that the governess found the girl a little later, her fingers dividing and dividing the doll's hair. She lifted Margaret's hands away gently. Girl, you harass my spirits. You are too old for your sisters' dolls, and what need has a healthy girl of wax toys when there is the wide world to play in? If she noticed the stiffness in

Margaret's legs, as they strolled in the orchard, she said nothing. They spoke of birds' nests, and poetry, and unrest among the French.

Mistress Mary had taken to writing a story in the bright July evenings. What was it about? Disappointment, she murmured, and would tell no more. Feeling neglected, Margaret became clumsier, tripping over shoes and toppling an inkstand. The governess forgave her everything. One morning Margaret found a double cherry hung over the handle of her wardrobe. She knew she had the power now. It brought her no joy.

I govern her completely, Mistress Mary wrote to her sister. She is a fine girl, and it only takes a cherry to win a smile from her. Her violence of temper remains deplorable, but I myself never feel the effects of it. She is wax in my hands. The truth is, this girl is the only consolation of my life in this backwater. How I look forward to my brief escape!

Nobody remembered to tell the children that their governess was spending two days with acquaintances in Tralee. Distracted by the details of mailcoaches and hats, Mistress Mary was gone before breakfast. Dot, passing the girl on the back staircase, had only time to whisper that the governess was gone.

Margaret stood in the middle of the empty bedchamber. Sure enough, Mistress Mary's travelling cloak was missing for the first time since October. Margaret was oddly calm. Her mind was busy wondering what she had done wrong, what brief immodesty or careless phrase would make her governess punish her so, by leaving without a word. She noticed that the writing case had been left behind. No reason not to, now: she wrenched it open and took a handful of pages. 'Pity is one of my prevailing passions', she read, and 'this world is a desert to me' at the top of another leaf. For a few moments the girl stood, savouring the grandeur of the phrases. But then they were dust in her mouth. All these words, and not an inch of warm skin left. As if Mistress Mary, who had never seemed too fond of having a body, had escaped in the form of a bird or a cloud.

The words were building up behind her tongue, making her gag. Nine months she has been living behind my hair, thought Margaret; that is as long as a baby. She parted her lips to breathe and a howl split her open.

After that she remembered nothing until her mother was standing over her.

Stop this fuss, her Ladyship advised. You are making a grand calamity out of nothing at all. Recollect yourself. Who are you?

I don't know.

You are Margaret King of . . .?

I don't remember.

The girl stood, at the rod's pleasure. It beat and beat and could not touch her.

Due to the excessive regret the girl had shown at the briefest of partings with her governess, her Ladyship explained to the household, she had decided that Mistress Mary would not be coming back. The black trunk was sent off before breakfast.

By August, Margaret was bleeding inside. Feeling herself seep away, she was not surprised. But Dot saw the red path down the girl's stocking; she took her into a closet and explained the business of the rags. Margaret nodded but did not believe her. She knew it was the first sign of the change. Blood had to trickle as the growth sped and the new freakish flesh pushed through. When she was three inches long, she would run away to Galway fair and show herself for sixpences. The pretend families would come with her, riding in the ropes of her hair.

A republican in 1798, Margaret would spit at bailiffs. An adulteress in Italy, she would meet her governess's daughter, who never knew her mother, and would tell her, I knew your mother.

BARONESS ORCZY

Two Good Patriots

ΩΩ

Being the deposition of citizeness Fanny Roussell, who was brought up, together with her husband, before the Tribunal of the Revolution on a charge of treason—both being subsequently acquitted.

My name is Fanny Roussell, and I am a respectable married woman, and as good a patriot as any of you sitting there.

Aye, and I'll say it with my dying breath, though you may send me to the guillotine . . . as you probably will, for you are all thieves and murderers, every one of you, and you have already made up your minds that I and my man are guilty of having sheltered that accursed Englishman whom they call the Scarlet Pimpernel . . . and of having helped him to escape.

But I'll tell you how it all happened, because, though you call me a traitor to the people of France, yet am I a true patriot and will prove it to you by telling you exactly how everything occurred, so that you may be on your guard against the cleverness of that man, who, I do believe, is a friend and confederate of the devil . . . else how could he have escaped that time?

Well! it was three days ago, and as bitterly cold as anything that my man and I can remember. We had no travellers staying in the house, for we are a good three leagues out of Calais, and too far for the folk who have business in or about the harbour. Only at midday the coffee-room would get full sometimes with people on their way to or from the port.

But in the evenings the place was quite deserted, and so lonely that at times we fancied that we could hear the wolves howling in the forest of St Pierre.

It was close on eight o'clock, and my man was putting up the shutters, when suddenly we heard the tramp of feet on the road outside, and then the quick word, 'Halt!'

The next moment there was a peremptory knock at the door. My man opened it, and there stood four men in the uniform of the 9th Regiment of the Line . . . the same that is quartered at Calais. The uniform, of course, I knew well, though I did not know the men by sight.

'In the name of the People and by the order of the Committee of

Public Safety!' said one of the men, who stood in the forefront, and who, I noticed, had a corporal's stripe on his left sleeve.

He held out a paper, which was covered with seals and with writing, but as neither my man nor I can read, it was no use our looking at it.

Hercule—that is my husband's name, citizens—asked the corporal what the Committee of Public Safety wanted with us poor *hôteliers* of a wayside inn.

'Only food and shelter for tonight for me and my men,' replied the corporal, quite civilly.

'You can rest here,' said Hercule, and he pointed to the benches in the coffee-room, 'and if there is any soup left in the stockpot, you are welcome to it.'

Hercule, you see, is a good patriot, and he had been a soldier in his day. . . . No! no . . . do not interrupt me any of you . . . you would only be saying that I ought to have known . . . but listen to the end.

'The soup we'll gladly eat,' said the corporal very pleasantly. 'As for shelter . . . well! I am afraid that this nice warm coffee-room will not exactly serve our purpose. We want a place where we can lie hidden, and at the same time keep a watch on the road. I noticed an outhouse as we came. By your leave we will sleep in there.'

'As you please,' said my man curtly.

He frowned as he said this, and it suddenly seemed as if some vague suspicion had crept into Hercule's mind.

The corporal, however, appeared unaware of this, for he went on quite cheerfully:

'Ah! that is excellent! *Entre nous*, citizen, my men and I have a desperate customer to deal with. I'll not mention his name, for I see you have guessed it already. A small red flower, what? . . . Well, we know that he must be making straight for the port of Calais, for he has been traced through St Omer and Ardres. But he cannot possibly enter Calais city tonight, for we are on the watch for him. He must seek shelter somewhere for himself and any other aristocrat he may have with him, and, bar this house, there is no other place between Ardres and Calais where he can get it. The night is bitterly cold, with a snow blizzard raging round. I and my men have been detailed to watch this road, other patrols are guarding those that lead toward Boulogne and to Gravelines; but I have an idea, citizen, that our fox is making for Calais, and that to me will fall the honour of handing that tiresome scarlet flower to the Public Prosecutor *en route* for Madame la Guillotine.'

Now I could not really tell you, citizens, what suspicions had by this time entered Hercule's head or mine; certainly what suspicions we did have were still very vague.

I prepared the soup for the men and they ate it heartily, after which my husband led the way to the outhouse where we sometimes stabled a traveller's horse when the need arose.

It is nice and dry, and always filled with warm, fresh straw. The entrance into it immediately faces the road; the corporal declared that nothing would suit him and his men better.

They retired to rest apparently, but we noticed that two men remained on the watch just inside the entrance, whilst the two others curled up in the straw.

Hercule put out the lights in the coffee-room, and then he and I went upstairs—not to bed, mind you—but to have a quiet talk together over the events of the past half-hour.

The result of our talk was that ten minutes later my man quietly stole downstairs and out of the house. He did not, however, go out by the front door, but through a back way which, leading through a cabbage-patch and then across a field, cuts into the main road some two hundred metres higher up.

Hercule and I had decided that he would walk the three leagues into Calais, despite the cold, which was intense, and the blizzard, which was nearly blinding, and that he would call at the post of *gendarmerie* at the city gates, and there see the officer in command and tell him the exact state of the case. It would then be for that officer to decide what was to be done; our responsibility as loyal citizens would be completely covered.

Hercule, you must know, had just emerged from our cabbage-patch on to the field when he was suddenly challenged:

'*Qui va là?*'

He gave his name. His certificate of citizenship was in his pocket; he had nothing to fear. Through the darkness and the veil of snow he had discerned a small group of men wearing the uniform of the 9th Regiment of the Line.

'Four men,' said the foremost of these, speaking quickly and commandingly, 'wearing the same uniform that I and my men are wearing . . . have you seen them?'

'Yes,' said Hercule hurriedly.

'Where are they?'

'In the outhouse close by.'

The other suppressed a cry of triumph.

'At them, my men!' he said in a whisper, 'and you, citizen, thank your stars that we have not come too late.'

'These men . . .' whispered Hercule. 'I had my suspicions.'

'Aristocrats, citizen,' rejoined the commander of the little party, 'and one of them is that cursed Englishman—the Scarlet Pimpernel.'

Already the soldiers, closely followed by Hercule, had made their way through our cabbage-patch back to the house.

The next moment they had made a bold dash for the barn. There was a great deal of shouting, a great deal of swearing and some firing, whilst Hercule and I, not a little frightened, remained in the coffee-room, anxiously awaiting events.

Presently the group of soldiers returned, not the ones who had first come, but the others. I noticed their leader, who seemed to be exceptionally tall.

He looked very cheerful, and laughed loudly as he entered the coffee-room. From the moment that I looked at his face I knew, somehow, that Hercule and I had been fooled, and that now, indeed, we stood eye to eye with that mysterious personage who is called the Scarlet Pimpernel.

I screamed, and Hercule made a dash for the door; but what could two humble and peaceful citizens do against this band of desperate men, who held their lives in their own hands? They were four and we were two, and I do believe that their leader has supernatural strength and power.

He treated us quite kindly, even though he ordered his followers to bind us down to our bed upstairs, and to tie a cloth round our mouths so that our cries could not be distinctly heard.

Neither my man nor I closed an eye all night, of course, but we heard the miscreants moving about in the coffee-room below. But they did no mischief, nor did they steal any of the food or wines.

At daybreak we heard them going out by the front door, and their footsteps disappearing toward Calais. We found their discarded uniforms lying in the coffee-room. They must have entered Calais by daylight, when the gates were opened—just like other peaceable citizens. No doubt they had forged passports, just as they had stolen uniforms.

Our maid-of-all-work released us from our terrible position in the course of the morning, and we released the soldiers of the 9th Regiment of the Line, whom we found bound and gagged, some of them wounded, in the outhouse.

That same afternoon we were arrested, and here we are, ready to die if we must, but I swear that I have told you the truth, and I ask you, in the name of justice, if we have done anything wrong, and if we did not act like loyal and true citizens, even though we were pitted against an emissary of the devil?

THOMAS HARDY

A Committee Man of 'The Terror'

ᘓᙣᘐ

We had been talking of the Georgian glories of our old-fashioned watering-place, which now, with its substantial russet-red and dun brick buildings in the style of the year eighteen hundred, looks like one side of a Soho or Bloomsbury Street transported to the shore, and draws a smile from the modern tourist who has no eye for solidity of build. The writer, quite a youth, was present merely as a listener. The conversation proceeded from general subjects to particular, until old Mrs H—, whose memory was as perfect at eighty as it had ever been in her life, interested us all by the obvious fidelity with which she repeated a story many times related to her by her mother when our aged friend was a girl—a domestic drama much affecting the life of an acquaintance of her said parent, one Mademoiselle V—, a teacher of French. The incidents occurred in the town during the heyday of its fortunes, at the time of our brief peace with France in 1802–3.

'I wrote it down in the shape of a story some years ago, just after my mother's death,' said Mrs H—. 'It is locked up in my desk there now.'

'Read it!' said we.

'No,' said she; 'the light is bad, and I can remember it well enough, word for word, flourishes and all.' We could not be choosers in the circumstances, and she began.

'There are two in it, of course, the man and the woman, and it was on an evening in September that she first got to know him. There had not been such a grand gathering on the Esplanade all the season. His Majesty King George the Third was present, with all the princesses and royal dukes, while upwards of three hundred of the general nobility and other persons of distinction were also in the town at the time. Carriages and other conveyances were arriving every minute from London and elsewhere; and when among the rest a shabby stage-coach came in by a by-route along the coast from Havenpool, and drew up at a second-rate tavern, it attracted comparatively little notice.

'From this dusty vehicle a man alighted, left his small quantity of luggage temporarily at the office, and walked along the street as if to look for lodgings.

'He was about forty-five—possibly fifty—and wore a long coat of faded superfine cloth, with a heavy collar, and a bunched-up neckcloth. He seemed to desire obscurity.

'But the display appeared presently to strike him, and he asked of a rustic he met in the street what was going on; his accent being that of one to whom English pronunciation was difficult.

'The countryman looked at him with a slight surprise, and said, "King Jarge is here and his royal Cwort."

'The stranger enquired if they were going to stay long.

'"Don't know, Sir. Same as they always do, I suppose."

'"How long is that?"

'"Till some time in October. They've come here every zummer since eighty-nine."

'The stranger moved onward down St Thomas Street, and approached the bridge over the harbour backwater, that then, as now, connected the old town with the more modern portion. The spot was swept with the rays of a low sun, which lit up the harbour lengthwise, and shone under the brim of the man's hat and into his eyes as he looked westward. Against the radiance figures were crossing in the opposite direction to his own; among them this lady of my mother's later acquaintance, Mademoiselle V—. She was the daughter of a good old French family, and at that date a pale woman, twenty-eight or thirty years of age, tall and elegant in figure, but plainly dressed and wearing that evening (she said) a small muslin shawl crossed over the bosom in the fashion of the time, and tied behind.

'At sight of his face, which, as she used to tell us, was unusually distinct in the peering sunlight, she could not help giving a little shriek of horror, for a terrible reason connected with her history, and after walking a few steps further, she sank down against the parapet of the bridge in a fainting fit.

'In his preoccupation the foreign gentleman had hardly noticed her, but her strange collapse immediately attracted his attention. He quickly crossed the carriageway, picked her up, and carried her into the first shop adjoining the bridge, explaining that she was a lady who had been taken ill outside.

'She soon revived; but, clearly much puzzled, her helper perceived that she still had a dread of him which was sufficient to hinder her complete recovery of self-command. She spoke in a quick and nervous way to the shopkeeper, asking him to call a coach.

'This the shopkeeper did, Mademoiselle V— and the stranger remaining in constrained silence while he was gone. The coach came up, and giving the man the address, she entered it and drove away.

'"Who is that lady?" said the newly arrived gentleman.

'"She's of your nation, as I should make bold to suppose," said the shopkeeper. And he told the other that she was Mademoiselle V—, governess at General Newbold's, in the same town.

'"You have many foreigners here?" the stranger enquired.

' "Yes, though mostly Hanoverians. But since the peace they are learning French a good deal in genteel society, and French instructors are rather in demand."

' "Yes, I teach it," said the visitor. "I am looking for a tutorship in an academy."

'The information given by the burgess to the Frenchman seemed to explain to the latter nothing of his countrywoman's conduct—which, indeed, was the case—and he left the shop, taking his course again over the bridge and along the south way to the Old Rooms Inn, where he engaged a bedchamber.

'Thoughts of the woman who had betrayed such agitation at sight of him lingered naturally enough with the newcomer. Though, as I stated, not much less than thirty years of age, Mademoiselle V—, one of his own nation, and of highly refined and delicate appearance, had kindled a singular interest in the middle-aged gentleman's breast, and her large dark eyes, as they had opened and shrunk from him, exhibited a pathetic beauty to which hardly any man could have been insensible.

'The next day, having written some letters, he went out and made known at the office of the town "Guide" and of the newspaper, that a teacher of French and calligraphy had arrived, leaving a card at the bookseller's to the same effect. He then walked on aimlessly, but at length enquired the way to General Newbold's. At the door, without giving his name, he asked to see Mademoiselle V—, and was shown into a little back parlour, where she came to him with a gaze of surprise.

'"My God! Why do you intrude here, Monsieur?" she gasped in French as soon as she saw his face.

'"You were taken ill yesterday. I helped you. You might have been run over if I had not picked you up. It was an act of simple humanity certainly; but I thought I might come to ask if you had recovered?"

'She had turned aside, and had scarcely heard a word of his speech. "I hate you, infamous man!" she said. "I cannot bear your helping me. Go away!"

'"But you are a stranger to me."

'"I know you too well!"

'"You have the advantage then, Mademoiselle. I am a newcomer here. I never have seen you before to my knowledge; and I certainly do not, could not, hate you."

' "Are you not Monsieur B—?"

'He flinched. "I am—in Paris," he said. "But here I am Monsieur G—."

' "That is trivial. You are the man I say you are."

' "How did you know my real name, Mademoiselle?"

' "I saw you in years gone by, when you did not see me. You were formerly Member of the Committee of Public Safety, under the Convention."

' "I was."

' "You guillotined my father, my brother, my uncle—all my family, nearly, and broke my mother's heart. They had done nothing but keep silence. Their sentiments were only guessed. Their headless corpses were thrown indiscriminately into the ditch of the Mousseaux Cemetry, and destroyed with lime."

'He nodded.

' "You left me without a friend, and here I am now, alone in a foreign land."

' "I am sorry for you," said he. "Sorry for the consequence, not for the intent. What I did was a matter of conscience, and, from a point of view indiscernible by you, I did right. I profited not a farthing. But I shall not argue this. You have the satisfaction of seeing me here an exile also, in poverty, betrayed by comrades, as friendless as yourself."

' "It is no satisfaction to me, Monsieur."

' "Well, things done cannot be altered. Now to the question: are you quite recovered?"

' "Not from dislike and dread of you—otherwise, yes."

' "Good morning, Mademoiselle."

' "Good morning."

'They did not meet again till one evening at the theatre (which my mother's friend was with great difficulty induced to frequent, to perfect herself in English pronunciation, the idea she entertained at that time being to become a teacher of English in her own country later on). She found him sitting next to her, and it made her pale and restless.

' "You are still afraid of me?"

' "I am. O cannot you understand!"

'He signified the affirmative.

' "I follow the play with difficulty," he said, presently.

' "So do I—*now*," said she.

'He regarded her long, and she was conscious of his look; and while she kept her eyes on the stage they filled with tears. Still she would not move, and the tears ran visibly down her cheek, though the play was a merry one, being no other than Mr Sheridan's comedy of *The Rivals*, with Mr S. Kemble as Captain Absolute. He saw her distress, and that her mind was elsewhere; and abruptly rising from his seat at candle-snuffing time he left the theatre.

'Though he lived in the old town, and she in the new, they frequently saw each other at a distance. One of these occasions was when she was on the north side of the harbour, by the ferry, waiting for the boat to take her across. He was standing by Cove Row, on the quay opposite. Instead of entering the boat when it arrived she stepped back from the quay; but looking to see if he remained she beheld him pointing with his finger to the ferry-boat.

' "Enter!" he said, in a voice loud enough to reach her.

'Mademoiselle V— stood still.

' "Enter!" he said, and, as she did not move, he repeated the word a third time.

'She had really been going to cross, and now approached and stepped down into the boat. Though she did not raise her eyes she knew that he was watching her over. At the landing steps she saw from under the brim of her hat a hand stretched down. The steps were steep and slippery.

' "No, Monsieur," she said. "Unless, indeed, you believe in God, and repent of your evil past!"

' "I am sorry you were made to suffer. But I only believe in the god called Reason, and I do not repent. I was the instrument of a national principle. Your friends were not sacrificed for any ends of mine."

'She thereupon withheld her hand, and clambered up unassisted. He went on, ascending the Look-out Hill, and disappearing over the brow. Her way was in the same direction, her errand being to bring home the two young girls under her charge, who had gone to the cliff for an airing. When she joined them at the top she saw his solitary figure at the further edge, standing motionless against the sea. All the while that she remained with her pupils he stood without turning, as if looking at the frigates in the roadstead, but more probably in meditation, unconscious where he was. In leaving the spot one of the children threw away half a sponge-biscuit that she had been eating. Passing near it he stooped, picked it up carefully, and put it in his pocket.

'Mademoiselle V— came homeward, asking herself, "Can he be starving?"

'From that day he was invisible for so long a time that she thought he had gone away altogether. But one evening a note came to her, and she opened it trembling. "I am here ill," it said, "and, as you know, alone. There are one or two little things I want done, in case my death should occur, and I should prefer not to ask the people here, if it could be avoided. Have you enough of the gift of charity to come and carry out my wishes before it is too late?"

'Now so it was that, since seeing him possess himself of the broken cake, she had insensibly begun to feel something that was more than curiosity, though perhaps less than anxiety, about this fellow-countryman of hers; and it was not in her nervous and sensitive heart to resist his appeal. She found his lodging (to which he had removed from the Old Rooms inn for economy) to be a room over a shop, half-way up the steep and narrow street of the old town, to which the fashionable visitors seldom penetrated. With some misgiving she entered the house, and was admitted to the chamber where he lay.

' "You are too good, too good," he murmured. And presently, "You need not shut the door. You will feel safer, and they will not understand what we say."

' "Are you in want, Monsieur? Can I give you—"

' "No, no. I merely want you to do a trifling thing or two that I have not strength enough to do myself. Nobody in the town but you knows who I really am—unless you have told?"

' "I have not told . . . I thought you *might* have acted from principle in those sad days, even—"

' "You are kind to concede that much. However, to the present. I was able to destroy my few papers before I became so weak. . . . But in the drawer there you will find some pieces of linen clothing—only two or three—marked with initials that may be recognized. Will you rip them out with a penknife?"

'She searched as bidden, found the garments, cut out the stitches of the lettering, and replaced the linen as before. A promise to post, in the event of his death, a letter he put in her hand, completed all that he required of her.

'He thanked her. "I think you seem sorry for me," he murmured. "And I am surprised. You are sorry?"

'She evaded the question. "Do you repent and believe?" she asked.

' "No."

'Contrary to her expectations and his own he recovered, though very slowly; and her manner grew more distant thenceforward, though his influence upon her was deeper than she knew. Weeks passed away, and the month of May arrived. One day at this time she met him walking slowly along the beach to the northward.

' "You know the news?" he said.

' "You mean of the rupture between France and England again?"

' "Yes; and the feeling of antagonism is stronger than it was in the last war, owing to Bonaparte's high-handed arrest of the innocent English who were travelling in our country for pleasure. I feel that the war will be long and bitter; and that my wish to live unknown in England will be frustrated. See here."

'He took from his pocket a piece of the single newspaper which circulated in the county in those days, and she read—

The magistrates acting under the Alien Act have been requested to direct a very scrutinizing eye to the Academies in our towns and other places, in which French tutors are employed, and to all of that nationality who profess to be teachers in this country. Many of them are known to be inveterate Enemies and Traitors to the nation among whose people they have found a livelihood and a home.

'He continued: "I have observed since the declaration of war a marked difference in the conduct of the rougher class of people here towards me. If a great battle were to occur—as it soon will, no doubt—feeling would grow to a pitch that would make it impossible for me, a disguised man of known occupation, to stay here. With you, whose duties and antecedents are known, it may be less difficult, but still unpleasant. Now I propose this. You have probably seen how my deep sympathy with you has quickened to a warm feeling; and what I say is, will you agree to give me a title to protect you by honouring me with your hand? I am older than you, it is true, but as husband and wife we can leave England together, and make the whole world our country. Though I would propose Quebec, in Canada, as the place which offers the best promise of a home."

' "My God! You surprise me!" said she.

' "But you accept my proposal?"

' "No, no!"

' "And yet I think you will, Mademoiselle, some day!"

' "I think not."

' "I won't distress you further now."

' "Much thanks. . . . I am glad to see you looking better, Monsieur; I mean you are looking better."

' "Ah, yes. I am improving. I walk in the sun every day."

'And almost every day she saw him—sometimes nodding stiffly only, sometimes exchanging formal civilities. "You are not gone yet," she said on one of these occasions.

' "No. At present I don't think of going without you."

' "But you find if uncomfortable here?"

' "Somewhat. So when will you have pity on me?"

'She shook her head and went on her way. Yet she was a little moved. "He did it on principle," she would murmur. "He had no animosity towards them, and profited nothing!"

'She wondered how he lived. It was evident that he could not be so poor as she had thought; his pretended poverty might be to escape notice. She could not tell, but she knew that she was dangerously interested in him.

'And he still mended, till his thin, pale face became more full and firm. As he mended she had to meet that request of his, advanced with even stronger insistency.

'The arrival of the King and Court for the season as usual brought matters to a climax for these two lonely exiles and fellow country-people. The King's awkward preference for a part of the coast in such dangerous proximity to France made it necessary that a strict military vigilance should be exercised to guard the royal residents. Half-a-dozen frigates were every night posted in a line across the bay, and two lines of sentinels, one at the water's edge and another behind the Esplanade, occupied the whole sea-front after eight every night. The watering-place was growing an inconvenient residence even for Mademoiselle V— herself, her friendship for this strange French tutor and writing-master who never had any pupils having been observed by many who slightly knew her. The General's wife, whose dependent she was, repeatedly warned her against the acquaintance; while the Hanoverian and other soldiers of the Foreign Legion, who had discovered the nationality of her friend, were more aggressive than the English military gallants who made it their business to notice her.

'In this tense state of affairs her answers became more agitated. "O Heaven, how can I marry you!" she would say.

' "You will; surely you will!" he answered again. "I don't leave without you, and I shall soon be interrogated before the magistrates if I stay here; probably imprisoned. You will come?"

'She felt her defences breaking down. Contrary to all reason and sense of family honour she was, by some abnormal craving, inclining to a tenderness for him that was founded on its opposite. Sometimes her warm sentiments burnt lower than at others, and then the enormity of her conduct showed itself in more staring hues.

'Shortly after this he came with a resigned look on his face. "It is as I expected," he said. "I have received a hint to go. In good sooth, I am no Bonapartist—I am no enemy to England; but the presence of the King made it impossible for a foreigner with no visible occupation, and who may be a spy, to remain at large in the town. The authorities are civil, but firm. They are no more than reasonable. Good. I must go. You must come also."

'She did not speak. But she nodded assent, her eyes drooping.

'On her way back to the house on the Esplanade she said to herself, "I am glad, I am glad! I could not do otherwise. It is rendering good for evil!" But she knew how she mocked herself in this, and that the moral principle had not operated one jot in her acceptance of him. In truth she had not realized till now the full presence of the emotion which had unconsciously grown up in her for this lonely and severe man, who, in her tradition, was vengeance and irreligion personified. He seemed to absorb her whole nature, and, absorbing, to control it.

'A day or two before the one fixed for the wedding there chanced to come to her a letter from the only acquaintance of her own sex and country she possessed in England, one to whom she had sent intelligence of her approaching marriage, without mentioning with whom. This friend's misfortunes had been somewhat similar to her own, which fact had been one cause of their intimacy; her friend's sister, a nun of the Abbey of Montmartre, having perished on the scaffold at the hands of the same Comité de Salut Public which had numbered Mademoiselle V—'s affianced among its members. The writer had felt her position much again of late, since the renewal of the war, she said; and the letter wound up with a fresh denunciation of the authors of their mutual bereavement and subsequent troubles.

'Coming just then, its contents produced upon Mademoiselle V— the effect of a pail of water upon a somnambulist. What had she been doing in betrothing herself to this man! Was she not making herself a parricide after the event? At this crisis in her feelings her lover called. He beheld her trembling, and, in reply to his question, she told him of her scruples with impulsive candour.

'She had not intended to do this, but his attitude of tender command coerced her into frankness. Thereupon he exhibited an agitation never before apparent in him. He said, "But all that is past. You are the symbol of Charity, and we are pledged to let bygones be."

'His words soothed her for the moment, but she was sadly silent, and he went away.

'That night she saw (as she firmly believed to the end of her life) a divinely sent vision. A procession of her lost relatives—father, brother, uncle, cousin—seemed to cross her chamber between her bed and the window, and when she endeavoured to trace their features she perceived them to be headless, and that she had recognized them by their familiar clothes only. In the morning she could not shake off the effects of this appearance on her nerves. All that day she saw nothing of her wooer, he being occupied in making arrangements for their departure. It grew towards evening—the marriage eve; but, in spite of his reassuring visit, her sense of family duty waxed stronger now that she was left alone. Yet, she asked herself, how could she, alone and unprotected, go at this eleventh hour and reassert to an affianced husband that she could not and would not marry him while admitting at the same time that she loved him? The situation dismayed her. She had relinquished her post as governess, and was staying temporarily in a room near the coach-office, where she expected him to call in the morning to carry out the business of their union and departure.

'Wisely or foolishly, Mademoiselle V——came to a resolution: that her only safety lay in flight. His contiguity influenced her too sensibly; she could not reason. So packing up her few possessions and placing on the table the small sum she owed, she went out privately, secured a last available seat in the London coach, and, almost before she had fully weighed her action, she was rolling out of the town in the dusk of the September evening.

'Having taken this startling step she began to reflect upon her reasons. He had been one of that tragic Committee the sound of whose name was a horror to the civilized world; yet he had been only *one* of several members, and, it seemed, not the most active. He had marked down names on principle, had felt no personal enmity against his victims, and had enriched himself not a sou out of the office he had held. Nothing could change the past. Meanwhile he loved her, and her heart inclined to as much of him as she could detach from the past. Why not, as he had suggested, bury memories, and inaugurate a new era by this union? In other words, why not indulge her tenderness, since its nullification could do no good.

'Thus she held self-communion in her seat in the coach, passing through Casterbridge and Shottsford, and on to the White Hart at Melchester, at which place the whole fabric of her recent intentions crumbled down. Better be staunch having got so far; let things take their course, and marry boldly the man who had so impressed her. How great he was; how small was she! And she had presumed to judge him! Abandoning her place in the coach with the precipitancy that had characterized her taking it, she waited till the vehicle had driven off, something in the departing shapes of the outside passengers against the starlit sky giving her a start, as she afterwards remembered. Presently the down coach, "The Morning Herald", entered the city, and she hastily obtained a place on the top.

' "I'll be firm—I'll be his—if it cost me my immortal soul!" she said. And with troubled breathings she journeyed back over the road she had just traced.

'She reached our royal watering-place by the time the day broke, and her first aim was to get back to the hired room in which her last few days had been spent. When the landlady appeared at the door in response to Mademoiselle V——'s summons, she explained her sudden departure and return as best she could; and no objection being offered to her re-engagement of the room for one day longer she ascended to the chamber and sat down panting. She was back once more, and her wild tergiversations were a secret from him whom alone they concerned.

'A sealed letter was on the mantlepiece. "Yes, it is directed to you, Mademoiselle," said the woman who had followed her. "But we were wondering what to do with it. A town messenger brought it after you had gone last night."

'When the landlady had left, Mademoiselle V—opened the letter and read—

MY DEAR AND HONOURED FRIEND.—You have been throughout our acquaintance absolutely candid concerning your misgivings. But I have been reserved concerning mine. That is the difference between us. You probably have not guessed that every qualm you have felt on the subject of our marriage has been paralleled in my heart to the full. Thus it happened that your involuntary outburst of remorse yesterday, though mechanically deprecated by me in your presence, was a last item in my own doubts on the wisdom of our union, giving them a force that I could no longer withstand. I came home; and, on reflection, much as I honour and adore you, I decide to set you free.

As one whose life has been devoted, and I may say sacrificed, to the cause of Liberty, I cannot allow your judgement (probably a permanent one) to be fettered beyond release by a feeling which may be transient only.

It would be no less than excruciating to both that I should announce this decision to you by word of mouth. I have therefore taken the less painful course of writing. Before you receive this I shall have left the town by the evening coach for London, on reaching which city my movements will be revealed to none.

Regard me, Mademoiselle, as dead, and accept my renewed assurances of respect, remembrance, and affection.

'When she had recovered from her shock of surprise and grief, she remembered that at the starting of the coach out of Melchester before dawn, the shape of a figure among the outside passengers against the starlit sky had caused her a momentary start, from its resemblance to that of her friend. Knowing nothing of each other's intentions, and screened from each other by the darkness, they had left the town by the same conveyance. "He, the greater, persevered; I, the smaller, returned!" she said.

'Recovering from her stupor, Mademoiselle V— bethought herself again of her employer, Mrs Newbold, whom recent events had estranged. To that lady she went with a full heart, and explained everything. Mrs Newbold kept to herself her opinion of the episode, and reinstalled the deserted bride in her old position as governess to the family.

'A governess she remained to the end of her days. After the final peace with France she became acquainted with my mother, to whom by degrees she imparted these experiences of hers. As her hair grew white, and her features pinched, Mademoiselle V— would wonder what nook of the world contained her lover, if he lived, and if by any chance she might see him again. But when, some time in the 'twenties, death came to her, at no great age, that outline against the stars of the morning remained as the last glimpse she ever obtained of her family's foe and her once affianced husband.'

GEORGETTE HEYER

Runaway Match

ЮХ

As the post-chaise and four entered the town of Stamford, young Mr
Morley, who had spent an uncomfortable night being jolted over the
road, remorselessly prodded his companion.

'We have reached Stamford,' he announced. 'We change horses here,
and whatever you may choose to do, *I* shall bespeak breakfast.'

Miss Paradise, snugly ensconced in her corner of the chaise, opened a
pair of dark eyes, blinked once or twice, yawned behind her feather muff,
and sat up.

'Oh!' said Miss Paradise, surveying the spring morning with enthusiasm.
'It is quite daylight! I have had the most delightful sleep.'

Mr Morley repeated his observation, not without a hint of pugnacity in
his voice.

Since the start of the elopement, rather more than nine hours before,
Miss Paradise, who was just eighteen, had been a trifle difficult to manage.
She had begun by taking strong exception to the ladder he had brought
for her escape from her bedroom window. Her remarks, delivered in an
indignant undertone as she had prepared to descend the ladder, might have
been thought to augur ill for the success of the runaway match; but Mr
Morley, who was also just eighteen, had quarelled with Miss Paradise from
the cradle, and thought her behaviour the most natural in the world. The
disposition she showed to take the management of the flight into her own
hands led to further wrangles, because, however much she might have been
in the habit of taking the lead in their past scrapes an elopement was a
very different matter, and called for a display of male initiative. But when he
had tried to point this out to Miss Paradise she had merely retorted, 'Stuff!
It was I who made the plan to elope. Now, Rupert, you know it was!'

This rejoinder was unanswerable, and Mr Morley, who had been arguing
in favour of putting up for the night at a respectable posting-house, had
allowed himself to be overruled. They had travelled swiftly northwards
by moonlight (a circumstance which had filled the romantic Miss Paradise
with rapture) with the result that a good deal of Mr Morley's zest for the
adventure had worn off by the time he made his announcement at eight
o'clock.

He was prepared to encounter opposition, but Miss Paradise, engaged in the task of tidying her tumbled curls, assented light-heartedly.

'To be sure, I am excessively hungry,' she said.

She picked up a chip hat from the seat and tied it on her head by its green gauze ribbons.

'I dare say I must look a positive fright,' she remarked; 'but you can have no notion how much I am enjoying myself.'

This buoyancy had the effect of making Mr Morley slightly morose.

'I can't imagine what there is to enjoy in being bumped about all night,' he said.

Miss Paradise turned her enchanting little face towards him, and exclaimed with considerable chagrin:

'Are you not enjoying it at all, Rupert? I must say I do think you need not get into a miff merely because of being bumped a trifle.'

'I am not in a miff!' said Mr Morley, 'but——'

'Oh, Rupert!' cried Miss Paradise, letting her muff fall. 'Don't, don't say that you do not want to elope with me, after all!'

'No, of course I do,' responded Mr Morley. 'The fact is, I didn't contrive to sleep above an hour or two. I shall be in better cue after breakfast.'

'Yes, I expect that's it,' nodded Miss Paradise, relieved. 'Only, I don't think we should waste very much time, you know, because when papa discovers our flight he is bound to pursue us.'

'I don't see that,' objected Mr Morley. 'He can't know where you have gone.'

'Yes, he can,' said Miss Paradise. 'I left a note on my pillow for him.'

'What!' ejaculated Mr Morley. 'Good heavens, Bab, why?'

'But he would be in a dreadful rout if I hadn't told him,' explained Miss Paradise. 'And even though he has behaved shockingly to me I don't want him to be anxious about me.'

Mr Morley retorted: 'If you think to have put an end to his anxiety by telling him you have eloped with me you very much mistake the matter.'

'No, but at least he will be sure that I am safe,' said Miss Paradise. 'You know that he likes you extremely, Rupert, even though he does not wish me to marry you. *That* is only because he says you are too young; and because he has this stupid notion that I must make a good match, of course,' she added candidly.

'Well, I think you must be mad,' said Mr Morley. 'I'll lay you a button he rides over immediately to tell my father. Then we shall have the pair of them at our heels, and a pretty pucker there will be.'

'I hadn't thought of that,' confessed Miss Paradise. 'But I dare say we shall have reached Gretna Green long before they come up with us.'

The chaise had arrived at the George Inn by this time, and had turned in under the archway to the courtyard. The steps were let down and the travellers alighted. Mr Morley felt stiff, but Miss Paradise gave her tiffany skirts a shake and tripped into the inn for all the world as though she had enjoyed a perfect night's rest.

There was not much sign of activity in the George at this early hour, but the landlord came out and led the way to a private parlour overlooking the street, and promised to serve breakfast in the shortest possible time. He betrayed no extraordinary curiosity, the extremely youthful appearance of his guests leading him to suppose them to be brother and sister.

Miss Paradise, realizing this, was disappointed, and commented on it to Mr Morley with considerable dissatisfaction.

'Well, thank Heaven for it,' said Mr Morley.

'Sometimes, Rupert,' said Miss Paradise, 'I think you are not romantic in the very least.'

'I never said that I was,' replied Mr Morley.

'You may not have said it, but you did say that you would rescue me from that odious Sir Roland, and if that is not——'

'Well, I *am* rescuing you,' interrupted Mr Morley, 'and I don't object to being romantic in reason. But when it comes to you wanting a rope-ladder to escape by,' he continued, last night's quarrel taking possession of his mind again, 'I call it the outside of enough.'

'Who ever heard of any other kind of ladder for an elopement?' demanded Miss Paradise scathingly.

'I don't know. But how was I to find such a thing? And now I come to think of it,' pursued Mr Morley, 'why the devil did you want a ladder at all? Your father and your aunt were both gone out, and you told me yourself the servants were all in bed.'

A disarming dimple peeped in Miss Paradise's cheek.

'Well, to tell you the truth, it wasn't *very* necessary,' she confessed. 'Only it seemed so much more exciting.'

The entrance of a serving-man with a tray prevented Mr Morley from uttering the indignant retort that sprang to his lips, and by the time the table was laid and the covers set on it the mingled aromas of coffee and broiled ham and ale had put all other thought than that of breakfast out of his head. He handed Miss Paradise to a chair, drew out one for himself, and was soon engaged in assuaging the first pangs of his hunger.

Miss Paradise, pursuing thoughts of her own, presently said:

'I dare say they won't have found my note yet.'

'I wish to Heaven you hadn't written it, Bab!'

'Well, so do I now,' admitted Miss Paradise. 'Because although I made certain that Aunt Albinia would not think of going to my room when she came home last night, it has all at once occurred to me that perhaps she might.'

Mr Morley, who was carving the cold sirloin, gave a groan.

'Why? If she never does——'

'Yes, but you see, I said I had the headache, and she might go to my room to see how I did. I had to say I had the headache, Rupert, because they would have forced me to go with them to the dinner-party if I hadn't.' Her brow darkened. 'To meet that odious old man,' she added broodingly.

'Sale?' enquired Mr Morley.

'Yes, of course.'

'He isn't as old as that, Bab. Hang it, he can't be much above thirty. And you don't know that he's odious after all.'

'Oh, yes, I do,' retorted Miss Paradise with strong feeling. 'He wrote to papa that he was perfectly willing to fulfil his obligations and marry me. I never heard of anything so odious in my life. He must be the most horrid creature imaginable, and as for papa, I am sorry to be obliged to say it, but he is very little better; in fact, I think, *worse*, because it was he who made this abominable plan to marry me to an Eligible Person with whom I am not even acquainted. And Sir Joseph Sale, too, of course, detestable old man that he was.'

'Gad, he was!' agreed Mr Morley. 'Do you remember——'

'No,' said Miss Paradise. 'At least, I'm not going to, because one should never speak ill of the dead. But you may depend upon it his nephew is just like him, and if papa thinks I am going to marry to oblige him he very much mistakes the matter. As though you and I had not said years and years ago that we would marry each other.'

'Parents are all alike,' said Mr Morley gloomily. 'However, this should show my father that I am not to be treated like a child any longer.'

'Yes,' said Miss Paradise, pouring out another cup of coffee. 'And if they don't like it on account of your being too young, I shall tell papa that it is all his fault, because if he hadn't made an arrangement for me to marry a man I've never clapped eyes on we shouldn't have thought of being married for a long time, should we?'

'No,' said Mr Morley. 'Not until I had come down from Oxford, at all events, and after that, I believe, there was a scheme for me to make the Grand Tour, which I must say I should have liked. I dare say we shouldn't have thought of being married for four or five years.'

Miss Paradise paused in the act of drinking her coffee and lowered the cup.

'Four or five years!' she repeated. 'But I should be twenty-two or three years old.'

'Well, so should I,' Mr Morley pointed out.

'But that is not at all the same thing,' said Miss Paradise indignantly.

'Oh, well, there's no sense in arguing about it,' replied Mr Morley, finishing his ale, and getting up. 'They have compelled us to elope, and there's an end to it. I had best tell them to have the horses put to at once, I think. We have no time to lose.'

Miss Paradise agreed to it, and engaged to be in readiness to resume the journey by the time Mr Morley had paid the reckoning and seen a fresh team harnessed to the chaise. He went out, and she was left to drink the last of her coffee, to tie on her becoming hat once more, and to straighten her tucker. This did not take long; she was ready before her swain, and was about to sally forth into the yard when the sound of a carriage being driven fast down the street made her run to the window.

It was not, however, a post-chaise, and no such dreaded sight as Sir John Paradise's face met her alarmed gaze. Instead, she saw a curricle and four driven by a gentleman in a very modish dress of dark blue with gold buttons. He wore a gold-laced tricorne on his own unpowdered hair, and a fringed cravat thrust through a gold buttonhole. A surtout with four laps on each side hung negligently open over his dress, and on his feet he had a pair of very highly polished top-boots. He was looking straight ahead, and so did not see Miss Paradise peeping at him over the short blind. She had a glimpse of a straight, rather haughty profile as the curricle passed the window; then the horses were checked, and the equipage swung round under the archway into the courtyard.

'Bab!' gasped Mr Morley, who had entered the room behind her. 'We are overtaken!'

Miss Paradise gave a shriek and dropped her muff.

'Mercy on me! Not papa?'

'No, I don't know who it can be, but a man has this instant driven into the yard——'

'Yes, yes, I saw him. But what in the world can he have to do with us?'

'I tell you I don't know, but he asked the landlord if he had seen anything of a young lady and gentleman. I did not wait for more, as you may imagine. What are we to do? Who in thunder can he be?'

A premonition had seized Miss Paradise. She took a step back, clasping her hands together in great agitation.

'Good heavens, Rupert! Could it be—Sir Roland?' Mr Morley stared at her.

'Sale? It can't be! How should he know of our elopement?'

'Papa must have brought him back with him last night. Oh, this is dreadful! I declare I am ready to sink!'

Mr Morley squared his shoulders.

'Well, if he is Sale, he shan't take you back, Bab. He has to reckon with me now.'

'But he is not in the least like Sir Joseph!' said Miss Paradise numbly. 'He is quite handsome!'

'What in the world has that to do with it?' demanded Mr Morley.

Miss Paradise turned scarlet.

'Nothing at all!' she replied. 'Whoever he is like he is odious. *Willing to fulfil*——But I never dreamed that *he* would follow us!'

At this moment the door was opened again, and a pleasant, slightly drawling voice said:

'So I have caught you, my children? I thought I might,' and the gentleman in the modish surtout walked into the room.

He paused on the threshold and raised his quizzing-glass. Miss Paradise, who had retreated to Mr Morley's side, blushed, and gave him back stare for stare.

'But I must humbly beg my apologies,' said the newcomer, a faintly quizzical smile in his grey eyes. 'I seem to have intruded. Madam——'

'Yes,' said Miss Paradise. 'You have intruded, Sir Roland!'

The quizzical smile lingered; one eyebrow went up.

'Now, I wonder how you knew me?' murmured the gentleman.

'I am well aware that you must be Sir Roland Sale,' said Miss Paradise, 'but I do *not* know you, and I do not desire to know you!'

Sir Roland laughed suddenly and shut the door.

'But are you not being a trifle hasty?' he enquired. 'Why don't you desire to know me?'

'I imagine you must know very well!' said Miss Paradise.

'Indeed I don't!' said Sir Roland. He came further into the room, and laid his hat and his elegant fringed gloves down on the table. He looked thoughtfully from one flushed countenance to the other, and said in a tone of amusement: 'Is it possible that you are running away from me?'

'Certainly not!' said Miss Paradise. 'But I think it only proper to tell you, sir, that this is the gentleman I am going to marry.'

Mr Morley tried to think of something dignified to add to this pronouncement, but under that ironic, not unkindly gaze, only succeeded in clearing his throat and turning redder than ever.

Sir Roland slid one hand into his pocket and drew out a snuff-box.

'But how romantic!' he remarked. 'Do, pray, present me!'

Mr Morley took a step forward.

'You must have guessed, sir, that my name is Morley. Miss Paradise has been promised to me these dozen years.'

Sir Roland bowed and offered his snuff-box.

'I felicitate you,' he said. 'But what part do I play in this charming—er—idyll?'

'None!' replied Miss Paradise.

Sir Roland, his snuff having been waved aside by Mr Morley, took a pinch and held it to one nostril. Then he fobbed his box with an expert flick of the finger and put it away again.

'I hesitate to contradict you, Miss Paradise,' he said, 'but I cannot allow myself to be thrust into the role of a mere onlooker.'

Miss Paradise replied, not quite so belligerently:

'I dare say you think you have a right to interfere, but you need not think that I will go back with you, for I won't!'

Mr Morley, feeling himself elbowed out of the discussion, said with some asperity:

'I wish you will leave this to me, Bab! Pray, do be quiet a moment!'

'Why should I be quiet?' demanded Miss Paradise. 'It is quite my own affair!'

'You always think you can manage everything,' said Mr Morley. 'But this is between men!'

'What nonsense!' said Miss Paradise scornfully. 'Pray, whom does he want to marry, you or me?'

'Lord, Bab, if you're going to talk like a fool I shall be sorry I ever said I'd elope with you!'

'Well, I'm sorry now!' said Miss Paradise instantly.

Mr Morley cast her a withering glance and turned once more to Sir Roland.

'Sir, no doubt you are armed with Sir John Paradise's authority, but——'

'Let me set your mind at rest at once,' interposed Sir Roland. 'I am here quite on my own authority.'

'Well, sir! Well, in that case——'

Miss Paradise entered into the conversation again.

'You can't pretend that you cared as much as that!' she said impetuously. 'You could not have wanted to marry me so very much when you had never so much as set eyes on me!'

'Of course not,' agreed Sir Roland. 'Until I set eyes on you I had not the least desire to marry you.'

'Then why did you write that odious letter to papa?' asked Miss Paradise reasonably.

'I never write odious letters,' replied Sir Roland calmly.

'I dare say you may think it was very civil and obliging of you,' said Miss Paradise; 'but for my part I have a very poor notion of a man allowing his marriage to be arranged for him, and when it comes to writing that you are willing to fulfil your——your *obligations*——'

A muscle quivered at the corner of Sir Roland's mouth.

'Did I write that?' he asked.

'You must know you did!'

'I am quite sure I wrote no such thing,' he said.

'Well, what did you write?' she demanded.

He walked forward till he stood quite close to her and held out his hand. He said, looking down at her:

'Does it signify what I wrote? After all, I had not seen you then. Now that we are acquainted I promise you I will not write or say anything to give you a disgust of me.'

She looked at him uncertainly. Even though his fine mouth was perfectly grave his eyes held a smile which one could hardly withstand. A little colour stole into her cheeks; the dimple peeped again; she put her hand shyly into his, and said:

'Well, perhaps it does not signify so *very* much. But I am going to marry Mr Morley, you know. That was all arranged between us years ago.'

Sir Roland still kept her hand clasped in his. 'Do you never change your mind, Miss Paradise?' he asked.

Mr Morley, who had begun in the presence of this polished gentleman to feel himself a mere schoolboy, interrupted at this moment and said hotly: 'Sir, I deny any right in you to interfere in Miss Paradise's affairs! She is under my protection, and will shortly be my wife. Bab, come with me! We should press on at once!'

'I suppose we should,' agreed Miss Paradise rather forlornly.

Mr Morley strode up to her and caught her wrist. Until the arrival of Sir Roland he had been regarding his approaching nuptials with mixed feelings, but to submit to a stranger's intervention, and to see his prospective bride in danger of being swayed by the undeniable charm of a man older, and far more at his ease than he was himself, was a little too much for him to stomach. There was a somewhat fiery light in his eyes as he said: 'Bab, you are promised to me! You know you are!'

Miss Paradise raised her eyes to Sir Roland's face. 'It is quite true,' she said with a faint sigh. 'I am promised to him, and one must keep one's word, you know.'

'Bab!' said Mr Morley sternly, 'you wanted to elope with me! It was your notion! Good heavens, you could not turn back now and go meekly home!'

'No, of course I couldn't,' said Miss Paradise, roused by this speech. 'I never heard of anything so flat!'

'I knew you would never fail!' said Mr Morley, casting a triumphant look at Sir Roland. 'Let us be on our way immediately.'

Sir Roland flicked a grain of snuff from his wide cuff. 'Not so fast, Mr Morley,' he said. 'I warned you, did I not, that I could not allow myself to be thrust into the role of mere onlooker?'

Mr Morley's eyes flashed. 'You have no right to interfere, sir!'

'My dear young man,' said Sir Roland, 'anyone has the right to do what he can to prevent two—er—young people from committing an act of the most unconscionable folly. You will not take Miss Paradise to Gretna today—or, in fact, any other day.'

There was a note of steel in the drawling voice. Miss Paradise, realizing that the adventure was becoming even more romantic than she had bargained for, clasped her hands in her muff and waited breathlessly.

Mr Morley laid a hand on his sword-hilt. 'Oh?' he said. 'Indeed, sir?'

Sir Roland, observing the gesture, raised his brows in some amusement.

Mr Morley said through his teeth: 'We shall do better to continue our discussion outside, sir, I believe.'

Miss Paradise caught her muff up to her chin, and over it looked imploringly at Sir Roland. He was not attending to her; he seemed to be considering Mr Morley. After a moment he said slowly: 'You are a little impetuous, are you not?'

'Sir,' said Mr Morley dramatically, 'if you want Bab you must fight for her!'

Miss Paradise's mouth formed an 'O' of mingled alarm and admiration.

There was a slight pause. Then Sir Roland smiled and said: 'Well, you have plenty of courage, at all events. I am perfectly prepared to fight for her.'

'Then follow me, sir, if you please!' said Mr Morley, striding to the door.

Miss Paradise gave a cry and sprang after him. 'Oh, Rupert, no!'

She was intercepted by Sir Roland, who laid a detaining hand on her arm. 'Don't be alarmed, Miss Paradise,' he said.

Miss Paradise said in an urgent undervoice: 'Oh, please don't! He can't fight you! He is only a boy. Sir Roland!'

Mr Morley, who was plainly enjoying himself at last, shut the door upon Miss Paradise, and demanded to know whether Sir Roland preferred swords or pistols. When Sir Roland unhesitatingly chose swords he bowed, and said that he believed there was a garden behind the inn which would serve their purpose.

He was right; there was a garden, with a small shrubbery screening part of it from the house. Sir Roland followed Mr Morley there and took off his coat and tossed it on to a wooden seat. 'This is damned irregular, you know.' he remarked, sitting down on the bench to pull off his boots. 'Are you very set on fighting me?'

'Yes, I am,' declared Mr Morley, removing his sword-belt. 'A pretty fellow I should be if I gave Bab—Miss Paradise—up to you for the mere asking!'

Sir Roland drew his sword from its sheath and bent the slender blade between his hands. 'You would be a still prettier fellow if you carried her off to Gretna,' he said dryly.

Mr Morley coloured. 'Well, I never wanted to elope,' he said defensively. 'It was all your doing that we were forced to!'

Sir Roland got up from the bench in his leisurely way, and stood waiting with his sword-point lightly resting on the ground. Mr Morley rolled up his sleeves, picked up his weapon, and announced that he was ready.

He had, of course, been taught to fence, and was by no means a dull pupil; but within ten seconds of engaging he was brought to a realization of the vast difference that lay between a friendly bout with foils and a duel with naked blades. He tried hard to remember all he had been taught, but the pace Sir Roland set was alarmingly swift, and made him feel singularly helpless and clumsy. It was all he could do to parry that flickering sword-point; several times he knew he had been too slow, and almost shut his eyes in the expectation of being run through. But, somehow, he always did seem to succeed in parrying the fatal lunge just in time, and once he managed to press Sir Roland hard with an attack in a high line. He was very soon dripping with sweat and quite out of breath, fighting gamely but with thudding pulses, and with a paralysing sensation of being pretty much at his opponent's mercy. And then, just as he had miraculously parried a thrust in *seconde*, Sir Roland executed a totally unexpected *volte*, and the next instant Mr Morley's sword was torn from his grasp and he had flung up his hands instinctively to guard his face.

'Mr Morley,' said Sir Roland, breathing a little fast, 'do you acknowledge yourself worsted?'

Mr Morley, sobbing for breath, could only nod.

'Then let us rejoin Miss Paradise,' said Sir Roland, giving him back his sword.

He moved towards the bench and began to pull on his boots again. Mr Morley presently followed his example, crestfallen and very much out of countenance.

'I suppose,' said Mr Morley disconsolately, 'you could have killed me if you had chosen?'

'Yes, certainly I could; but then, you see, I am a very good swordsman,' said Sir Roland, smiling. 'Don't look so downcast. I think, one day, you may be a very good swordsman, too.'

Considerably cheered, Mr Morley followed him back to the inn parlour. Miss Paradise, who was looking pale and frightened, sprang up at their entrance and gave a gasp of relief.

'Oh, you haven't killed each other!' she cried thankfully.

'No; it was much too fine a morning for anything of that nature,' said Sir Roland. 'Instead, we have decided that it will be best if I take you back to your papa, Miss Paradise. These Gretna marriages are not quite the thing, you know.'

Miss Paradise seemed undecided, and looked towards Mr Morley for support.

'We shall have to give it up, Bab,' he said gloomily.

Miss Paradise sighed.

'I suppose we shall, though it does seem horridly flat to go home without any adventure at all.'

'Well, I've fought my first duel,' pointed out Mr Morley.

'Yes, but I haven't done anything!' objected Miss Paradise.

'On the contrary,' put in Sir Roland tactfully, 'you were the whole cause of the duel.'

'So I was!' said Miss Paradise, brightening. She gave Sir Roland one of her frank smiles. 'You are not at all what I thought you would be,' she confided. 'I didn't suppose you were the sort of person who would come after us so—so romantically!'

Sir Roland looked down at her with a rueful twinkle in his eyes.

'Miss Paradise, I must make a confession. I did not come after you.'

'You did not? But—but what did you come for, then?' she asked, considerably astonished.

'I came to meet my sister and my young brother,' said Sir Roland.

'Sister! Brother!' echoed Miss Paradise. 'I did not know you had any. How can this be? Did you not see my father last night? There must be some mistake!'

'I have never met your father in my life,' said Sir Roland.

Light broke in on Mr Morley. He cried out:

'Oh, good heavens! *Are* you Sir Roland Sale?'

'No,' said the other. 'I am only one Philip Devereux, who got up early to meet his sister on the last stage of her return from Scotland, and stumbled upon an adventure.'

Miss Paradise gave a choked cry.

'Oh, how *could* you?' she said, in a suffocating voice.

Mr Morley, quite pale with excitement, waved her aside.

'Not—not *the* Devereux?' he faltered. 'Not—oh, not *Viscount* Devereux of Frensham?'

'Well, yes, I am afraid so,' replied his lordship apologetically.

'Bab!' ejaculated Mr Morley. 'Do you hear that? I have actually crossed swords with one of the first swordsmen in Europe! Only think of it!'

Miss Paradise showed no desire to think of it. She turned her head away.

The viscount said: 'Do you think you could go and see what has been done with your chaise and my curricle, Mr Morley?'

'Oh, yes, certainly!' said Mr Morley. 'I'll go now, shall I?'

'If you please,' said his lordship, his eyes on Miss Paradise's profile. He waited until the door was shut behind Mr Morley, and then said gently: 'Forgive me, Miss Paradise!'

'You let me say—you let me believe you were the man papa says is going to marry me, and I——'

She stopped, for he had taken her hands and was looking down at her in a way that made her heart beat suddenly fast.

'I haven't the least idea what papa will say, but I can assure you that I am the man who is going to marry you,' said his lordship, with complete composure.

ARTHUR QUILLER-COUCH

The Singular Adventure of a Small Free-Trader

ᘒᘒ

The events which are to be narrated happened in the spring of 1803, and just before the rupture of the Peace of Amiens between our country and France; but were related to my grandfather in 1841 by one Yann, or Jean, Riel, a Breton 'merchant', alias smuggler— whether or not a descendant of the famous Hervé of that name, I do not know. He chanced to fall ill while visiting some friends in the small Cornish fishing-town, of which my grandfather was the only doctor; and this is one of a number of adventures recounted by him during his convalescence. I take it from my grandfather's MSS, but am not able, at this distance of time, to learn how closely it follows the actual words of the narrator.

Smuggling in 1841 was scotched, but certainly not extinct, and the visit of M. Riel to his old customers was, as likely as not, connected with business.—Q.

'Item, of the Cognac 25 degrees above proof, according to sample in the little green flask, *144* ankers at *4* gallons per anker, at *5*s. *6*d. per gallon, the said ankers to be ready slung for horse-carriage.'

'Now may the mischief fly away with these English!' cried my father, to whom my mother was reading the letter aloud. 'It costs a man a working day, with their gallons and sixpences, to find out of how much they mean to rob him at the end of it.'

'Item, *2* ankers of colouring stuff at *4* gallons per anker, price as usual. The place to be as before, under Rope Hauen, east side of Blackhead, unless warned: and a straight run. Come close in, any wind but easterly, and can load up horses alongside. *March 24th or 25th* will be best, night tides suiting, and no moon. Horses will be there: two fenced lights, pilchard-store and beach, showing S$\frac{1}{4}$E to ESE. Get them in line. Usual pay for freighting, and crew *17*l. per man, being a straight run.'

'And little enough,' was my father's comment.

'Item, *15* little wooden dolls, jointed at the knees and elbows, the same as tante Yvonne used to sell for two sols at Saint Pol de Léon——.'

' "Fifteen little wooden dolls"! "Fifteen little woo——" ' My father dropped into his chair, and sat speechless, opening and shutting his mouth like a fish.

'It is here in black and white,' said my mother. I found the letter, years after, in her kist. It was written, as were all the letters we received from this Cornish venturer, in a woman's hand, small and delicate, with up-strokes like spider's thread; written in French, too, quite easy and careless. My mother held it close to the window. '"Fifteen little wooden dolls,"' she repeated, '"jointed at the knees and elbows."'

'Well, I've gone to sea with all sorts, from Admiral Brueys upwards; but fifteen little wooden dolls—jointed—at—the—knees!'

'I know the sort,' I put in from the hearth, where my mother had set me to watch the *bouillon*. 'You can get as many as you like in the very next street, and at two sols apiece. I will look to that part of the cargo.'

'You, for example . . .?'

'Yes, I; since you promised to take me on the very next voyage after I was twelve.'

'But that's impossible. This is a straight run, as they call it, and not a mere matter of sinking the crop.'

'And next time,' I muttered bitterly, 'we shall be at war with England again, and then it will be the danger of privateers—always one excuse or another!'

My mother sighed as she looked out of window towards the Isle de Batz. I had been coaxing her half the morning, and she had promised me to say nothing.

Well, the result was that I went. My father's lugger carried twelve hands—I counted myself, of course; and indeed my father did the same when it came to charging for the crew. Still, twelve was not an out-of-the-way number, since in these *chassemarées* one must lower and rehoist the big sails at every fresh tack. As it happened, however, we had a fair wind right across from Roscoff, and made a good landfall of the Dodman at four in the afternoon, just twenty hours after starting. This was a trifle too early for us; so we dowsed sail, to escape notice, and waited for night-fall. As soon as it grew dark, we lowered the two tub-boats we carried—one on davits and the other inboard—and loaded them up and started to pull for shore, leaving two men behind on the lugger. My father steered the first boat, and I the other, keeping close in his wake—and a proud night that was for me! We had three good miles between us and shore; but the boats were mere shells and pulled light even with the tubs in them. So the men took it easy. I reckon that it was well past midnight before we saw the two lights which the letter had promised.

After this everything went easily. The beach at Rope Hauen is steep-to; and with the light breeze there was hardly a ripple on it. On a rising tide we ran the boats in straight upon the shingle; and in less than a minute the kegs were being hove out. By the light of the lantern on the beach I could see the shifting faces of the crowd, and the troop of horses standing behind, quite quiet, shoulder to shoulder, shaved from forelock to tail, all smooth and shining with grease. I had heard of these Cornish horses, and how closely they were clipped; but these beat all I had ever imagined. I could see no hair on them; and I saw them quite close; for in the hurry each horse, as his turn came, was run out alongside the boat; the man who led him standing knee-deep until the kegs were slung across by the single girth. As soon as this was done, a slap on the rump sent the beast shoreward, and the man scrambled out after him. There was scarcely any talk, and no noise except that caused by the wading of men and horses.

Now all this time I carried my parcel of little dolls in a satchel slung at my shoulder, and was wondering to whom I ought to deliver it. I knew a word or two of English, picked up from the smugglers that used to be common as skate at Roscoff in those days; so I made shift to ask one of the men alongside where the freighter might be. As well as I could make out, he said that the freighter was not on the beach; but he pointed to a tall man standing beside the lantern and gave me to understand that this was the 'deputy'. So I slipped over the gunwale and waded ashore towards him.

As I came near, the man moved out of the light, and strolled away into the darkness to the left, I don't know upon what errand. I ran after him, as I thought, but missed him. I stood still to listen. This side of the track was quite deserted, but the noise of the runners behind me, though not loud, was enough to confuse the sound of his footsteps. After a moment, though, I heard a slight scraping of shingle, and ran forward again—plump against the warm body of some living thing.

It was a black mare, standing here close under the cliff, with the kegs ready strapped upon her. I saw the dark forms of other horses behind, and while I patted the mare's shoulder, and she turned her head to sniff and nuzzle me, another horse came up laden from the water and joined the troop behind, no man leading or following. The queer thing about my mare, though, was that her coat had no grease on it like the others, but was close and smooth as satin, and her mane as long as a colt's. She seemed so friendly that I, who had never sat astride a horse in my life, took a sudden desire to try what it felt like. So I walked round, and finding a low rock on the other side, I mounted it and laid my hands on her mane.

On this she backed a foot or two and seemed uneasy, then turned her muzzle and sniffed at my leg. 'I suppose,' thought I, 'a Cornish horse won't understand my language.' But I whispered to her to be quiet, and quiet she was at once. I found that the tubs, being slung high, made quite a little cradle between them. 'Just a moment,' I told myself, 'and then I'll slip off and run back to the boat'; and twining the fingers of my left hand in her mane, I took a spring and landed my small person prone between the two kegs, with no more damage than a barked shin-bone.

And at that very instant I heard a shrill whistle and many sudden cries of alarm; and a noise of shouting and galloping across the beach; and was raising my head to look when the mare rose too, upon her hind-legs, and with the fling of her neck caught me a blow on the nose that made me see stars. And then long jets of fire seemed to mingle with the stars, and I heard the *pop-pop* of pistol-shots and more shouting.

But before this we were off and away—I still flat on the mare's back, with a hand in her mane and my knees wedged against the tubs; away and galloping for the head of the beach, with the whole troop of laden horses pounding at our heels. I could see nothing but the loom of the cliff ahead and the white shingle underfoot; and I thought of nothing but to hold on—and well it was that I did, for else the horses behind had certainly trampled me flat in the darkness. But all the while I heard shouting, louder and louder, and now came more pounding of hoofs alongside, or a little ahead, and a tall man on horseback sprang out of the night, and, cannoning against the mare's shoulder, reached out a hand to catch her by rein, mane, or bridle. I should say that we raced in this way, side by side, for ten seconds or so. I could see the gilt buttons twinkling on his sleeve as he reached past my nose, and finding neither bit nor rein, laid his hand at length right on top of mine. I believe that, till then, the riding-officer—it was he, for the next time I saw a riding-officer I recognized the buttons—had no guess of anyone's being on the mare's back. But instead of the oath that I expected, he gave a shrill scream, and his arm dropped, for the mare had turned and caught it in her teeth, just above the elbow. The next moment she picked up her stride again, and forged past him. As he dropped back, a bullet or two sang over us, and one went *ping*! into the right-hand keg. But I had no time to be afraid, for the mare's neck rose again and caught me another sad knock on the nose as she heaved herself up the cliff-track, and now I had work to grip the edge of the keg, and twine my left hand tighter in her mane to prevent myself slipping back over her tail, and on to those deadly hoofs. Up we went, the loose stones flying behind us into the bushes right and left. Further behind I heard the scrambling of many hoofs, but whether of the tub-carriers or the troopers' horses it was not for me to guess. The mare knew, however, for as the slope grew easier, she whinnied and slackened her pace to give them time to come up. This also gave me a chance to shift my seat a bit, for the edges of the kegs were nipping my calves cruelly. The beach below us was like the wicked place in a priest's sermon

—black as pitch and full of cursing—and by this time all alive with lanterns; but they showed us nothing. There was no more firing, though, and I saw no lights out at sea, so I hoped my father had managed to push off and make for the lugger.

We were now on a grassy down at the head of the cliff, and my mare, after starting again at a canter which rattled me abominably, passed into an easy gallop. I declare that except for my fears—and now, as the chill of the wind bit me, I began to be horribly afraid—it was like swinging in a hammock to the pitch of a weatherly ship. I was not in dread of falling, either; for her heels fell so lightly on the turf that they persuaded all fear of broken bones out of the thought of falling; but I *was* in desperate dread of those thundering tub-carriers just behind, who seemed to come down like a black racing wave right on top of us, and to miss us again and again by a foot or less. The *weight* of them on this wide, empty down—that was the nightmare we seemed to be running from.

We passed through an open gate, then another; then out upon hard road for half a mile or so (but I can tell you nothing of the actual distance or the pace), and then through a third gate. All the gates stood open; had been left so on purpose, of course; and the grey granite side-posts were my only milestones throughout the journey. Every mortal thing was strange as mortal thing could be. Here I was, in a foreign land I had never seen in my life, and could not see now; on horseback for the first time in my life; and going the dickens knew whither, at the dickens knew what pace; in much certain and more possible danger; alone, and without speech to explain myself when—as I supposed must happen sooner or later—my runaway fate should shoot me among human folk. And overhead—this seemed the oddest thing of all—shone the very same stars that were used to look in at my bedroom window over Roscoff quay. My mother had told me once that these were millions of miles away, and that people lived in them; and it came into my head as a monstrous queer thing that these people should be keeping me in view, and my own folk so far away and lost to me.

But the stars, too, began to grow faint; and little by little the fields and country took shape around us—plough, and grass, and plough again; then hard road, and a steep dip into a valley where branches met over the lane and scratched the back of my head as I ducked it; then a moorland rising straight in front, and rounded hills with the daylight on them. And as I saw this, we were dashing over a granite bridge and through a whitewashed street, our hoofs drumming the villagers up from their beds. Faces looked out of windows and were gone, like scraps of a dream. But just beyond the village we passed an old labourer trudging to his work, and he jumped into the hedge and grinned as we went by.

We were climbing the moor now, at a lopping gallop that set the packet of dolls bob-bobbing on my back to a sort of tune. The horses behind were nearly spent, and the sweat had worked their soaped hides into a complete lather. But the mare generalled them all the while; and striking on a cart-track beyond the second rise of the moor, slowed down to a walk, wheeled round and scanned the troop. As they struggled up she whinnied loudly. A whistle answered her far down the lane, and at the sound of it she was off again like a bird.

The track led down into a hollow, some acres broad, like a saucer scooped between two slopes of the moor; and in the middle of it—just low enough to be hidden from the valley beneath—stood a whitewashed farmhouse, with a courtlege in front and green-painted gate; and by this gate three persons watched us as we came—a man and two women.

The man by his dress was plainly a farmer; and catching sight of me, he called out something I could not understand, and turned towards the woman beside him, whom I took to be his wife. But the other woman, who stood some paces away, was a very different person—tall and slight, like a lady; grey-haired, and yet not seeming old; with long white hands and tiny high-heeled shoes, and dressed in black silk, with a lace shawl crossed over her shoulders, and a silver whistle hanging from her neck. She came forward, holding out a handful of sugar, and spoke to the mare, if you'll believe me, in my very own Breton.

'Good Lilith!' said she. 'Ah, what a mess for me to groom! See what a coat! Good Lilith!' Then, as Lilith munched the sugar—'Who are you, little boy? I never saw you before. Explain yourself, kindly, little boy.'

'My name is Yann,' said I; 'Yann Riel. I am from Roscoff, and—oh, how tired, madame!'

'He is Breton! He speaks the Breton!' She clapped her hands, drew me down from my seat, and kissed me on both cheeks.

'Yann, you shall sleep now—this instant. Tell me only how you came— a word or two—that I may repeat to the farmer.'

So I did my best, and told her about the run, and the dragoons on the beach, and how I came on Lilith's back.

'Wonderful, wonderful! But how came she to allow you?'

'That I know not, madame. But when I spoke to her she was quiet at once.'

'In the Breton—you spoke in the Breton? Yes, yes, that explains—*I* taught her. Dear Lilith!' She patted the mare's neck, and broke off to clap her hands again and interpret the tale to the farmer and his wife; and the farmer growled a bit, and then they all began to laugh.

'He says you are a "rumgo", and you had better be put to bed. But the packet on your back—your nightshirt, I suppose? You have managed it all so complete, Yann!' And she laughed merrily.

'It holds fifteen little wooden dolls,' said I, 'jointed at the knees and elbows; and they cost two sols apiece.'

'My little dolls—you clever boy! Oh, you clever little boy!' She kissed me twice again. 'Come, and you shall sleep, and then, when you wake, you shall see.'

She took me by the hand and hurried me into the house, and upstairs to a great bedroom with a large oaken four-post bed in it, and a narrow wooden bed beside, and a fire lit, and an armchair by the hearth. The four-post bed had curtains of green damask, all closely pinned around it, and a green valence. But she went to the little bed, which was hung with pink dimity, and pulled the white sheets out of it and replaced them with others from a great wardrobe sunk in the wall. And while I sat in the chair by the fire, munching a crust of bread and feeling half inclined to cry and more than half inclined to sleep, she left me, and returned with a can of hot water and a vast night-shirt of the farmer's, and bade me good-night.

'Be quick and undress, little one.' She turned at the door. 'The tubs are all in hiding by this time. Good night, Yann.'

I believe I slept as soon as my head touched the sweet-smelling pillow; and I must have slept the round of the clock before I opened my eyes, for the room was now bright with candles, and in the armchair by the fire sat the Breton lady sewing as if for dear life.

But the wonder of her was that she now wore a short plain dress such as girls wear in the convent schools in Brittany, and her grey hair was tied just like a girl's. One little foot rested on the brass fender, and the fire-light played on its silver shoe-buckle.

I coughed, to let her know that I was awake, and she looked across and nodded.

'Almost ten o'clock, Yann, and time for you to rise and have supper. And after supper—are you sorry?—another journey for you. At midnight you start in the gig with Farmer Ellory, who will drive you to the coast, to a town called Fowey, where some friends of his "in the trade" are starting for Roscoff. In six hours you will be aboard ship again; and in another twenty, perhaps, you will see your mother—and your father too, if he escaped clear away. In little more than a day you will be back in Brittany. But first you must lie quite still, and I will show you something.'

'To be sure I will, madame.'

'You must not call me that. I am the Demoiselle Héloïse Kéranguin. You know St Pol de Léon, Yann?'

'Almost as well as my own town, mademoiselle.'

'And the Convent of the Grey Nuns, on the road to Morlaix, a little beyond the town?'

It was on my tongue to tell her that fire and soldiery had wiped it even with the ground, during the 'Terror'. But she interrupted me. Setting down her work-basket, which was heaped high with reels and parti-coloured rags of silk, she pushed a small table over to the big bed and loaded it with candlesticks. There were three candles already alight in the room, but she lit others and set them in line—brass candlesticks, plated candlesticks, candlesticks of chinaware—fourteen candlesticks in all, and fresh candles in each. Laying a finger on her lip, she stepped to the big bed and unfastened the corking-pins which held the green curtains together. As she pushed the curtains back I lifted myself on an elbow.

It was into a real theatre that I looked. She had transformed the whole level of the bed into a miniature stage, with buildings of cardboard, cleverly painted, and gardens cut out of silk and velvet and laid down, and rose-trees gummed on little sticks, and a fishpond and brook of looking-glass, with embroidered flowers stuck along their edges, and along the paths (of real sand) a score of little dolls walking, all dressed in the uniform of the Grey Nuns. I declare it was so real, you could almost hear the fountain playing, with its *jet d'eau* of transparent beads strung on an invisible wire.

'But how pretty, mademoiselle!' I cried.

She clasped her hands nervously. 'But is it *like*, Yann? It is so long ago that I may have forgotten. Tell me if it is like; or if there is anything wrong. I promise not to be offended.'

'It is exactly like, mademoiselle.'

'See, here is the Mother Superior; and this is Sœur Gabrielle. I have to make the dresses full and stiff, or they wouldn't stand up. And that, with the blue eyes, is Sœur Hyacinthe. She walks with me—this is I—as she always did. And what do you think? With the fifteen dolls that you have brought I am going to have a real Pardon, and townspeople and fisher-people to stand and worship at the altar of the Virgin, there in the corner. I made it of wax, and stamped the face with a seal that Charles gave me. He was to have been my husband when I left the school.'

'Indeed, mademoiselle?'

'Yes, but the soldiers burnt his house. It was but a week after I left the school, and the Château Sant-Ervoan lay but a mile from my mother's house. He fled to us, wounded; and we carried him to the coast—there was a price on his head, and we, too, had to flee—and escaped over to England. He died on this bed, Yann. Look——'

She lifted a candle, and there on the bed's ledge I read, in gilt lettering, some words I have never forgotten, though it was not until years after that I got a priest to explain them to me. They were 'C. DE. R. COMES ET ECSUL. MDCCXCIII.'

While I stared, she set the candle down again and gently drew the curtains round the bed.

'Rise now and dress, dear child, or your supper will be cold and the farmer impatient. You have done me good. Although I have written the farmer's letters for him, it never seemed to me that I wrote to living people: for all I used to know in Brittany, ten years ago, are dead. For the future I shall write to you.'

She turned at the door as she said this, and that was the last I ever saw of her. For when I passed out of the room, dressed and ready for my journey, it was quite dark on the landing, where she met and kissed me. Then she slipped a little packet into my hand.

'For the dolls,' she said.

In the kitchen I slipped it out of my pocket and examined it under the table's edge. It was a little silver crucifix, and I have kept it to this day.

How the Brigadier Played For a Kingdom

៛∞ଽ

It has sometimes struck me that some of you, when you have heard me tell these little adventures of mine, may have gone away with the impression that I was conceited. There could not be a greater mistake than this, for I have always observed that really fine soldiers are free from this failing. It is true that I have had to depict myself sometimes as brave, sometimes as full of resource, always as interesting; but, then, it really was so, and I had to take the facts as I found them. It would be an unworthy affectation if I were to pretend that my career has been anything but a fine one. The incident which I will tell you tonight, however, is one which you will understand that only a modest man would describe. After all, when one has attained such a position as mine, one can afford to speak of what an ordinary man might be tempted to conceal.

You must know, then, that after the Russian campaign the remains of our poor army were quartered along the western bank of the Elbe, where they might thaw their frozen blood and try, with the help of the good German beer, to put a little between their skin and their bones. There were some things which we could not hope to regain, for I dare say that three large commissariat fourgons would not have sufficed to carry the fingers and the toes which the army had shed during that retreat. Still, lean and crippled as we were, we had much to be thankful for when we thought of our poor comrades whom we had left behind, and of the snowfields—the horrible, horrible snowfields. To this day, my friends, I do not care to see red and white together. Even my red cap thrown down upon my white counterpane has given me dreams in which I have seen those monstrous plains, the reeling, tortured army, and the crimson smears which glared upon the snow behind them. You will coax no story out of me about that business, for the thought of it is enough to turn my wine to vinegar and my tobacco to straw.

Of the half-million who crossed the Elbe in the autumn of the year '12, about forty thousand infantry were left in the spring of '13. But they were terrible men, these forty thousand: men of iron, eaters of horses, and sleepers in the snow; filled, too, with rage and bitterness against the Russians. They would hold the Elbe until the great army of conscripts, which the Emperor was raising in France, should be ready to help them to cross it once more.

But the cavalry was in a deplorable condition. My own hussars were at Borna, and when I paraded them first, I burst into tears at the sight of them. My fine men and my beautiful horses—it broke my heart to see the state to which they were reduced. 'But, courage,' I thought, 'they have lost much, but their Colonel is still left to them.' I set to work, therefore, to repair their disasters, and had already constructed two good squadrons, when an order came that all colonels of cavalry should repair instantly to the depots of the regiments in France to organize the recruits and the remounts for the coming campaign.

You will think, doubtless, that I was overjoyed at this chance of visiting home once more. I will not deny that it was a pleasure to me to know that I should see my mother again, and there were a few girls who would be very glad at the news; but there were others in the army who had a stronger claim. I would have given my place to any who had wives and children whom they might not see again. However, there is no arguing when the blue paper with the little red seal arrives, so within an hour I was off upon my great ride from the Elbe to the Vosges. At last I was to have a period of quiet. War lay behind my mare's tail and peace in front of her nostrils. So I thought, as the sound of the bugles died in the distance, and the long, white road curled away in front of me through plain and forest and mountain, with France somewhere beyond the blue haze which lay upon the horizon.

It is interesting, but it is also fatiguing, to ride in the rear of an army. In the harvest time our soldiers could do without supplies, for they had been trained to pluck the grain in the fields as they passed, and to grind it for themselves in their bivouacs. It was at that time of year, therefore, that those swift marches were performed which were the wonder and the despair of Europe. But now the starving men had to be made robust once more, and I was forced to draw into the ditch continually as the Coburg sheep and the Bavarian bullocks came streaming past with waggon loads of Berlin beer and good French cognac. Sometimes, too, I would hear the dry rattle of the drums and the shrill whistle of the fifes, and long columns of our good little infantry men would swing past me with the white dust lying thick upon their blue tunics. These were old soldiers drawn from the garrisons of our German fortresses, for it was not until May that the new conscripts began to arrive from France.

Well, I was rather tired of this eternal stopping and dodging, so that I was not sorry when I came to Altenburg to find that the road divided, and that I could take the southern and quieter branch. There were few wayfarers between there and Greiz, and the road wound through groves of oaks and beeches, which shot their branches across the path. You will think it strange that a Colonel of hussars should again and again pull up his horse in order to admire the beauty of the feathery branches and the little, green, new-budded leaves, but if you had spent six months among the fir trees of Russia you would be able to understand me.

There was something, however, which pleased me very much less than the beauty of the forests, and that was the words and looks of the folk who lived in the woodland villages. We had always been excellent friends with the Germans, and during the last six years they had never seemed to bear us any malice for having made a little free with their country. We had shown kindnesses to the men and received them from the women, so that good, comfortable Germany was a second home to all of us. But now there was something which I could not understand in the behaviour of the people. The travellers made no answer to my salute; the foresters turned their heads away to avoid seeing me; and in the villages the folk would gather into knots in the roadway and would scowl at me as I passed. Even women would do this, and it was something new for me in those days to see anything but a smile in a woman's eyes when they were turned upon me.

It was in the hamlet of Schmolin, just ten miles out of Altenburg, that the thing became most marked. I had stopped at the little inn there just to damp my moustache and to wash the dust out of poor Violette's throat. It was my way to give some little compliment, or possibly a kiss, to the maid who served me; but this one would have neither the one nor the other, but darted a glance at me like a bayonet-thrust. Then when I raised my glass to the folk who drank their beer by the door they turned their backs on me, save only one fellow, who cried, 'Here's a toast for you, boys! Here's to the letter T!' At that they all emptied their beer mugs and laughed; but it was not a laugh that had good-fellowship in it.

I was turning this over in my head and wondering what their boorish conduct could mean, when I saw, as I rode from the village, a great T new carved upon a tree. I had already seen more than one in my morning's ride, but I had given no thought to them until the words of the beer-drinker gave them an importance. It chanced that a respectable-looking person was riding past me at the moment, so I turned to him for information.

'Can you tell me, sir,' said I, 'what this letter T is?'

He looked at it and then at me in the most singular fashion. 'Young man,' said he, 'it is not the letter N.' Then before I could ask further he clapped his spurs into his horse's ribs and rode, stomach to earth, upon his way.

At first his words had no particular significance in my mind, but as I trotted onwards Violette chanced to half turn her dainty head, and my eyes were caught by the gleam of the brazen N's at the end of the bridle-chain. It was the Emperor's mark. And those T's meant something which was opposite to it. Things had been happening in Germany, then, during our absence, and the giant sleeper had begun to stir. I thought of the mutinous faces that I had seen, and I felt that if I could only have looked into the hearts of these people I might have had some strange news to bring into France with me. It made me the more eager to get my remounts, and to see ten strong squadrons behind my kettledrums once more.

While these thoughts were passing through my head I had been alternately walking and trotting, as a man should who has a long journey before, and a willing horse beneath him. The woods were very open at this point, and beside the road there lay a great heap of faggots. As I passed there came a sharp sound from among them, and, glancing round, I saw a face looking out at me—a hot, red face, like that of a man who is beside himself with excitement and anxiety. A second glance told me that it was the very person with whom I had talked an hour before in the village.

'Come nearer!' he hissed. 'Nearer still! Now dismount and pretend to be mending the stirrup leather. Spies may be watching us, and it means death to me if I am seen helping you.'

'Death!' I whispered. 'From whom?'

'From the Tugendbund. From Lutzow's night-riders. You Frenchmen are living on a powder magazine, and the match has been struck that will fire it.'

'But this is all strange to me,' said I, still fumbling at the leathers of my horse. 'What is this Tugendbund?'

'It is the secret society which has planned the great rising which is to drive you out of Germany, just as you have been driven out of Russia.'

'And these T's stand for it?'

'They are the signal. I should have told you all this in the village, but I dared not be seen speaking with you. I galloped through the woods to cut you off, and concealed both my horse and myself.'

'I am very much indebted to you,' said I, 'and the more so as you are the only German that I have met today from whom I have had common civility.'

'All that I possess I have gained through contracting for the French armies,' said he. 'Your Emperor has been a good friend to me. But I beg that you will ride on now, for we have talked long enough. Beware only of Lutzow's night-riders!'

'Banditti?' I asked.

'All that is best in Germany,' said he. 'But for God's sake ride forwards, for I have risked my life and exposed my good name in order to carry you this warning.'

Well, if I had been heavy with thought before, you can think how I felt after my strange talk with the man among the faggots. What came home to me even more than his words was his shivering, broken voice, his twitching face, and his eyes glancing swiftly to right and left, and opening in horror whenever a branch cracked upon a tree. It was clear that he was in the last extremity of terror, and it is possible that he had cause, for shortly after I had left him I heard a distant gunshot and a shouting from somewhere behind me. It may have been some sportsman halloaing to his dogs, but I never again hear of or saw the man who had given me my warning.

I kept a good look-out after this, riding swiftly where the country was open, and slowly where there might be an ambuscade. It was serious for me, since 500 good miles of German soil lay in front of me; but somehow I did not take it very much to heart, for the Germans had always seemed to me to be a kindly, gentle people, whose hands closed more readily round a pipe-stem than a sword-hilt—not out of want of valour, you understand, but because they are genial, open souls, who would rather be on good terms with all men. I did not know then that beneath that homely surface there lurks a devilry as fierce as, and far more persistent than, that of the Castilian or the Italian.

And it was not long before I had shown to me that there was something more serious abroad than rough words and hard looks. I had come to a spot where the road runs upwards through a wild tract of heathland and vanishes into an oak wood. I may have been half-way up the hill when, looking forward, I saw something gleaming under the shadow of the tree-trunks, and a man came out with a coat which was so slashed and spangled with gold that he blazed like a fire in the sunlight. He appeared to be very drunk, for he reeled and staggered as he came towards me. One of his hands was held up to his ear and clutched a great red handkerchief, which was fixed to his neck.

I had reined up the mare and was looking at him with some disgust, for it seemed strange to me that one who wore so gorgeous a uniform should show himself in such a state in broad daylight. For his part, he looked hard in my direction and came slowly onwards, stopping from time to time and swaying about as he gazed at me. Suddenly, as I again advanced, he screamed out his thanks to Christ, and, lurching forwards, he fell with a crash upon the dusty road. His hands flew forward with the fall, and I saw that what I had taken for a red cloth was a monstrous wound, which had left a great gap in his neck, from which a dark blood-clot hung, like an epaulette upon his shoulder.

'My God!' I cried, as I sprang to his aid. 'And I thought that you were drunk!'

'Not drunk, but dying,' said he. 'But thank Heaven that I have seen a French officer while I have still strength to speak.'

I laid him among the heather and poured some brandy down his throat. All round us was the vast countryside, green and peaceful, with nothing living in sight save only the mutilated man beside me.

'Who has done this?' I asked, 'and what are you? You are French, and yet the uniform is strange to me.'

'It is that of the Emperor's new guard of honour. I am the Marquis of Château St Arnaud, and I am the ninth of my blood who has died in the service of France. I have been pursued and wounded by the night-riders of Lutzow, but I hid among the brushwood yonder, and waited in the hope that a Frenchman might pass. I could not be sure at first if you were friend or foe, but I felt that death was very near, and that I must take the chance.'

'Keep your heart up, comrade,' said I; 'I have seen a man with a worse wound who has lived to boast of it.'

'No, no,' he whispered; 'I am going fast.' He laid his hand upon mine as he spoke, and I saw that his finger-nails were already blue. 'But I have papers here in my tunic which you must carry at once to the Prince of Saxe-Felstein, at his Castle of Hof. He is still true to us, but the Princess is our deadly enemy. She is striving to make him declare against us. If he does so, it will determine all those who are wavering, for the King of Prussia is his uncle and the King of Bavaria his cousin. These papers will hold him to us if they can only reach him before he takes the last step. Place them in his hands tonight, and, perhaps, you will have saved all Germany for the Emperor. Had my horse not been shot, I might, wounded as I am——' he choked, and the cold hand tightened into a grip, which left mine as bloodless as itself. Then, with a groan, his head jerked back, and it was all over with him.

Here was a fine start for my journey home. I was left with a commission of which I knew little, which would lead me to delay the pressing needs of my hussars, and which at the same time was of such importance that it was impossible for me to avoid it. I opened the Marquis's tunic, the brilliance of which had been devised by the Emperor in order to attract those young aristocrats from whom he hoped to raise these new regiments of his Guard. It was a small packet of papers which I drew out, tied up with silk, and addressed to the Prince of Saxe-Felstein. In the corner, in a sprawling, untidy hand, which I knew to be the Emperor's own, was written: 'Pressing and most important.' It was an order to me, those four words—an order as clear as if it had come straight from the firm lips with the cold grey eyes looking into mine. My troopers might wait for their horses, the dead Marquis might lie where I had laid him amongst the heather, but if the mare and her rider had a breath left in them the papers should reach the Prince that night.

I should not have feared to ride by the road through the wood, for I have learned in Spain that the safest time to pass through a guerrilla country is after an outrage, and that the moment of danger is when all is peaceful. When I came to look upon my map, however, I saw that Hof lay further to the south of me, and that I might reach it more directly by keeping to the moors. Off I set, therefore, and had not gone fifty yards before two carbine shots rang out of the brushwood and a bullet hummed past me like a bee. It was clear that the night-riders were bolder in their ways than the brigands of Spain, and that my mission would have ended where it had begun if I had kept to the road.

It was a mad ride, that—a ride with a loose rein, girth-deep in heather and in gorse, plunging through bushes, flying down hillsides, with my neck at the mercy of my dear little Violette. But she—she never slipped, she never faltered, as swift and as surefooted as if she knew that her rider carried the fate of all Germany beneath the buttons of his pelisse. And I—I had long borne the name of being the best horseman in the six brigades of light cavalry, but I never rode as I rode then. My friend the Bart has told me of how they hunt the fox in England, but the swiftest fox would have been captured by me that day. The wild pigeons which flew overhead did not take a straighter course than Violette and I below. As an officer, I have always been ready to sacrifice myself for my men, though the Emperor would not have thanked me for it, for he had many men, but only one—well, cavalry leaders of the first class are rare.

But here I had an object which was indeed worth a sacrifice, and I thought no more of my life than of the clods of earth that flew from my darling's heels.

We struck the road once more as the light was failing, and galloped into the little village of Lobenstein. But we had hardly got upon the cobblestones when off came one of the mare's shoes, and I had to lead her to the village smithy. His fire was low, and his day's work done, so that it would be an hour at the least before I could hope to push on to Hof. Cursing at the delay, I strode into the village inn and ordered a cold chicken and some wine to be served for my dinner. It was but a few miles to Hof, and I had every hope that I might deliver my papers to the Prince on that very night, and be on my way for France next morning with dispatches for the Emperor in my bosom. I will tell you now what befell me in the inn of Lobenstein.

The chicken had been served and the wine drawn, and I had turned upon both as a man may who has ridden such a ride, when I was aware of a murmur and a scuffling in the hall outside my door. At first I thought that it was some brawl between peasants in their cups, and I left them to settle their own affairs. But of a sudden there broke from among the low, sullen growl of the voices such a sound as would send Étienne Gerard leaping from his deathbed. It was the whimpering cry of a woman in pain. Down clattered my knife and my fork, and in an instant I was in the thick of the crowd which had gathered outside my door.

The heavy-cheeked landlord was there and his flaxen-haired wife, the two men from the stables, a chambermaid and two or three villagers. All of them, women and men, were flushed and angry, while there in the centre of them, with pale cheeks and terror in her eyes, stood the loveliest woman that ever a soldier would wish to look upon. With her queenly head thrown back, and a touch of defiance mingled with her fear, she looked as she gazed round her like a creature of a different race from the vile, coarse-featured crew who surrounded her. I had not taken two steps from my door before she sprang to meet me, her hand resting upon my arm and her blue eyes sparkling with joy and triumph.

'A French soldier and gentleman!' she cried. 'Now at last I am safe.'

'Yes, madam, you are safe,' said I, and I could not resist taking her hand in mine in order that I might reassure her. 'You have only to command me,' I added, kissing the hand as a sign that I meant what I was saying.

'I am Polish,' she cried; 'the Countess Palotta is my name. They abuse me because I love the French. I do not know what they might have done to me had Heaven not sent you to my help.'

I kissed her hand again lest she should doubt my intentions. Then I turned upon the crew with such an expression as I know how to assume. In an instant the hall was empty.

'Countess,' said I, 'you are now under my protection. You are faint, and a glass of wine is necessary to restore you.' I offered her my arm and escorted her into my room, where she sat by my side at the table and took the refreshment which I offered her.

How she blossomed out in my presence, this woman, like a flower before the sun! She lit up the room with her beauty. She must have read my admiration in my eyes, and it seemed to me that I also could see something of the sort in her own. Ah! my friends, I was no ordinary-looking man when I was in my thirtieth year. In the whole light cavalry it would have been hard to find a finer pair of whiskers. Murat's may have been a shade longer, but the best judges are agreed that Murat's were a shade too long. And then I had a manner. Some women are to be approached in one way and some in another, just as a siege is an affair of fascines and gabions in hard weather and of trenches in soft. But the man who can mix daring with timidity, who can be outrageous with an air of humility, and presumptuous with a tone of deference, that is the man whom mothers have to fear. For myself, I felt that I was the guardian of this lonely lady, and knowing what a dangerous man I had to deal with, I kept strict watch upon myself. Still, even a guardian has his privileges, and I did not neglect them.

But her talk was as charming as her face. In a few words she explained that she was travelling to Poland, and that her brother who had been her escort had fallen ill upon the way. She had more than once met with ill-treatment from the country folk because she could not conceal her good-will towards the French. Then turning from her own affairs she questioned me about the army, and so came round to myself and my own exploits. They were familiar to her, she said, for she knew several of Poniatowski's officers, and they had spoken of my doings. Yet she would be glad to hear them from my own lips. Never have I had so delightful a conversation. Most women make the mistake of talking rather too much about their own affairs, but this one listened to my tales just as you are listening now, ever asking for more and more and more. The hours slipped rapidly by, and it was with horror that I heard the village clock strike eleven, and so learned that for four hours I had forgotten the Emperor's business.

'Pardon me, my dear lady,' I cried, springing to my feet, 'but I must on instantly to Hof.'

She rose also, and looked at me with a pale, reproachful face. 'And me?' she said. 'What is to become of me?'

'It is the Emperor's affair. I have already stayed far too long. My duty calls me, and I must go.'

'You must go? And I must be abandoned alone to these savages? Oh, why did I ever meet you? Why did you ever teach me to rely upon your strength?' Her eyes glazed over, and in an instant she was sobbing upon my bosom.

Here was a trying moment for a guardian! Here was a time when he had to keep a watch upon a forward young officer. But I was equal to it. I smoothed her rich brown hair and whispered such consolations as I could think of in her ear, with one arm round her, it is true, but that was to hold her lest she should faint. She turned her tear-stained face to mine. 'Water,' she whispered. 'For God's sake, water!'

I saw that in another moment she would be senseless. I laid the drooping head upon the sofa, and then rushed furiously from the room, hunting from chamber to chamber for a carafe. It was some minutes before I could get one and hurry back with it. You can imagine my feelings to find the room empty and the lady gone.

Not only was she gone, but her cap and silver-mounted riding switch which had lain upon the table were gone also. I rushed out and roared for the landlord. He knew nothing of the matter, had never seen the woman before, and did not care if he never saw her again. Had the peasants at the door seen anyone ride away? No, they had seen nobody. I searched here and searched there, until at last I chanced to find myself in front of a mirror, where I stood with my eyes staring and my jaw as far dropped as the chin-strap of my shako would allow.

Four buttons of my pelisse were open, and it did not need me to put my hand up to know that my precious papers were gone. Oh! the depth of cunning that lurks in a woman's heart. She had robbed me, this creature, robbed me as she clung to my breast. Even while I smoothed her hair, and whispered kind words into her ear, her hands had been at work beneath my dolman. And here I was, at the very last step of my journey, without the power of carrying out this mission which had already deprived one good man of his life, and was likely to rob another one of his credit. What would the Emperor say when he heard that I had lost his dispatches? Would the army believe it of Étienne Gerard? And when they heard that a woman's hand had coaxed them from me, what laughter there would be at mess-table and at camp-fire! I could have rolled upon the ground in my despair.

But one thing was certain—all this affair of the fracas in the hall and the persecution of the so-called Countess was a piece of acting from the beginning. This villainous innkeeper must be in the plot. From him I might learn who she was and where my papers had gone. I snatched my sabre from the table and rushed out in search of him. But the scoundrel had guessed what I would do, and had made his preparations for me. It was in the corner of the yard that I found him, a blunderbuss in his hands and a mastiff held upon a leash by his son. The two stable-hands, with pitchforks, stood upon either side, and the wife held a great lantern behind him, so as to guide his aim.

'Ride away, sir, ride away!' he cried, with a crackling voice. 'Your horse is at the door, and no one will meddle with you if you go your way; but if you come against us, you are alone against three brave men.'

I had only the dog to fear, for the two forks and the blunderbuss were shaking about like branches in a wind. Still, I considered that, though I might force an answer with my sword-point at the throat of this fat rascal, still I should have no means of knowing whether that answer was the truth. It would be a struggle, then, with much to lose and nothing certain to gain. I looked them up and down, therefore, in a way that set their foolish weapons shaking worse than ever, and then, throwing myself upon my mare, I galloped away with the shrill laughter of the landlady jarring upon my ears.

I had already formed my resolution. Although I had lost my papers, I could make a very good guess as to what their contents would be, and this I would say from my own lips to the Prince of Saxe-Felstein, as though the Emperor had commissioned me to convey it in that way. It was a bold stroke and a dangerous one, but if I went too far I could afterwards be disavowed. It was that or nothing, and when all Germany hung on the balance the game should not be lost if the nerve of one man could save it.

It was midnight when I rode into Hof, but every window was blazing, which was enough in itself, in that sleepy country, to tell the ferment of excitement in which the people were. There was hooting and jeering as I rode through the crowded streets, and once a stone sang past my head, but I kept upon my way, neither slowing nor quickening my pace, until I came to the palace. It was lit from base to battlement, and the dark shadows, coming and going against the yellow glare, spoke of the turmoil within. For my part, I handed my mare to a groom at the gate, and striding in I demanded, in such a voice as an ambassador should have, to see the Prince instantly, upon business which would brook no delay.

The hall was dark, but I was conscious as I entered of a buzz of innumerable voices, which hushed into silence as I loudly proclaimed my mission. Some great meeting was being held then—a meeting which, as my instincts told me, was to decide this very question of war and peace. It was possible that I might still be in time to turn the scale for the Emperor and for France. As to the major-domo, he looked blackly at me, and showing me into a small antechamber he left me. A minute later he returned to say that the Prince could not be disturbed at present, but that the Princess would take my message.

The Princess! What use was there in giving it to her? Had I not been warned that she was German in heart and soul, and that it was she who was turning her husband and her State against us?

'It is the Prince that I must see,' said I.

'Nay, it is the Princess,' said a voice at the door, and a woman swept into the chamber. 'Von Rosen, you had best stay with us. Now, sir, what is it that you have to say to either Prince or Princess of Saxe-Felstein?'

At the first sound of the voice I had sprung to my feet. At the first glance I had thrilled with anger. Not twice in a lifetime does one meet that noble figure, that queenly head, and those eyes as blue as the Garonne, and as chilling as her winter waters.

'Time presses, sir!' she cried, with an impatient tap of her foot. 'What have you to say to me?'

'What have I to say to you?' I cried. 'What can I say, save that you have taught me never to trust a woman more? You have ruined and dishonoured me for ever.'

She looked with arched brows at her attendant.

'It this the raving of fever, or does it come from some less innocent cause?' said she. 'Perhaps a little blood-letting——'

'Ah, you can act!' I cried. 'You have shown me that already.'

'Do you mean that we have met before?'

'I mean that you have robbed me within the last two hours.'

'This is past all bearing,' she cried, with an admirable affectation of anger. 'You claim, as I understand, to be an ambassador, but there are limits to the privileges which such an office brings with it.'

'You brazen it admirably,' said I. 'Your Highness will not make a fool of me twice in one night.' I sprang forward and, stooping down, caught up the hem of her dress. 'You would have done well to change it after you had ridden so far and so fast,' said I.

It was like the dawn upon a snow-peak to see her ivory cheeks flush suddenly to crimson.

'Insolent!' she cried, 'Call the foresters and have him thrust from the palace!'

'I will see the Prince first.'

'You will never see the Prince. Ah! Hold him, Von Rosen, hold him!'

She had forgotten the man with whom she had to deal—was it likely that I would wait until they could bring their rascals? She had shown me her cards too soon. Her game was to stand between me and her husband. Mine was to speak face to face with him at any cost. One spring took me out of the chamber. In another I had crossed the hall. An instant later I had burst into the great room from which the murmur of the meeting had come. At the far end I saw a figure upon a high chair under a dais. Beneath him was a line of high dignitaries, and then on every side I saw vaguely the heads of a vast assembly. Into the centre of the room I strode, my sabre clanking, my shako under my arm.

'I am the messenger of the Emperor,' I shouted. 'I bear his message to His Highness the Prince of Saxe-Felstein.'

The man beneath the dais raised his head, and I saw that his face was thin and wan, and that his back was bowed as though some huge burden was balanced between his shoulders.

'Your name, sir?' he asked.

'Colonel Étienne Gerard, of the Third Hussars.'

Every face in the gathering was turned upon me, and I heard the rustle of the innumerable necks and saw countless eyes without meeting one friendly one amongst them. The woman had swept past me, and was whispering, with many shakes of her head and dartings of her hands, into the Prince's ear. For my own part I threw out my chest and curled my moustache, glancing round in my own debonair fashion at the assembly. They were men, all of them, professors from the college, a sprinkling of their students, soldiers, gentlemen, artisans, all very silent and serious. In one corner there sat a group of men in black, with riding-coats drawn over their shoulders. They leaned their heads to each other, whispering under their breath, and with every movement I caught the clank of their sabres or the clink of their spurs.

'The Emperor's private letter to me informs me that it is the Marquis Château St Arnaud who is bearing his dispatches,' said the Prince.

'The Marquis has been foully murdered,' I answered, and a buzz rose up from the people as I spoke. Many heads were turned, I noticed, towards the dark men in the cloaks.

'Where are your papers?' asked the Prince.

'I have none.'

A fierce clamour rose instantly around me. 'He is a spy! He plays a part!' they cried. 'Hang him!' roared a deep voice from the corner, and a dozen others took up the shout. For my part, I drew out my handkerchief and flicked the dust from the fur of my pelisse. The Prince held out his thin hands, and the tumult died away.

'Where, then, are your credentials, and what is your message?'

'My uniform is my credential, and my message is for your private ear.'

He passed his hand over his forehead with the gesture of a weak man who is at his wits' end what to do. The Princess stood beside him with her hand upon his throne, and again whispered in his ear.

'We are here in council together, some of my trusty subjects and my-self,' said he. 'I have no secrets from them, and whatever message the Emperor may send to me at such a time concerns their interests no less than mine.'

There was a hum of applause at this, and every eye was turned once more upon me. My faith, it was an awkward position in which I found myself, for it is one thing to address eight hundred hussars, and another to speak to such an audience on such a subject. But I fixed my eyes upon the Prince, and tried to say just what I should have said if we had been alone, shouting it out, too, as though I had my regiment on parade.

'You have often expressed friendship for the Emperor,' I cried. 'It is now at last that this friendship is about to be tried. If you will stand firm, he will reward you as only he can reward. It is an easy thing for him to turn a Prince into a King and a province into a power. His eyes are fixed upon you, and though you can do little to harm him, you can ruin your-self. At this moment he is crossing the Rhine with two hundred thousand men. Every fortress in the country is in his hands. He will be upon you in a week, and if you have played him false, God help both you and your people. You think that he is weakened because a few of us got the chil-blains last winter. Look there!' I cried, pointing to a great star which blazed through the window above the Prince's head. 'That is the Emperor's star. When it wanes, he will wane—but not before.'

You would have been proud of me, my friends, if you could have seen and heard me, for I clashed my sabre as I spoke, and swung my dolman as though my regiment was picketed outside in the courtyard. They listened to me in silence, but the back of the Prince bowed more and more as though the burden which weighed upon it was greater than his strength. He looked round with haggard eyes.

'We have heard a Frenchman speak for France,' said he. 'Let us have a German speak for Germany.'

The folk glanced at each other, and whispered to their neighbours. My speech, as I think, had its effect, and no man wished to be the first to commit himself in the eyes of the Emperor. The Princess looked round her with blazing eyes, and her clear voice broke the silence.

'Is a woman to give this Frenchman his answer?' she cried. 'Is it possible, then, that among the night-riders of Lutzow, there is none who can use his tongue as well as his sabre?'

Over went a table with a crash, and a young man had bounded upon one of the chairs. He had the face of one inspired—pale, eager, with wild hawk eyes, and tangled hair. His sword hung straight from his side, and his riding-boots were brown with mire.

'It is Korner!' the people cried. 'It is young Korner, the poet! Ah, he will sing, he will sing.'

And he sang! It was soft, at first, and dreamy, telling of old Germany, the mother of nations, of the rich, warm plains, and the grey cities, and the fame of dead heroes. But then verse after verse rang like a trumpet-call. It was of the Germany of now, the Germany which had been taken unawares and overthrown, but which was up again, and snapping the bonds upon her giant limbs. What was life that one should covet it? What was glorious death that one should shun it? The mother, the great mother, was calling. Her sigh was in the night wind. She was crying to her own children for help. Would they come? Would they come? Would they come?

Ah, that terrible song, the spirit face and the ringing voice! Where were I, and France, and the Emperor? They did not shout, these people—they howled. They were up on the chairs and the tables. They were raving, sobbing, the tears running down their faces. Korner had sprung from the chair, and his comrades were round him with their sabres in the air. A flush had come into the pale face of the Prince, and he rose from his throne.

'Colonel Gerard,' said he, 'you have heard the answer which you are to carry to your Emperor. The die is cast, my children. Your Prince and you must stand or fall together.'

He bowed to show that all was over, and the people with a shout made for the door to carry the tidings into the town. For my own part, I had done all that a brave man might, and so I was not sorry to be carried out amid the stream. Why should I linger in the palace? I had had my answer and must carry it, such as it was. I wished neither to see Hof nor its people again until I entered it at the head of a vanguard. I turned from the throng, then, and walked silently and sadly in the direction in which they had led the mare.

It was dark down there by the stables, and I was peering round for the ostler, when suddenly my two arms were seized from behind. There were hands at my wrists and at my throat, and I felt the cold muzzle of a pistol under my ear.

'Keep your lips closed, you French dog,' whispered a fierce voice. 'We have him, captain.'

'Have you the bridle?'

'Here it is.'

'Sling it over his head.'

I felt the cold coil of leather tighten round my neck. An ostler with a stable lantern had come out and was gazing upon the scene. In its dim light I saw stern faces breaking everywhere through the gloom, with the black caps and dark cloaks of the night-riders.

'What would you do with him, captain?' cried a voice.

'Hang him at the palace gate.'

'An ambassador?'

'An ambassador without papers.'

'But the Prince?'

'Tut, man, do you not see that the Prince will then be committed to our side? He will be beyond all hope of forgiveness. At present he may swing round tomorrow as he has done before. He may eat his words, but a dead hussar is more than he can explain.'

'No, no, Von Strelitz, we cannot do it,' said another voice.

'Can we not? I shall show you that!' and there came a jerk on the bridle which nearly pulled me to the ground. At the same instant a sword flashed and the leather was cut through within two inches of my neck.

'By Heaven, Korner, this is rank mutiny,' cried the captain. 'You may hang yourself before you are through with it.'

'I have drawn my sword as a soldier and not as a brigand,' said the young poet. 'Blood may dim its blade, but never dishonour. Comrades, will you stand by and see this gentleman mishandled?'

A dozen sabres flew from their sheaths, and it was evident that my friends and my foes were about equally balanced. But the angry voices and the gleam of steel had brought the folk running from all parts.

'The Princess!' they cried. 'The Princess is coming!'

And even as they spoke I saw her in front of us, her sweet face framed in the darkness. I had cause to hate her, for she had cheated and befooled me, and yet it thrilled me then and thrills me now to think that my arms have embraced her, and that I have felt the scent of her hair in my nostrils. I know not whether she lies under her German earth, or whether she still lingers, a grey-haired woman in her Castle of Hof, but she lives ever, young and lovely, in the heart and memory of Étienne Gerard.

'For shame!' she cried, sweeping up to me, and tearing with her own hands the noose from my neck. 'You are fighting in God's own quarrel, and yet you would begin with such a devil's deed as this. This man is mine, and he who touches a hair of his head will answer for it to me.'

They were glad enough to slink off into the darkness before those scornful eyes. Then she turned once more to me.

'You can follow me, Colonel Gerard,' she said. 'I have a word that I would speak to you.'

I walked behind her to the chamber into which I had originally been shown. She closed the door, and then looked at me with the archest twinkle in her eyes.

'Is it not confiding of me to trust myself with you?' said she. 'You will remember that it is the Princess of Saxe-Felstein and not the poor Countess Palotta of Poland.'

'Be the name what it might,' I answered, 'I helped a lady whom I believed to be in distress, and I have been robbed of my papers and almost of my honour as a reward.'

'Colonel Gerard,' said she, 'we have been playing a game, you and I, and the stake was a heavy one. You have shown by delivering a message which was never given to you that you would stand at nothing in the cause of your country. My heart is German and yours is French, and I also would go all lengths, even to deceit and to theft, if at this crisis I could help my suffering fatherland. You see how frank I am.'

'You tell me nothing that I have not seen.'

'But now that the game is played and won, why should we bear malice? I will say this, that if ever I were in such a plight as that which I pretended in the inn of Lobenstein, I should never wish to meet a more gallant protector or a truer-hearted gentleman than Colonel Étienne Gerard. I had never thought that I could feel for a Frenchmen as I felt for you when I slipped the papers from your breast.'

'But you took them, none the less.'

'They were necessary to me and to Germany. I knew the arguments which they contained and the effect which they would have upon the Prince. If they had reached him all would have been lost.'

'Why should your Highness descend to such expedients when a score of these brigands, who wished to hang me at your castle gate, would have done the work as well?'

'They are not brigands, but the best blood of Germany,' she cried, hotly. 'If you have been roughly used, you will remember the indignities to which every German has been subjected, from the Queen of Prussia downwards. As to why I did not have you waylaid upon the road, I may say that I had parties out on all sides, and that I was waiting at Lobenstein to hear of their success. When instead of their news you yourself arrived I was in despair, for there was only the one weak woman betwixt you and my husband. You see the straits to which I was driven before I used the weapon of my sex.'

'I confess that you have conquered me, your Highness, and it only remains for me to leave you in possession of the field.'

'But you will take your papers with you.' She held them out to me as she spoke. 'The Prince has crossed the Rubicon now, and nothing can bring him back. You can return these to the Emperor, and tell him that we refused to receive them. No one can accuse you then of having lost your dispatches. Goodbye, Colonel Gerard, and the best I can wish you is that when you reach France you may remain there. In a year's time there will be no place for a Frenchman upon this side of the Rhine.'

And thus it was that I played the Princess of Saxe-Felstein with all Germany for a stake, and lost my game to her. I had much to think of as I walked my poor, tired Violette along the highway which leads westward from Hof. But amid all the thoughts there came back to me always the proud, beautiful face of the German woman, and the voice of the soldier-poet as he sang from the chair. And I understood then that there was something terrible in this strong, patient Germany—this mother root of nations—and I saw that such a land, so old and so beloved, never could be conquered. And as I rode I saw that the dawn was breaking, and that the great star at which I had pointed through the palace window was dim and pale in the western sky.

Catherine Carr

∞

The ninth day of August 1814, was hot and sultry on the Connecticut coast. Ever since morning, old Madam Carr, Catherine's grandmother, had been presaging a thunder-tempest. Sitting on the north porch, knitting an interminable length of silk stocking, she had sniffed the charred dust of the field-grass consuming beneath the fierce sun, as in a very *auto-de-fé* of nature, and turned keen old eyes of suspicion upon the northwest horizon, where the thunderstorms were wont to brew. All day the northwest had been vague, as with dregs of colour instead of cloud, for all the purple-blue of the intense sky seemed to have settled there. 'There will be a thunder-tempest before night,' said Madam Carr, in her deep voice, almost a bass, coming unexpectedly from a slender old throat swathed in folds of delicate lace.

'Yes, grandmother, I think so,' responded Catherine.

She was dressed in some thin stuff of an indeterminate pattern of purple stripes and garlands of pink flowers, faded and darned profusely, as was her lace tucker. Everything in the old Carr mansion was only held from the collapse of age by the timely stitches of thrift. Madam Carr's black satin was actually embroidered by her own cunning fingers in a rose pattern to conceal the threadbare places, and her black lace mittens were almost gone as to the original weave. Madam Carr always wore her black lace mittens of an afternoon, as ladies had been accustomed to do when she was young. She had come of a fine old English family which had not sought the New World at all until the colonies were nearly fledged for independent flight. There had been wealth and state in the daily life of Madam Carr, although now she lived in actual poverty with her granddaughter Catherine. The Carr estate had dribbled away through improvident fingers, until there was little left except the old mansion on the outskirts of the village, and a few acres of scanty pasturage, and a woodlot or two. Catherine had a brother, two years older than herself, who might retrieve the family fortunes, but he was fighting in Perry's squadron against the British, and his grandmother had little to say of him. Madam Carr was a Tory born and bred, and, if she had her way, would have set up the throne of England at the Capitol with no delay. She prayed in her closet for the English arms, and never asked if Catherine had heard from young Harry Carr, though she loved him well, and though she knew that his scanty pay, sent home as regularly as might be, kept her and her household alive. Catherine herself, from her early training, was not as true a daughter of the republic as she might have been in her principles, and had striven as best she might to dissuade her brother Harry. 'Go fight for the one true country and the one true government,' said she, with that gentle imperiousness which folk said was like her grandmother's, and became her well, but Harry would not listen.

'They have ground us into the dust,' he declared, fiercely, 'and now we will arise and let them see the stuff we are made of.'

''Tis British stuff,' argued Catherine.

'British stuff grafted on to a tree of the West, till 'tis the noblest fruit of the whole earth,' said Harry Carr, and was off in his blue uniform.

Catherine wrote to him, though her disapproval was still active. Catherine looked as her grandmother had looked when she was a girl. She was small and fair, with a face as round and innocent as a child's; but she bore herself like a queen, and had, at times, a severity of manner which no one, not even her grandmother, gainsaid. Catherine Carr had been sought in marriage by one Captain Miles Wadsworth, of the blood of the Wadsworth who had hidden the charter of Connecticut in the hollow tree.

Catherine had accepted him, and he had for some six months visited her as her acknowledged lover; then there had been high words, none but they two knew over what, and Captain Miles had sailed away in his ship, *The Commonwealth*, to Liverpool. That was during the year before the war, and Catherine had not heard a word from her lover since. If she grieved for him, nobody knew. Now and then a swain with courting intent rode over from Stonington to see this damsel with a fair face and good blood in her veins, though no gold in her purse, but she would have none of him.

'Catherine will never be wedded,' said Madam Carr, 'but no one need scorn her for it, since 'tis not for lack of chances. 'Tis an honorable estate, when it comes with choice, and has been held by many a woman of our family.' If the truth had been told, Madam Carr somewhat regretted that she had not held that estate herself. She had been a true and faithful wife to her husband, now dead some twenty years, but she had not been in love with him, and he was in her memory but an additional shadow cast by her fleshly life. She had been a Carr herself, having married her own cousin. She looked with favour on Catherine remaining in her single blessedness. 'There is freedom and dignity in it, child, let them say what they will,' said she, 'and the old place will feed you as long as you live.'

Catherine Carr led her life of peaceful monotony of industry, and seemed content enough, though she was young and there was warm blood in her veins. The days for three years had dawned and dimmed with no change, except their own variety of eternity, but now this ninth day of August was to hold enough to spice a lifetime.

Catherine, when she waked that morning, had felt a strange quickening of her pulses and a turmoil of her whole mind, both as to its memories and its imagery of the future. An electrical ferment of storms and up-heavals of present harmonies seemed to be in her little sphere of life, as well as in the greatness of nature. 'There will be a thunder-tempest be-fore night,' said Madam Carr. 'Something will happen before night,' said Catherine Carr, but only to herself. All the forenoon she and old black Sylvy had been putting up little pots of currant jelly and jam. Her finger-tips were still rosy with the currant juice, though she had rubbed them well. She sat that afternoon with her grandmother on the porch and darned a petticoat of fine red damask which had belonged to her mother.

'I remember well when your mother first wore that; she had just turned sixteen,' said Madam Carr. 'It is a handsome colour, though not for this time of year.'

'The people of this country like it not overwell, at this time of year or any other,' said Catherine, with a laugh, referring to the red coats of the British soldiers. Catherine's laugh had a nervous ring, her delicate face was strained and tense, her blue eyes almost black. She kept turning an ear this way and that, as if listening; she glanced often at the road, visible through the lilac shrubbery in the front yard; all at once Catherine's whole face lightened and sharpened with attention like a hunting dog's. She heard a noise like the feet of a swift runner; then she saw one man, then another, and another, come pelting up the road; then a stout woman labouring behind with heavy joltings of hips like the panniers of a donkey. Catherine recognized them for the landlord, his wife and son and man-servant, from the tavern a half-mile below. The tavern was a small hostelry of none too good repute, serving mainly as a resting and carousing place for sailors, for there was a good anchorage at that point. Catherine ran out around the corner of the house, through the lilacs, to the road, and her grandmother, knitting-work in hand, went after her, nearly as agile as she.

'What is it?' called Catherine, and her face was pale; for those were troublous times on the coast, with an enemy on the seas. The woman answered her, though she was scantest of breath.

'The British! The British!' she gasped out.

Catherine ran after her, grasping her arm and keeping alongside easily. 'Are they come? Are the British come?'

'Their ships are off yonder, four of 'em, and a boat from the biggest—a-comin'—a-comin' ashore. Let go! let go!'

'Are the men going to Stonington to warn the town?'

'No; the town will be all in a light blaze in a half-hour. We be goin' to the swamp, where they can't find us. Best you and your folks go too, Catherine Carr. Get your teaspoons an' come. I have my half-dozen in my pocket. Let go! let go!'

With that the woman, the wife of John Tabb the tavern-keeper, wrenched away her arm, and was on with the rushing men.

'Oh, go to Stonington! go to Stonington! to warn the town!' cried Catherine, after them; but John Tabb pelted on with his stubborn face as if he did not hear.

'Go to Stonington! go to Stonington, Mr Tabb!' shrieked Catherine. But John Tabb went on with a steady rush, like a panic-stricken ox, and they were all out of sight.

Catherine ran back to her grandmother, standing pale, but with a certain air of tremulous triumph, at the house gate. 'We have nothing to fear, Catherine,' said she. 'There is no need for women of the Carr family to fear anything from British soldiers.'

'I am not afraid,' said Catherine, proudly, 'but the people in Stonington will have no time to defend the town. Grandmother, I am going to warn them.'

'You will not go one step,' declared Madam Carr. 'The British will give them notice in time to remove the women and children; the men will stay and fight. What more would you?'

'There will be no time to strengthen the earthworks.'

'Let them fall, then! What you are thinking of, Catherine Carr, is Miles Wadsworth's father and mother, and sister Pamela.'

Catherine went rosy red, and ran, with her grandmother calling vainly after her, through the house to the garden. It was in her mind to cross to the fields from the garden, and then to run through them to a point further up the road, that being considered a shorter way.

Catherine hurried through the old garden, overrun with box in sinuous green windings, under the arches of ancient rose-trees. Black Sylvy's eyes rolled white with childlike wonder and terror from the shadowy kitchen window. Catherine ran down the main garden path, between the humping rows of box. She climbed the wall at the rear, and had just set foot on the sun-baked grass of the field when she gave a choking cry and started back. Captain Miles Wadsworth was coming towards her, running weakly, as if he were about to fall, flinging out with uncertain knees like a drunken man. His face was darkly flushed, though it had grown so thin that one who did not know it as well as her own might not have recognized it, and he kept one hand pressed hard against his side.

For a second Catherine felt as if she were in a dream, and as if it were her eyes awaver with sleep which gave this dream man his wavering gait;

then she sprang forward. 'Miles! Miles!' she cried out, and caught him by the arm; and he leaned against her shoulder, and would have fallen except for her. 'How came you here? What is the matter? Oh, Miles!' she cried, trembling and supporting him, and forgetting that there had ever been enmity between them.

'The British! Alarm the town!' gasped out Captain Miles Wadsworth.

'Miles, have you escaped from the British ships?'

'Yes. Sweetheart, hurry; I can go no further. I swam ashore, and the sun heat—and—I have had a wound in my side. Go, go!'

'Miles, if they find you, you will be shot.'

'Never mind me, Catherine. Run to the town!'

'And a boatload is coming to the anchorage at the tavern,' cried Catherine in a frenzy. 'They will find you here; they will stop and search the house; 'tis you they are after, Miles.'

'No, I—overheard—they will—bombard Stonington in an hour. Go, Catherine.'

'They will stop and search, and if they find you, they will kill you, Miles.'

'As well one time as another, Catherine,' replied Captain Miles Wadsworth, faintly, with a little bitterness of sadness. 'I have no sweetheart and no ship, and my old father and mother have another son and will not miss me. I must die for my country, which alone has any need of me, and as well now. Run to Stonington, Catherine, for ten minutes on the earthworks may mean victory and the salvation of the town. Leave me where I am, and go.'

But Catherine Carr, with her mouth set hard, was already urging the young man towards the house, though he continued to beg her not, almost piteously. 'Catherine, for your life leave me and go,' he begged, stammering and stumbling, for the pain and dizziness in his head were great, and his feet met the ground with strange shocks. Captain Miles Wadsworth was nearly spent with exhaustion from the heat and the stress and anxiety of his escape, being also weakened in health by anxiety of mind and a bullet wound received some time since in his side. His good ship *The Commonwealth* had been captured by the British on the high seas before the war had begun, and he himself impressed into service in the British navy. That morning, having discovered that messengers were to be sent to warn Stonington and give her an hour to remove the inhabitants, he had jumped overboard and swum ashore; but on reaching the land he had sunk down exhausted, and lain there he knew not how long.

Catherine half dragged, half bore him into the house, to the bedroom out of the cool north parlour, and forced him gently upon the bed, on which he sank, gasping faintly, for his strength was almost spent.

Madam Carr stood in the entry as they passed, and black Sylvy's eyes rolled white over her shoulder. 'Who—who—?' demanded Madam Carr, and her face was as pale as her darned laces. She did not know Miles Wadsworth with his emaciated figure and his thin flushed face.

Catherine answered not a word till she had laid Miles on the bed and come back panting. 'Fetch cold water from the well, and a towel, quick,' she ordered Sylvy, imperatively, then ran to the cupboard in the south room for a glass of cordial; but Madam Carr caught her by the arm. 'Who is it, Catherine Carr?' she demanded.

'Miles—it is Miles; let me go.'

'How comes he here?'

'He has escaped from the British. Let me go, grandmother.'

'They will shoot him if they find him here.'

'They shall not find him.'

'I saw the gleam of their bayonets down the road before I came in. They are almost here. They will shoot him if they find him, Catherine Carr.'

Catherine gazed at her grandmother with a face of pale desperation; then the colour came back, her eyes flashed with sudden resolution and fun, and she gave one leap towards the north porch, where lay the red damask petticoat which she had been mending.

'Catherine, Catherine Carr, what are you going to do?'

Catherine made a swift notch with her scissors, then, with a fierce twitch of her strong young arm, rent the petticoat from belt to hem.

'Catherine, what are you doing? Are you gone mad?'

Catherine made no reply; she rushed to the fireplace, caught up the poker, and thrust it through the band of the damask petticoat, then scuttled up the winding spiral of the front stair, and hung out the red flag which she had improvised from the second-storey window over the front door.

Madam Carr, all in a tremulous whir of silken skirts, stood at the foot of the stair; she thought that her granddaughter had lost her wits over the peril of her lover. 'Catherine,' she called, faintly, and as she called caught a scarlet gleam through the side-lights of the front door, heard the tramp of heavy feet, and the jingle of swords, then a great clatter of the knocker. Then she heard the voice of Catherine, who was waving the red flag from the window overhead—

'What do you want, gentlemen?'

A voice of enquiry sounded from without, and it was imperious, though somewhat tempered by admiration for Catherine's fair face, and respect for her gentle, fine-bred manners.

Then Madam Carr heard her granddaughter's voice again, gentle as ever, yet with a tone in it which she had never heard, and which seemed to show in a flash the girl to her as a stranger: 'It is the flag of the small-pox, gentlemen. You can enter if you please, but you will do so at your peril.'

There was a commotion among the redcoats; then the man's voice sounded again, with a quavery of consternation in it, 'Who is ill, madam?'

And Catherine Carr answered, unfaltering over the first lie of her life, 'My brother, gentlemen.'

'A sailor has escaped from our ship this morning. Has one come here seeking shelter?'

'No, sir.'

There was a confused murmur of voices; then the leading one was heard again: 'I am dispatched by the British commodore to give notice to the inhabitants of Stonington that the bombardment by his Majesty's ships will begin in an hour's time, and to warn them to remove to places of safety.'

'Oh, I pray you, gentlemen,' cried Catherine Carr—and if she was guilty of a bit of malicious humour in the midst of her terror and danger they did not perceive it—'I pray you, gentlemen, to assist me in removing my poor brother, who is in the worst stage of the disease, to some place of safety before the bombardment begins.'

Then the young officer, who had a face as fair and rosy as the ladies of whom he was a dear lover, answered in great haste: 'There is no cause for alarm, madam; you are well out of range of our fire. There is no occasion to remove your brother, and it might do him great harm, since he is in such a state. I bid you good-day, and may the sad affliction which is upon your house spare your roses!'

Then Madam Carr heard the heavy tramp of retreating feet, and saw the vanishing swing of scarlet shoulders, and Catherine came downstairs, laughing.

Her grandmother, pale with anxiety and bewilderment, yet looked at her with asperity. 'Laugh, if you can, at falsehood and disloyalty against your rightful country and ruler,' said she, 'but you know not if they will not return and shoot your lover in there, and you know not if he be not lying dead now with a sunstroke. You would laugh on the verge of the grave.'

Catherine turned pale, and hastened into the north bedroom, where Captain Miles lay, with black Sylvy wetting his head with cloths wet in cold water, and crooning the while with a half-savage murmur of sympathy and love. Miles did not open his eyes as Catherine bent over him, and Sylvy shook her head warningly. In spite of his peril and anxiety, Miles Wadsworth had fallen into the repose of utter exhaustion.

Madam Carr beckoned Catherine out into the parlour. 'What if they suspect, and stop on their way back—what then?' she whispered.

'They will not stop when they are bound on such an errand,' Catherine whispered back, but her forehead was contracted a little. Then suddenly her whole face lighted with the flash of her ready wit, and her mouth twitched, for so full of spirit was she that she had presence of mind to love the jest which sometimes rides abreast with danger. Out she ran to the kitchen, got a jar of the currant jelly which she had just made, and the porringer of paste which she had used for fastening the little circles of letter-paper over the jars, and was back in the bedroom. Then, while the black woman watched her in consternation, though she ceased not her cold-water applications, she made with soft and deft touches Captain Miles Wadsworth's handsome face resemble most hideously that of a smallpox patient.

When Catherine turned, after finishing her work, she saw her grandmother standing in the doorway, watching her with an expression which resembled her own. Full of perturbation and anxiety was Madam Carr's face, and yet there was a lift at her mouth corners which denoted something else. Suddenly she glided swiftly to the cupboard beside the chimney in the parlour, where they kept a small stock of domestic remedies, and displayed, with a shrewd narrowing of her eyes, a small vial to Catherine. 'It is assafœtida, and, in case they return, it can be spilled on the floor for the evil smell of the disease,' said she, and Catherine nodded, laughing again; but at that moment came a clattering knock on the front door, which caused them both to start and pale.

'They have come back,' gasped Catherine, losing for the moment her courage; but her grandmother stood firm. 'Go to the door,' said she, 'and fetch them in if they have a mind!' Then she went herself into the bedroom, ordered Sylvy away, and took her place at Captain Miles's head.

Catherine ran up the stairs, flung open the window, and brandished her ominous flag. Only one man stood below—a grizzled old king's veteran, with a square jaw, small eyes of stubborn defiance, and a face deeply dinted with the smallpox. In doubt and suspicion as to whether the deserter, or a body of the enemy and arms, might not be concealed in the house, this man, John Busby by name, had volunteered to return and face the pestilence.

'You will enter at your peril, sir,' declared Catherine's sweet voice from the window. She had quite recovered herself; her eyes flashed, and her cheeks were as red as her flag.

When the caller declared gruffly his intention to enter, Catherine ran down at once and opened the door, and led the way to the bedroom wherein Captain Miles Wadsworth lay.

'I beg you, sir, to be quiet, and not disturb my brother, or his life may answer for it,' said Catherine, softly; and the soldier made no answer, but trod on tiptoe, and barely stepped over the bedroom threshold. Indeed, he found one look and one breath enough. Madam Carr had spilled all the assafœtida as she proposed, and Captain Wadsworth's face in the darkened room was a hideous sight. Madam Carr had slipped the wet towel over his forehead and eyes, but the lower part of his face was exposed, and it looked, as John Busby reported, for all the world like a Christmas pudding. And as for the disease being smallpox—'Go there and draw one long breath,' said John Busby.

John Busby retreated, fairly routed, though he knew it not, by the wit of two women, sturdy soldier though he was, and the gleam of his scarlet coat was soon out of sight on the Stonington road.

After he was gone, Captain Miles Wadsworth lay in such a peace of rest and slumber that seemed almost like that of death, and Madam Carr ordered that he be left to himself. 'He will do now,' said she, 'but we will leave his face as 'tis lest they come again, and he be shot, and we convicted of perjury.'

She went out with her stately glide, but Catherine lingered, gave a swift glance around to make sure that she was not seen, then bent over the sleeping man and kissed him softly. Then she slipped out, all glowing with blushes, and looked at the tall clock on the stair landing. It was about five o'clock in the afternoon. Catherine reckoned, standing there. 'In about half an hour they will have reached Stonington and given warning,' she reflected; 'then they have to return to their ship. That will take near an hour longer, as they have computed. The bombardment will begin about the half-hour after six.'

Suddenly the thought came to Catherine Carr, 'If—if those messengers could be delayed two hours—one hour—only one half-hour—the people of Stonington might be able to save the town.' For she remembered what her lover had said as to the value of even fifteen minutes for the strengthening of defences. Catherine Carr, a slim young figure, in her gown of faded purples and roses, stood with her fair alert face fronting the face of the clock as if it were the veritable one of time and eternity, in which she could find the eyelight of wisdom.

Then she threw back her head, her white throat swelled, and she gave a short laugh, which was to her gay undaunted temper the expression of resolution.

Catherine hurried out on the porch, caught up the remaining half of her red damask petticoat, and raced down the road.

Catherine hastened towards the tavern, which had been so incontinently deserted, down the stretch of dusty road between the powdered thickets of gold and purple midsummer bloom. The sun rays beat over from the westward like the fierce arrows of retreating hunters, there were broadsides of heat from earth and sky, and every leaf and blade and flower had seemingly its own shot for the weary eye that beheld it, for it was one of the hottest days of the year. But Catherine had no thought of the heat as she sped along, being fully possessed with fiercest love and loyalty, not for her country—for as to that she was divided—but for her lover. She remembered Miles Wadsworth's anxiety for the safety of Stonington, and how he had risked his own life for that end, and rather than fire against his countrymen.

'Stonington shall be saved, Miles,' she cried, exultantly, as she hurried along. She thought of the good folk whom she knew in the town, and how Miles Wadsworth's father and mother and sister dwelt there.

When she reached the tavern she saw the boat from the British frigate rocking on the blue swell, and to her great relief it was empty. Her fear had been lest a sailor or two were left in charge. She did not know that such was the case, and that they had been lured from their allegiance by the abandoned barrels of New England rum in the tavern, and were then therein, forgetful of friend and foe alike.

Catherine got a hatchet from the tavern woodpile, and ran down the beach to the boat. She had a strong arm and wrist; she had soon cut a hole in the bottom of the boat, and severed the rope. All the time she was in deadly fear lest someone on the ships which swarmed aloof in the haze of the heat might spy her out with a glass. She had, to guard against that, brought the severed half of her damask petticoat and pinned it over her shoulders, that any watcher might take it for the red coat of England.

When her work was done, and she was hurrying up the road, she cast a glance back, to see the dim ships moving in a line towards a position more directly opposite the town, all except one, which lingered for the return of the boat.

'That boat will never return,' thought Catherine Carr, 'and they cannot bombard until they are sure that the first shot will not kill their own men.'

Then she reflected—suppose she had scuttled the boat, how long would it take for the soldiers to signal the ship, and for another boat to be sent to them? It would take some time, certainly, but could she not delay them longer still? A spirit of utter daring and mischief seized the girl, and imaginations as extravagant as those of a child possessed her. She was of an exceeding nervous temperament, as finely responsive to all new conditions as a bird which slants its wings to every change of wind. Moreover, the intense heat of the day, instead of depressing her with slumbrous languor, had stung her to freer life. She had risen with higher and higher understanding and capacity to every new emergency of danger, until now she was in a fairly abnormal condition of activity, and barriers were as nothing to the leaping powers of her mind.

'I will delay the British ships longer still,' said Catherine Carr.

Just before she reached her own house she saw a thick cloud of dust ahead, then horns tossed through it, followed by bounding red flanks and lashing tails. Then she heard the shrill scream of a fife, and knew that little Johnny Purdy was driving home the cows, playing the while on the fife on which his grandfather had played in the Revolution, and in which his soul delighted.

Catherine bounded forward and caught the boy by the arm. He stared at her in bewilderment, his fife still at his pouting lips. He and his widowed mother, living off the main road, had heard naught of the approach of the British.

Catherine talked fast; the boy's face kindled as he listened. He was a sunburnt lad of twelve, with ready blushes, but steady eyes.

When Catherine released him, bidding him hurry, he went forward with a leap, driving the cows at a furious rate, but his fife was still.

Catherine ran towards the house, and her grandmother met her at the door. 'He is still asleep,' she said. 'Where have you been, Catherine Carr?'

'I have been to the British boat and scuttled it.'

'They will shoot you for a traitor, and you are one,' said Madam Carr, hotly, but her eyes snapped.

Catherine made no reply, but hurried up to the garret and got the two old muskets which her grandfather Carr had once used against the colonists.

'What are you going to do with the muskets? Is your brain turned?' demanded Madam Carr, when Catherine came down.

'Go to the door and watch, grandmother, and you will see,' answered Catherine, running out with a musket over each shoulder, though they were of prodigious weight and length, and she could scarcely walk under them.

The country road was bordered with stone walls. Catherine laid her two muskets one over each wall, a little above the house, with their muzzles pointing up the road at a slight slant. Then back she went for some pokers and tongs and shovels, and even warming-pans with long brass-bound handles, which, seen in an uncertain light of gathering dusk, and also of storm, for that which had threatened all day was rolling high in the north-west, might well deceive the eyes of suspicion in a hostile land. All these she ranged over the stone walls, with the muskets at the ends nearest the town; and then came little Johnny Purdy, dragging painfully the enormous gun used by his ancestor of Plymouth against the Indians.

That, when placed, seemed to project half across the road with its furious old muzzle; and though it could not be fired without such labour and preparation as were out of the question, yet its look was an intimation of a far-reaching death.

Then Johnny Purdy, who was pale under his freckles, for his ancestral blood was surging high in his heart, went, according to instructions, his fife in hand, to a point some distance away in the fields, whence he could fly easily to cover of the woods; and Catherine Carr, with a tin pan and two sticks for a drum, stationed herself aloof on the other side of the road.

When the men from the British ship, with the young officer at their head, came down the road on their way to take boat, having warned the town of Stonington according to instructions, they saw to their dismay a gleaming file as of muskets projecting over either wall just before they reached the Carr house. The thunderclouds were gathering fast, and there was a sallow twilight over the whole country. They doubted not that they saw at least twelve good muskets, six on either side of the road, forming a short but certain land of death for the King's men to pass. They halted, and suddenly from the right came the shrill scream of a fife in the tune of 'Yankee Doodle'; then the triumphant beat of a drum responded from the left. That meant, to the British, American reinforcements in both quarters, and the impossibility of taking to the fields and skirting the muskets.

The men were brave soldiers as any in his Majesty's service, but a little pause before what seemed a certain march on death did them no discredit. Some time was spent by them in conference as to what course to pursue, some holding it was best to retreat on the town and seize a boat, some holding that since their flag did not protect them here it would not there, and they would either be cut in pieces, or held prisoners to prevent the bombardment.

And all the time the Stonington men were working for their lives at the earthworks, while the ancient, useless muskets and the pokers, tongs, and warming-pans pointed fiercely over the stone walls, above the Carr house, at the enemy; and the fife played on the right nearer and nearer, now this way, now that; while the drum sounded bravely on the left.

Finally the men from the British ship placed themselves shoulder to shoulder, levelled their guns, and made a rushing charge, as brave in spirit as that of the six hundred, though it was against a powerless enemy, and their bravery seemed to recoil upon themselves with a shock of mortification and ridicule.

After the bloodless passage between the bristling walls, there was a halt and hoarse shout, half of mirth, half of rage.

Catherine had fled like a deer through the garden into the house, and little Johnny Purdy had dropped into a misty hollow, where they might have searched all night for him in vain. There he lay low, shaking with exultant laughter, though the thunder shower burst over him, drenching him to the skin, and pelting him as with bullets of water.

As Catherine entered the house there came a blinding blue flash, and a report like a whole broadside of artillery. Catherine rushed through into the south room, where her grandmother was. She was laughing, but her grandmother's look checked her.

'See what you have done!' cried Madam Carr, panting. 'They are coming back here!'

Catherine looked out of the window, and there were the redcoats coming on a run. Their charge had carried them past, but they had turned as soon as they had collected their wits, and made for the Carr house. They hesitated no longer before that flying ensign of a dread pestilence. Either they, having just experienced such a scurvy trick, suspected the subterfuge, or, being so wroth with those who had mocked them, feared no danger, or else were minded to seek one moment's shelter from the fury of the storm, though that would have scarcely been to their credit, being on the King's service; but whatever may have been the reason, they were coming.

Catherine Carr's brave heart and gay daring failed her utterly. 'They— will—find him!' she gasped. This contingency had not entered her head, which was, after all, that of a child's, without enough retrospective distance to give her the best foresight.

Then it was that Madam Carr arose. 'Be quiet, Catherine Carr, and do as I bid you, and I will save him,' said she, and went straight across to the north parlour, with Catherine at her heels, though her knees shook so that she could scarcely step.

In the north parlour was a carved chest of precious Indian wood. It had been one of Madam Carr's wedding treasures, and held some of her most precious belongings, her bridal gown and veil and satin shoes, and her dead husband's knee-buckles among them.

Madam Carr flung open the chest, and drew therefrom a parcel securely folded in a fine linen cloth, laid well under the others, for it was a contraband thing, and had been kept for years at the risk of disparagement to Madam Carr's loyalty.

She unfolded the parcel while the knocks of the British soldiers resounded through the house, and shook out the British ensign, which had been struck on a British man-of-war in the Revolution, and had fallen into her husband's hands, to be held sacred by him, and his widow after him.

In the bedroom, dark with the storm, Captain Miles Wadsworth was stirring in faint bewildered wise, awakened by the roll of the thunder and the clatter at the door. Madam Carr, holding the flag behind her, went into the room and up to the bed, Catherine following.

'The British are about to enter this house,' said Madam Carr. 'They will shoot you for a deserter if they find you here. Shut your eyes and lie still, and hold your breath, for your life.'

Miles tried to speak.

'Do as I bid you, for your life,' ordered Madame Carr. 'Shut your eyes and hold your breath.'

Captain Miles Wadsworth, borne down in his weakness by this on-slaught of maternal will, closed his eyes. Then Madam Carr flung the British ensign over him as he lay there, and stood at his side, her head bowed as in pale and tearless grief, when the redcoats rushed into the room.

There was a sudden hush as they saw the still shape on the bed under the British flag, the stony grief of the stately old woman, and Catherine, who was weeping in good earnest, for her nerves had given way.

They all uncovered, and then the young officer gave the word to leave the room; but the old soldier John Busby spoke up in surly defiance of authority. 'Sir,' said he, 'they have just served us one trick; this may be another.'

'Silence!' cried the officer, with a side glance towards the weeping Catherine, and maybe an uneasy consciousness of the evil odour in the room—'Silence! The man under my command who respects not the dead covered by the British flag falls himself. March!'

The young officer gave one more admiring and passionate glance at Catherine. If he suspected aught wrong, he hid his suspicions from himself, and as he spoke he heard a signal-gun from his ship, and knew that there was no time to waste upon such play as this. Then he went out of the room, his men following.

The signal-gun from the ship was fired again as the redcoats hurried down the road, with the storm driving hard at their heels, and the lightning playing upon their scarlet backs like whips of fire.

The bombardment of Stonington was a good three hours late that night, and though continued until midnight, and with intervals for three days, and some property demolished, the town was saved, and finally the British fleet sailed away, leaving the American flag still floating over the ramparts.

To Catherine Carr the safety of the town, though she had striven so hard for it, was as nothing to that of her lover; but his rejoicing was somewhat tempered with mortification when he learned that he had been saved by the flag which was hostile to his country. 'A traitor and a coward might have availed himself of such a means of safety,' he said once, hotly;

but Madam Carr faced him with stern indignation: 'English you were born, and your forebears. English you will live and die, and those who come after you,' declared she. 'Blood is blood, and race is race, and you cannot change it, though you deny your King, and crown all your own unworthy heads. You will speak with an English tongue, and look askance at the rest of the world with English eyes, as long as you live. The English flag covered those who came before you from death and insult, and will cover those who come after you. It has saved your life now, and you have cause for honour and pride, and not shame.' With that Madam Carr went out of the room, her head as high and stiff as if she bore the majesty of England upon it, and left Catherine and her lover alone in the north parlour. That was a year later, when peace was established and Miles's health quite restored, and he and Catherine were to be married the next day. Even then, Miles's sister Pamela, and Catherine's brother Harry, and some of her mates, were out in the fields gathering wild flowers to trim the old Carr house for the wedding.

Catherine Carr, who was strong in natural feminine wiles for the soothing of man, and understood well how to inveigle him sweetly away from a bone of argument rather than allow him to pause to wrangle over it, simply remarked to Miles that breathing New England air for three-score years and ten had not changed grandmother Carr's old England heart, and old trees must lean their own way, and went on to tell him gayly of little Johnny Purdy's trepidation over bringing home the wedding cake from Stonington, whither it had been taken to be iced, on his wheelbarrow, and how he had sighed with relief when he landed it safe at the door, having neither spilled it in the dust nor had it tasted by wasp or bee or boy.

Miles Wadsworth looked at Catherine's beautiful, laughing face, and thought no more then of how his life had been saved, since saved it was for so great happiness; but Catherine, with her head on her lover's shoulder, cast a mindful glance at the old carved chest in the corner which held the British flag.

Wash

Sutpen stood above the pallet bed on which the mother and child lay. Between the shrunken planking of the wall the early sunlight fell in long pencil strokes, breaking upon his straddled legs and upon the riding whip in his hand, and lay across the still shape of the mother, who lay looking up at him from still, inscrutable, sullen eyes, the child at her side wrapped in a piece of dingy though clean cloth. Behind them an old negro woman squatted beside the rough hearth where a meagre fire smouldered.

'Well, Milly,' Sutpen said, 'too bad you're not a mare. Then I could give you a decent stall in the stable.'

Still the girl on the pallet did not move. She merely continued to look up at him without expression, with a young, sullen, inscrutable face still pale from recent travail. Sutpen moved, bringing into the splintered pencils of sunlight the face of a man of sixty. He said quietly to the squatting negress, 'Griselda foaled this morning.'

'Horse or mare?' the negress said.

'A horse. A damned fine colt. . . . What's this?' He indicated the pallet with the hand which held the whip.

'That un's a mare, I reckon.'

'Hah,' Sutpen said. 'A damned fine colt. Going to be the spit and image of old Rob Roy when I rode him North in '61. Do you remember?'

'Yes, Marster.'

'Hah.' He glanced back towards the pallet. None could have said if the girl still watched him or not. Again his whip hand indicated the pallet. 'Do whatever they need with whatever we've got to do it with.' He went out, passing out the crazy doorway and stepping down into the rank weeds (there yet leaned rusting against the corner of the porch the scythe which Wash had borrowed from him three months ago to cut them with) where his horse waited, where Wash stood holding the reins.

When Colonel Sutpen rode away to fight the Yankees, Wash did not go. 'I'm looking after the Kernel's place and niggers,' he would tell all who asked him and some who had not asked—a gaunt, malaria-ridden man with pale, questioning eyes, who looked about thirty-five, though it was known that he had not only a daughter but an eight-year-old grand-daughter as well. This was a lie, as most of them—the few remaining men between eighteen and fifty—to whom he told it, knew, though there were some who believed that he himself really believed it, though even these believed that he had better sense than to put it to the test with Mrs Sutpen or the Sutpen slaves. Knew better or was just too lazy and shiftless to try it, they said, knowing that his sole connection with the Sutpen plantation lay in the fact that for years now Colonel Sutpen had allowed him to squat in a crazy shack on a slough on the river bottom on the Sutpen place, which Sutpen had built for a fishing lodge in his bachelor days and which had since fallen in dilapidation from disuse, so that now it looked like an aged or sick wild beast crawled terrifically there to drink in the act of dying.

The Sutpen slaves themselves heard of his statement. They laughed. It was not the first time they had laughed at him, calling him white trash behind his back. They began to ask him themselves, in groups, meeting him in the faint road which led up from the slough and the old fish camp, 'Why ain't you at de war, white man?'

Pausing, he would look about the ring of black faces and white eyes and teeth behind which derision lurked. 'Because I got a daughter and family to keep,' he said. 'Git out of my road, niggers.'

'Niggers?' they repeated; 'niggers?' laughing now. 'Who him, calling us niggers?'

'Yes,' he said. 'I ain't got no niggers to look after my folks if I was gone.'

'Nor nothing else but dat shack down yon dat Cunnel wouldn't *let* none of us live in.'

Now he cursed them; sometimes he rushed at them, snatching up a stick from the ground while they scattered before him, yet seeming to surround him still with that black laughing, derisive, evasive, inescapable, leaving him panting and impotent and raging. Once it happened in the very back yard of the big house itself. This was after bitter news had come down from the Tennessee mountains and from Vicksburg, and Sherman had passed through the plantation, and most of the negroes had followed him. Almost everything else had gone with the Federal troops, and Mrs Sutpen had sent word to Wash that he could have the scuppernongs ripening in the arbor in the backyard. This time it was a house servant, one of the few negroes who remained; this time the negress had to retreat up the kitchen steps, where she turned. 'Stop right dar, white man. Stop right whar you is. You ain't never crossed dese steps whilst Cunnel here, and you ain't ghy' do hit now.'

This was true. But there was this of a kind of pride: he had never tried to enter the big house, even though he believed that if he had, Sutpen would have received him, permitted him. 'But I ain't going to give no black nigger the chance to tell me I can't go nowhere,' he said to himself. 'I ain't even going to give Kernel the chance to have to cuss a nigger on my account.' This, though he and Sutpen had spent more than one afternoon together on those rare Sundays when there would be no company in the house. Perhaps his mind knew that it was because Sutpen had nothing else to do, being a man who could not bear his own company. Yet the fact remained that the two of them would spend whole afternoons in the scuppernong arbor, Sutpen in the hammock and Wash squatting against a post, a pail of cistern water between them, taking drink for drink from the same demijohn. Meanwhile on weekdays he would see the fine figure of the man—they were the same age almost to a day, though neither of them (perhaps because Wash had a grandchild while Sutpen's son was a youth in school) ever thought of himself as being so—on the fine figure of the black stallion, galloping about the plantation. For that moment his heart would be quiet and proud. It would seem to him that that world in which negroes, whom the Bible told him had been created and cursed by God to be brute and vassal to all men of white skin, were better found and housed and even clothed than he and his; that world in which he sensed always about him mocking echoes of black laughter was but a dream and an illusion, and that the actual world was this one across which his own lonely apotheosis seemed to gallop on the black thoroughbred, thinking how the Book said also that all men were created in the image of God and hence all men made the same image in God's eyes at least; so that he could say, as though speaking of himself, 'A fine proud man. If God Himself was to come down and ride the natural earth, that's what He would aim to look like.'

Sutpen returned in 1865, on the black stallion. He seemed to have aged ten years. His son had been killed in action the same winter in which his wife had died. He returned with his citation for gallantry from the hand of General Lee to a ruined plantation, where for a year now his daughter had subsisted partially on the meagre bounty of the man to whom fifteen years ago he had granted permission to live in that tumbledown fishing camp whose very existence he had at the time forgotten. Wash was there to meet him, unchanged: still gaunt, still ageless, with his pale, questioning gaze, his air diffident, a little servile, a little familiar. 'Well, Kernel,' Wash said, 'they kilt us but they ain't whupped us yit, air they?'

That was the tenor of their conversation for the next five years. It was inferior whisky which they drank now together from a stoneware jug, and it was not in the scuppernong arbor. It was in the rear of the little store which Sutpen managed to set up on the highroad: a frame shelved room where, with Wash for clerk and porter, he dispensed kerosene and staple foodstuffs and stale gaudy candy and cheap beads and ribbons to negroes or poor whites of Wash's own kind, who came afoot or on gaunt mules to haggle tediously for dimes and quarters with a man who at one time could gallop (the black stallion was still alive; the stable in which his jealous get lived was in better repair than the house where the master himself lived) for ten miles across his own fertile land and who had led troops gallantly in battle; until Sutpen in fury would empty the store, close and lock the doors from the inside. Then he and Wash would repair to the rear and the jug. But the talk would not be quiet now, as when Sutpen lay in the hammock, delivering an arrogant monologue while Wash squatted guffawing against his post. They both sat now, though Sutpen had the single chair while Wash used whatever box or keg was handy, and even this for just a little while, because soon Sutpen would reach that stage of impotent and furious undefeat in which he would rise, swaying and plunging, and declare again that he would take his pistol and the black stallion and ride single-handed into Washington and kill Lincoln, dead now, and Sherman, now a private citizen. 'Kill them!' he would shout. 'Shoot them down like the dogs they are—'

'Sho, Kernel; sho, Kernel,' Wash would say, catching Sutpen as he fell. Then he would commandeer the first passing wagon or, lacking that, he would walk the mile to the nearest neighbour and borrow one and return and carry Sutpen home. He entered the house now. He had been doing so for a long time, taking Sutpen home in whatever borrowed wagon might be, talking him into locomotion with cajoling murmurs as though he were a horse, a stallion himself. The daughter would meet them and hold open the door without a word. He would carry his burden through the once white formal entrance, surmounted by a fanlight imported piece by piece from Europe and with a board now nailed over a missing pane, across a velvet carpet from which all nap was now gone, and up a formal stairs, now but a fading ghost of bare boards between two strips of fading paint, and into the bedroom. It would be dusk by now, and he would let his burden sprawl on to the bed and undress it and then he would sit quietly in a chair beside. After a time the daughter would come to the door. 'We're all right now,' he would tell her. 'Don't you worry none, Miss Judith.'

Then it would become dark, and after a while he would lie down on the floor beside the bed, though not to sleep, because after a time— sometimes before midnight—the man on the bed would stir and groan and then speak. 'Wash?'

'Hyer I am, Kernel. You go back to sleep. We ain't whupped yit, air we? Me and you kin do hit.'

Even then he had already seen the ribbon about his granddaughter's waist. She was now fifteen, already mature, after the early way of her kind. He knew where the ribbon came from; he had been seeing it and its kind daily for three years, even if she had lied about where she got it, which she did not, at once bold, sullen, and fearful. 'Sho now,' he said. 'Ef Kernel wants to give hit to you, I hope you minded to thank him.'

His heart was quiet, even when he saw the dress, watching her secret, defiant, frightened face when she told him that Miss Judith, the daughter, had helped her to make it. But he was quite grave when he approached Sutpen after they closed the store that afternoon, following the other to the rear.

'Get the jug,' Sutpen directed.

'Wait,' Wash said. 'Not yit for a minute.'

Neither did Sutpen deny the dress. 'What about it?' he said.

But Wash met his arrogant stare; he spoke quietly. 'I've knowed you for going on twenty years. I ain't never yit denied to do what you told me to do. And I'm a man nigh sixty. And she ain't nothing but a fifteen-year-old gal.'

'Meaning that I'd harm a girl? I, a man as old as you are?'

'If you was ara other man, I'd say you was as old as me. And old or no old, I wouldn't let her keep that dress nor nothing else that come from your hand. But you are different.'

'How different?' But Wash merely looked at him with his pale, questioning, sober eyes. 'So that's why you are afraid of me?'

Now Wash's gaze no longer questioned. It was tranquil, serene. 'I ain't afraid. Because you air brave. It ain't that you were a brave man at one minute or day of your life and got a paper to show hit from General Lee. But you air brave, the same as you air alive and breathing. That's where hit's different. Hit don't need no ticket from nobody to tell me that. And I know that whatever you handle or tech, whether hit's a regiment of men or a ignorant gal or just a hound dog, that you will make hit right.'

Now it was Sutpen who looked away, turning suddenly, brusquely. 'Get the jug,' he said sharply.

'Sho, Kernel,' Wash said.

So on that Sunday dawn two years later, having watched the negro midwife, which he had walked three miles to fetch, enter the crazy door beyond which his granddaughter lay wailing, his heart was still quiet though concerned. He knew what they had been saying—the negroes in cabins about the land, the white men who loafed all day long about the store, watching quietly the three of them: Sutpen, himself, his granddaughter with her air of brazen and shrinking defiance as her condition became daily more and more obvious, like three actors that came and went upon a stage. 'I know what they say to one another,' he thought. 'I can almost hyear them: *Wash Jones has fixed old Sutpen at last. Hit taken him twenty years, but he has done hit at last.*'

It would be dawn after a while, though not yet. From the house, where the lamp shone dim beyond the warped doorframe, his granddaughter's voice came steadily as though run by a clock, while thinking went slowly and terrifically, fumbling, involved somehow with a sound of galloping hooves, until there broke suddenly free in midgallop the fine proud figure of the man on the fine proud stallion, galloping; and then that at which thinking fumbled, broke free too and quite clear, not in justification nor even explanation, but as the apotheosis, lonely, explicable, beyond all fouling by human touch: 'He is bigger than all them Yankees that kilt his son and his wife and taken his niggers and ruined his land, bigger than this hyer durn country that he fit for and that has denied him into keeping a little country store; bigger than the denial which hit helt to his lips like the bitter cup in the Book. And how could I have lived this nigh to him for twenty years without being teched and changed by him? Maybe I ain't as big as him and maybe I ain't done none of the galloping. But at least I done been drug along. Me and him kin do hit, if so be he will show me what he aims for me to do.'

Then it was dawn. Suddenly he could see the house, and the old negress in the door looking at him. Then he realized that his granddaughter's voice had ceased. 'It's a girl,' the negress said. 'You can go tell him if you want to.' She re-entered the house.

'A girl,' he repeated; 'a girl'; in astonishment, hearing the galloping hooves, seeing the proud galloping figure emerge again. He seemed to watch it pass, galloping through avatars which marked the accumulation of years, time, to the climax where it galloped beneath a brandished sabre and a shot-torn flag rushing down a sky in colour like thunderous sulphur, thinking for the first time in his life that perhaps Sutpen was an old man like himself. 'Gittin a gal,' he thought in that astonishment; then he thought with the pleased surprise of a child: 'Yes, sir. Be dawg if I ain't lived to be a great-grandpaw after all.'

He entered the house. He moved clumsily, on tiptoe, as if he no longer lived there, as if the infant which had just drawn breath and cried in light had dispossessed him, be it of his own blood too though it might. But even above the pallet he could see little save the blur of his granddaughter's exhausted face. Then the negress squatting at the hearth spoke, 'You better gawn tell him if you going to. Hit's daylight now.'

But this was not necessary. He had no more than turned the corner of the porch where the scythe leaned which he had borrowed three months ago to clear away the weeds through which he walked, when Sutpen himself rode up on the old stallion. He did not wonder how Sutpen had got the word. He took it for granted that this was what had brought the other out at this hour on Sunday morning, and he stood while the other dismounted, and he took the reins from Sutpen's hand, an expression on his gaunt face almost imbecile with a kind of weary triumph, saying, 'Hit's a gal, Kernel. I be dawg if you ain't as old as I am—' until Sutpen passed him and entered the house. He stood there with the reins in his hand and heard Sutpen cross the floor to the pallet. He heard what Sutpen said, and something seemed to stop dead in him before going on.

The sun was now up, the swift sun of Mississippi latitudes, and it seemed to him that he stood beneath a strange sky, in a strange scene, familiar only as things are familiar in dream, like the dreams of falling to one who has never climbed. 'I kain't have heard what I thought I heard,' he thought quietly. 'I know I kain't.' Yet the voice, the familiar voice which had said the words was still speaking, talking now to the old negress about a colt foaled that morning. 'That's why he was up so early,' he thought. 'That was hit. Hit ain't me and mine. Hit ain't even hisn that got him outen bed.'

Sutpen emerged. He descended into the weeds, moving with that heavy deliberation which would have been haste when he was younger. He had not yet looked full at Wash. He said, 'Dicey will stay and tend to her. You better—' Then he seemed to see Wash facing him and paused. 'What?' he said.

'You said—' To his own ears Wash's voice sounded flat and ducklike, like a deaf man's. 'You said if she was a mare, you could give her a good stall in the stable.'

'Well?' Sutpen said. His eyes widened and narrowed, almost like a man's fists flexing and shutting, as Wash began to advance towards him, stooping a little. Very astonishment kept Sutpen still for the moment, watching that man whom in twenty years he had no more known to make any motion save at command than he had the horse which he rode. Again his eyes narrowed and widened; without moving he seemed to rear suddenly upright. 'Stand back,' he said suddenly and sharply. 'Don't you touch me.'

'I'm going to tech you, Kernel,' Wash said in that flat, quiet, almost soft voice, advancing.

Sutpen raised the hand which held the riding whip; the old negress peered around the crazy door with her black gargoyle face of a worn gnome. 'Stand back, Wash,' Sutpen said. Then he struck. The old negress leaped down into the weeds with the agility of a goat and fled. Sutpen slashed Wash again across the face with the whip, striking him to his knees. When Wash rose and advanced once more he held in his hands the scythe which he had borrowed from Sutpen three months ago and which Sutpen would never need again.

When he re-entered the house his granddaughter stirred on the pallet bed and called his name fretfully. 'What was that?' she said.

'What was what, honey?'

'That ere racket out there.'

''Twarn't nothing,' he said gently. He knelt and touched her hot forehead clumsily. 'Do you want ara thing?'

'I want a sup of water,' she said querulously. 'I been laying here wanting a sup of water a long time, but don't nobody care enough to pay me no mind.'

'Sho now,' he said soothingly. He rose stiffly and fetched the dipper of water and raised her head to drink and laid her back and watched her turn to the child with an absolute stonelike face. But a moment later he saw that she was crying quietly. 'Now, now,' he said, 'I wouldn't do that. Old Dicey says hit's a right fine gal. Hit's all right now. Hit's all over now. Hit ain't no need to cry now.'

But she continued to cry quietly, almost sullenly, and he rose again and stood uncomfortably above the pallet for a time, thinking as he had thought when his own wife lay so and then his daughter in turn: 'Women. Hit's a mystry to me. They seem to want em, and yit when they git em they cry about hit. Hit's mystry to me. To ara man.' Then he moved away and drew a chair up to the window and sat down.

Through all that long, bright, sunny forenoon he sat at the window, waiting. Now and then he rose and tiptoed to the pallet. But his granddaughter slept now, her face sullen and calm and weary, the child in the crook of her arm. Then he returned to the chair and sat again, waiting, wondering why it took them so long, until he remembered that it was Sunday. He was sitting there at mid-afternoon when a half-grown white boy came around the corner of the house upon the body and gave a choked cry and looked up and glared for a mesmerized instant at Wash in the window before he turned and fled. Then Wash rose and tiptoed again to the pallet.

The granddaughter was awake now, wakened perhaps by the boy's cry without hearing it. 'Milly,' he said, 'air you hungry?' She didn't answer, turning her face away. He built up the fire on the hearth and cooked the food which he had brought home the day before: fatback it was, and cold corn pone; he poured water into the stale coffee pot and heated it. But she would not eat when he carried the plate to her, so he ate himself, quietly, alone, and left the dishes as they were and returned to the window.

Now he seemed to sense, feel, the men who would be gathering with horses and guns and dogs—the curious, and the vengeful: men of Sutpen's own kind, who had made the company about Sutpen's table in the time when Wash himself had yet to approach nearer to the house than the scuppernong arbour—men who had also shown the lesser ones how to fight in battle, who maybe also had signed papers from the generals saying that they were among the first of the brave; who had also galloped in the old days arrogant and proud on the fine horses across the fine plantations—symbols also of admiration and hope; instruments too of despair and grief.

That was who they would expect him to run from. It seemed to him that he had no more to run from than he had to run to. If he ran, he would merely be fleeing one set of bragging and evil shadows for another just like them, since they were all of a kind throughout all the earth which he knew, and he was old, too old to flee far even if he were to flee. He could never escape them, no matter how much or how far he ran: a man going on sixty could not run that far. Not far enough to escape beyond the boundaries of earth where such men lived, set the order and the rule of living. It seemed to him that he now saw for the first time, after five years, how it was that Yankees or any other living armies had managed to whip them: the gallant, the proud, the brave; the acknowledged and chosen best among them all to carry courage and honour and pride. Maybe if he had gone to the war with them he would have discovered them sooner. But if he had discovered them sooner, what would he have done with his life since? How could he have borne to remember for five years what his life had been before?

Now it was getting towards sunset. The child had been crying; when he went to the pallet he saw his granddaughter nursing it, her face still bemused, sullen, inscrutable. 'Air you hungry yit?' he said.

'I don't want nothing.'

'You ought to eat.'

This time she did not answer at all, looking down at the child. He returned to his chair and found that the sun had set. 'Hit kain't be much longer,' he thought. He could feel them quite near now, the curious and the vengeful. He could even seem to hear what they were saying about him, the undercurrent of believing beyond the immediate fury: *Old Wash Jones he come a tumble at last. He thought he had Sutpen, but Sutpen fooled him. He thought he had Kernel where he would have to marry the gal or pay up. And Kernel refused.* 'But I never expected that, Kernel!' he cried aloud, catching himself at the sound of his own voice, glancing quickly back to find his granddaughter watching him.

'Who you talking to now?' she said.

'Hit ain't nothing. I was just thinking and talked out before I knowed hit.'

Her face was becoming indistinct again, again a sullen blur in the twilight. 'I reckon so. I reckon you'll have to holler louder than that before he'll hear you, up yonder at that house. And I reckon you'll need to do more than holler before you get him down here too.'

'Sho now,' he said. 'Don't you worry none.' But already thinking was going smoothly on: 'You know I never. You know how I ain't never expected or asked nothing from ara living man but what I expected from you. And I never asked that. I didn't think hit would need. I said, *I don't need to. What need has a fellow like Wash Jones to question or doubt the man that General Lee himself says in a handwrote ticket that he was brave?* Brave,' he thought. 'Better if nara one of them had never rid back home in '65'; thinking *Better if his kind and mine too had never drawn the breath of life on this earth. Better that all who remain of us be blasted from the face of earth than that another Wash Jones should see his whole life shredded from him and shrivel away like a dried shuck thrown on to the fire.*

He ceased, became still. He heard the horses, suddenly and plainly; presently he saw the lantern and the movement of men, the glint of gun barrels, in its moving light. Yet he did not stir. It was quite dark now, and he listened to the voices and the sounds of underbrush as they surrounded the house. The lantern itself came on; its light fell upon the quiet body in the weeds and stopped, the horses tall and shadowy. A man descended and stooped in the lantern light, above the body. He held a pistol; he rose and faced the house. 'Jones,' he said.

'I'm here,' Wash said quietly from the window. 'That you, Major?'

'Come out.'

'Sho,' he said quietly. 'I just want to see to my granddaughter.'

'We'll see to her. Come on out.'

'Sho, Major. Just a minute.'

'Show a light. Light your lamp.'

'Sho. In just a minute.' They could hear his voice retreat into the house, though they could not see him as he went swiftly to the crack in the chimney where he kept the butcher knife: the one thing in his slovenly life and house in which he took pride, since it was razor sharp. He approached the pallet, his granddaughter's voice:

'Who is it? Light the lamp, grandpaw.'

'Hit won't need no light, honey. Hit won't take but a minute,' he said, kneeling, fumbling towards her voice, whispering now. 'Where air you?'

'Right here,' she said fretfully. 'Where would I be? What is . . .' His hand touched her face. 'What is . . . Grandpaw! Grand . . .'

'Jones!' the sheriff said. 'Come out of there!'

'In just a minute, Major,' he said. Now he rose and moved swiftly. He knew where in the dark the can of kerosene was, just as he knew that it was full, since it was not two days ago that he had filled it at the store and held it there till he got a ride home with it, since the five gallons were heavy. There were still coals on the hearth; besides the crazy building itself was like tinder: the coals, the hearth, the walls exploding in a single blue glare. Against it the waiting men saw him in a wild instant springing towards them with the lifted scythe before the horses reared and whirled. They checked the horses and turned them back toward the glare, yet still in wild relief against it the gaunt figure ran towards them with the lifted scythe.

'Jones!' the sheriff shouted; 'stop! Stop, or I'll shoot. Jones! *Jones!*' Yet still the gaunt, furious figure came on against the glare and roar of the flames. With the scythe lifted, it bore down upon them, upon the wild glaring eyes of the horses and the swinging glints of gun barrels, without any cry, any sound.

E. M. DELAFIELD

The Way We Lived Then

∞

When young Edward Ivey married Mary Laidloe, they took a small house in Chester Terrace and lived there very happily. Edward went every day to the City, and dealt with bonded tea, as his father had done before him, and Mary very often drove there to fetch him in the little one-horse brougham that had been a wedding present from her parents—and both of them spent a good deal of time with their children. For a new baby arrived punctually, every two years or so. By the time they had been married ten years, they had six children—Lilias, Edward, Adelaide, Henry, Isabel, and Constance—and the Chester Terrace house was much too small to accommodate them at all comfortably. Edward had long given up his dressing-room to Lilias and Adelaide, the three bedrooms on the top-floor were filled to overflowing with little Edward, Henry, Isabel, the nurse, and the nursery-maid, and the baby's cot had had to be squeezed into the front bedroom, occupied by Mary and Edward themselves.

Then Edward Ivey's father had died of an apoplectic stroke, towards the end of a big City banquet, and the young people—as they still felt themselves to be, although Edward was now approaching forty, and Mary not much younger—took up their residence in the Cavendish Square house originally bought by Edward's grandfather.

Old Mrs Ivey, who knew very well that ladies did not understand business, was very glad that the house should have been left to Edward, especially as she had every intention of continuing to live in it herself. She had a comfortable jointure, and Edward was her only child; she was quite fond of Mary, and devoted to her six grandchildren. Everything was arranged in the most friendly and comfortable way possible, and Mary had the drawing-room repapered in blue with a pattern of gold fern-leaves, and Edward installed a bathroom.

At the time of the move to Cavendish Square, Lilias, the Iveys' eldest child, was thirteen years old; very pretty and intelligent, and the favourite of her parents. All the other children were well aware of this partiality, and resented it very much indeed. They were not deceived by their mother's glib assurances, that it was because Lilias was the eldest, that she was to come out for a drive in the brougham, to be sent for when there were visitors in the drawing-room, or to accompany her parents to the Crystal Palace.

In due course, Edward and Henry went to school—and came back quite different, for their first holidays, and were somehow never the same any more—but Lilias, Isabel, Adelaide, and Constance had a governess at home, because ladylike little girls, said their mother, did not go to school to mix with all sorts of children and perhaps pick up slang expressions and bad manners.

The governess gave them lessons—including music, drawing, German, and calisthenics—and took them to McPherson's Gymnasium in Sloane Street twice a week, and to walk in Kensington Gardens or beside the railings of Rotten Row, every morning. They also attended dancing-classes at the Albert Hall, where they met a very select company of other little girls. If they wanted to ask any of these to tea in Cavendish Square, it was first of all necessary to find out whether their respective parents were known to one another. If so, Lilias would petition her mother, who would probably consent, and the invitation might be given. If not, Mrs Ivey said: 'I'm very sorry, darling, but I don't know anything about this child's mother, and so you'll have to be content with speaking to her nicely and politely at the dancing-class.' And nothing more happened.

So the years went on, very peacefully and prosperously. Grandmama was sometimes a little bit shocked by the modern freedom that was allowed to the children, and by the fact that none of them had ever been whipped in their nursery days, and she was really indignant when Lilias was allowed to hang up a picture of a woman without any clothes on, called *La Source*, in her bedroom. But grandmama was getting old, and kept a good deal to her own sitting-room, on the ground-floor, with her cross and devoted maid Louisa, who had been with her more than thirty years.

Indeed, the servants were something of a problem, in Cavendish Square, for they were all old, and yet had been in the family's service for so very many years that it was impossible to think of dismissing any of them. Only old Nurse—who had brought up Mary Laidloe—was given a pension, and went to live in a cottage at Plumstead, when Lilias was eighteen and about to be presented, and even little Constance, the youngest, had passed her tenth birthday, and could no longer be said to require a nurse.

Lilias had heard a great deal, in the last year or two, about her 'coming-out'. She looked forward to it with mingled excitement and fear. It would be exciting to be grown-up, to have no more lessons, and to have late dinner downstairs every night—but it would be terrifying to go to balls and dinner-parties with one's mother, and run the fearful risk of perhaps not being a success after all.

To be admired by young men was to be a success.

Lilias had a secret dread that young men might not admire her. The only ones she had ever had anything to do with were her two brothers, and nothing was more certain than that Edward and Henry did not admire her.

Why, indeed, should they, Lilias sometimes wondered rather despondently, looking at her reflection in the glass. But that was before she put her hair up.

After this epoch-making operation—performed by her mother's maid until Lilias should have learned how to 'manage' for herself—she had not nearly so many doubts.

For she really was very pretty indeed.

Her skin was pale and delicate, her eyes large and hazel, and her features belonged to the type called classical. Her dark hair, straight and fine, was parted in the middle, carefully waved with the curling-tongs on either side of the parting, and drawn to the back of her head, where it was rolled into a number of little sausages, each of them secured by a stout hairpin at either end, the whole being firmly fastened on to a large pad.

It was a matter of regret to Lilias that she was tall—most people said *very* tall, for her slenderness made her five-foot seven inches seem more— and she instinctively stooped a little, always, in an endeavour to reduce her height, because it made her feel so self-conscious.

The night that she went to her first ball, Lilias wore a white satin dress, carefully chosen by her mother and made by her mother's dressmaker, her grandmother's pearls, and white shoes and stockings. Her long white kid gloves reached nearly to her shoulders, and she carried a little lace fan with mother-of-pearl sticks and handle.

The servants were allowed to come in and look at her, after she had been inspected by her mother, her mother's maid, and the young ladies' maid.

The two boys were at Eton, but the three younger girls stood around, half-envious and half-admiring, in their schoolroom frocks of grey-and-red shepherd's plaid, all three made exactly alike, their hair brushed away from their foreheads, and tied with two bows of black ribbon—one on the top of their heads, and the other behind their ears.

'Now, darling,' said Mrs Ivey, 'you must come in and show yourself to grandmama.'

Grandmama admired the white satin dress, told Lilias to hold herself properly, readjusted the pearl necklace in accordance with some mysterious requirement of her own, and then said:

'Not more than two dances with any one partner, my dear child, and go back to your mother after every dance.'

The carriage was announced, Lilias bent down to kiss her grandmother respectfully, and then followed her mother downstairs.

'Goodbye, father. I wish you were coming too.'

'You must tell me all about it at breakfast tomorrow,' replied her father, very kindly coming down the steps to hold open the carriage-door for them.

Lilias smiled and nodded, preoccupied with an anxious fear that the summer evening breeze might make her hair untidy. As a matter of fact, her mother had already told her that she was to stay in bed next morning until she was allowed to get up.

Both of them were almost equally nervous as the brougham stopped in front of the house in Belgrave Square where the ball was taking place. There was an awning, and a red carpet across the pavement. A small crowd had gathered to watch the arrivals. Lilias would have liked to scuttle into the shelter of the house, but her mother walked slowly, looking neither to the right nor to the left, and she was obliged to follow.

In a very few minutes they had left their cloaks in a room that was already heaped with evening wraps, and were following a footman up the stairs. Trails and trails of smilax, with pots of azaleas and hydrangeas, formed a lane all the way up . . .

At the head of the stairs, the butler awaited them. 'Mrs and Miss *Ivey*!' he bawled, into an open doorway through which came strains of dance-music, and an enormous babble of talk.

Lilias, for a moment, wished herself back again in the familiar security of the schoolroom at home.

Then her hostess, wearing a dress that flashed and sparkled with sequins, an emerald tiara, and carrying a bouquet of red carnations, greeted her and her mother.

'*So* glad you've brought your girl, Mary—though I don't know how to believe you've got a grown-up daughter. How do you like being grown-up, Lilias?'

'Very much, thank you,' Lilias shyly returned the answer that she knew was expected.

'That's right. My girl must find you some partners. Molly—where are you, Molly?'

'I'm here, mother,' replied a rather breathless voice, and a girl a year or two older than Lilias detached herself from a small group of young people standing just inside the room.

'Don't move away like that again,' said her mother rather sharply. 'You're to stay *here*, by me, for the present. This is Lilias Ivey, and it's her first ball—(such a compliment to us, Mary, to let it be ours!) Find her some partners, Molly, before the next dance begins. And Molly—introduce her to some girls as well.'

As Lilias followed the daughter of the house, she caught a murmur from her hostess to Mrs Ivey:

'An older girl, who's been out a year or two, can be such a help . . .'

Lilias knew what was meant. A help in getting to know young men.

Molly Stanford certainly did her best. She brought up at least seven young men, all of whom asked for the pleasure of a dance with Miss Ivey. One of them, who looked exceedingly young, asked if he might have two. Lilias felt flattered, until it occurred to her that perhaps he, like herself, did not know many people in the room. Then Molly introduced her to two girls, one of whom, in her turn, made Lilias known to a cousin.

Finally, just as the third waltz began, Lilias enjoyed the supreme glory of acquiring a partner already known to her—a Mr Christopher Carswell, next to whom she had sat at dinner, only a week earlier, at her first grown-up dinner-party.

Christopher Carswell was large, and very fair, with a deep, booming voice, and Lilias thought that he must be at least thirty. He was very easy to talk to, and complimented Lilias on her waltzing. When the dance was ended, he asked her to give him another one, and to let him take her in to supper.

'Are you enjoying yourself, darling?' said Mrs Ivey anxiously, as Lilias returned to her after each dance.

'Oh yes, *very* much.' Lilias always replied, and it was true.

Thanks to Molly, and Molly's friends, she was engaged for every dance up to the last supper-extra, so that her mind was freed from the dread of being a wallflower.

None of her other partners proved to be as entertaining as Christopher Carswell. Most of them said that the band was good, and that the floor was good, and that it was rather crowded. One of them told her that he had not once been to bed before three o'clock throughout the week, and another one said that he had to go back to Oxford next morning.

Lilias made suitable replies, but was not very much interested.

When Christopher Carswell came to fetch her to take her into the supper-room, she felt so pleased to see him that it was like meeting an old friend.

'Are you enjoying your first ball?' he asked.

'Yes, very, very much.'

'I'm so glad,' he said, looking at her as though he really meant it.

Lilias blushed.

They sat at their little table for a long while, and if Christopher Carswell made an excellent meal off iced consommé, quail, fruit salad, and champagne, Lilias—dawdling with an ice and a preserved apricot—was not aware of it.

They were talking all the time. Quite serious talk.

Christopher Carswell wanted to know all about Lilias—what she read, and what she most enjoyed doing, and how many balls she was going to that summer. Two of her prospective hostesses were also to be his, and Christopher Carswell asked her to keep some dances for him.

Then the conversation became more intimate.

'I suppose you know heaps of people here tonight?'

'Oh, yes,' said Lilias innocently. 'Miss Stanford introduced me to a good many.'

'Have any of your partners told you that you're the prettiest girl here?'

Lilias blushed more deeply than before. She had always been taught that personal remarks were not polite—but perhaps that no longer applied to people when they were grown-up?

She felt very grown-up indeed.

'You mustn't talk like that. You *know* it isn't true,' she admonished him gravely.

'But *I* think it is.'

'I'm not so silly as to believe you,' Lilias replied, without very much conviction.

When she suddenly woke up to the fact that at least three dances must have gone by whilst they had been sitting at their little table, Lilias felt very guilty.

What *would* her mother say?

Christopher was looking at her, and smiling, almost as though he guessed what was in her mind.

'Let's finish out this dance,' he suggested, as they reached the door of the ballroom.

It seemed rather deceitful, but Lilias quite saw that this would be better than going straight to her mother, who would then perceive that she had only just come upstairs again.

'Keep the last waltz for me,' murmured Christopher, when he bowed, and left her at the corner where Mrs Ivey, with half-a-dozen mothers, sat on a little gilt chair, every now and then raising her fan to conceal the involuntary beginnings of a yawn.

She smiled very kindly at her daughter.

'How many more dances are you engaged for, darling?'

'Two more and—and the last waltz.'

'We shan't be here for the last waltz,' Mrs Ivey replied placidly. 'I'm going after the next two, darling.'

'Oh, mother!'

Her mother laughed comfortably.

'I'm glad you've enjoyed yourself so much, my pet.'

'But can't we stay for the last waltz?'

'No darling, certainly not. It's quite late enough as it is.'

Then there was nothing more to be said, and Lilias knew it. She was terribly disappointed, and besides, what *would* Christopher think of her, for cutting his dance?

She looked at every couple that passed, through the next two dances, but could nowhere see Christopher's broad shoulders and fair head towering above other people's.

Probably he was sitting out somewhere with his partner. Lilias felt a rather queer sensation of annoyance as she wondered with whom he was talking, and whether he was saying the same kind of things that he had said to her.

'Now, darling——'

Lilias followed her mother to where their hostess was standing shaking hands with departing guests.

'I hope you've enjoyed your first ball?' said Mrs Stanford kindly.

'Oh, yes, very much indeed, thank you. I danced every dance.'

'Your girl was too sweet, about finding partners for Lilias,' interposed Mrs Ivey quickly.

'Oh, but Molly would simply *love* to do anything——'

'Won't she come and have tea with Lilias one day next week?'

'She'd be delighted.'

'Lilias shall write and suggest something then. Good-night—thank you again for a *most* delightful evening . . .'

With a last despairing look round, that nowhere revealed Christopher Carswell, Lilias was obliged to go downstairs and get her cloak. She stood by her mother at the foot of the stairs, whilst the carriage was being called.

Dance-music was still going on, and fortunate couples for whom the ball was not yet over were sitting on the stairs, or politely passing up and down them.

Suddenly, Lilias saw him.

Darting from her mother's side, she spoke breathlessly.

'I'm so sorry—I'm being taken away now——'

'What a shame—I was so looking forward to our dance.'

'So was I.'

'But I'll see you on Thursday, at the Ritz.'

'Oh, yes—Good-night.'

'Good-night.'

His look, and smile, seemed to plunge down through her eyes, as they met his in full, to some mysterious recess of her inmost being.

That moment was worth the severe rebuke that Lilias received from her mother as they drove home through the silent streets and squares, to Cavendish Square.

'Who was that, Lilias?'

'Who was who?' said Lilias feebly.

'You know perfectly well what I mean. The man you rushed up to in that ill-bred, schoolgirl way, just as I was ready to start.'

'His name is—is Mr Carswell. He sat next to me at dinner at Lady Crale's and I met him again tonight and danced with him.'

'Isn't he the man who took you down to supper?'

'Yes—I—I think he was.'

'Well, I'm sure I don't know what he'll think of you, after this. Young men like girls to be quiet, and well-bred and not run after them like hoydens. I was quite ashamed of you.'

'I'm very sorry, mother.'

'Very well, darling, we'll say no more about it. Only remember that another time, when mother is waiting for you——' Mrs Ivey continued to say more about it until the carriage stopped.

She did not, however, tell Lilias's father, which was what the girl had most feared, and indeed, from Edward Ivey's pleasant rallying of her the following day, Lilias gathered that she was considered to have achieved quite a success at her first ball.

Long before the end of her first season, Lilias was in love with Christopher Carswell.

Mrs Ivey was the first person to make this discovery, with the possible exception of Christopher Carswell himself—and she took immediate steps to find out whether or not she might rejoice in it. Naturally, she wanted her daughter to fall in love and marry—the sooner the better, with three sisters younger than herself—but it would have to be a really *good* marriage.

Was Christopher Carswell a desirable *parti*?

On this point, Mrs Ivey asked the opinion of half-a-dozen London hostesses, with some slender pretence of doing so in confidence.

The result was universally, and terribly, unsatisfactory.

Christopher Carswell was not a desirable young man at all. He hadn't a penny. He had already proposed to, and been refused by, the daughter of a very rich Jewish financier. Moreover, he was a flirt, and always ready to make any pretty girl conspicuous.

'My dear,' said Mrs Ivey's friends in chorus, 'don't let your girl have anything to do with him. It's the greatest pity that he was ever introduced to her.'

'It will ruin her chances with other men, pretty as she is, if it gets about that he's a friend of hers.'

'But indeed he isn't!' exclaimed the horrified Mrs Ivey. 'Naturally,

the child danced with him when he asked her to, but that wasn't her
fault——'

'No, indeed.'

'And I shall see that it's put a stop to, *at* once.'

Mrs Ivey carried her resolution into immediate effect.

'By the way, Lilias, I don't want you to have anything more to do with
that Mr Carswell. Don't be rude, of course, just say you're engaged for
every dance, next time he asks you. He'll soon understand what that
means.'

'Mother, why?' gasped Lilias.

'Because I say so, my darling.' Then Mrs Ivey relented a little. 'I've
heard one or two things about him that I don't quite like. You trust
mother, don't you?'

'Yes, of course I do. But I—I—I'm afraid he'll think it's that I'm not—
that I——'

'It doesn't in the least matter what he thinks,' said Mrs Ivey serenely.
'Run and get your things, dear, I'm taking you to pay some calls with me
this afternoon.'

Lilias, in a state of consternation, was obliged to obey. Anything else
than obedience, did not so much as occur to her.

Until Christopher Carswell put it into her head.

This he did on the very next occasion that they met. It was at
Hurlingham. The Stanfords had taken Lilias with them, to see the polo.
Molly Stanford, a soldier cousin, and Lilias, were walking about, when
they met Christopher. Molly, who had been told nothing, and was besides
much interested in her cousin, appeared to take it for granted that
Christopher and Lilias should move away, side by side, as though to see
something of the game.

Lilias was so much agitated, so anxious to explain what she feared must
happen, and so certain that it would be wrong to do so, that she could not
speak. Christopher, looking down at her flushed, troubled face, at once
asked what was the matter.

In a trembling voice, with many hesitations, Lilias murmured that she
was afraid he would think—she didn't know how to explain—it wasn't
always very easy—Mother was most terribly particular.

'Have you been told not to have anything to do with me?' asked
Christopher. He was actually smiling!

'Yes,' Lilias answered, very low, and with a dreadful break in the
middle of the word.

There was a silence.

Then Christopher, in a very earnest way, began to tell her about himself, and how unjust people were because, when he was younger, he had

sometimes behaved rather foolishly. He added that if he had, in those days, known a girl like Lilias Ivey, things might have been very different. The very first time he met her, he had felt that she could influence him—it was almost, he thought, as though they had met in a former life.

Strangely thrilled, Lilias admitted that exactly the same idea—about the former life—had crossed her own mind, more than once.

Christopher, at that, caught her hand, in its white kid glove, and gave it a moment's pressure, that frightened her dreadfully, but also filled her with bliss.

She faltered something about going back to Mrs Stanford.

'I know,' said Christopher. 'It's impossible to talk, like this, but there are things I *must* say to you. I suppose you're not allowed to go out anywhere by yourself, are you?'

'Oh, no.'

'But with your maid?'

'Yes—sometimes.'

'Could you——' he hesitated, 'come to the National Gallery, tomorrow morning, about eleven?'

'It wouldn't be right——'

'It's only for this once. If you're not to have anything more to do with me, I *must* speak to you once more. There's something I must say to you,' he urged.

Between her own inclinations, curiosity as to what he had to say to her, and Christopher's importunity, Lilias gave in.

She promised to meet him the following morning. Her maid was an extraordinarily stupid girl, and it was very easy to make her believe that Lilias had been seized with a sudden wish to study the pictures in Gallery IV.

'Go and look at anything in any of the other rooms, if you'd like to, Norris.' Lilias told her in a shaking voice. 'I'll sit here. I—I want to look at that big picture over there rather carefully.'

It turned out, after all, that Christopher had no special communication to make, but in the guilty rapture of hearing what he *did* say, Lilias scarcely noticed that.

She was in love, and was being made love to, and it was all clandestine, and exciting, and unheard of.

For the affair naturally did not stop at that meeting in the National Gallery. It was unfair and ridiculous, Christopher assured Lilias, that a girl of her age and judgement should not be allowed to choose her own friends. If her parents would not trust her, then she was entirely justified in withholding her confidence from them.

Although she did not really believe this, Lilias was far too much in love not to let herself be overpersuaded.

It proved possible to enlist the sympathy of Norris, the maid, who revealed an unexpected passion for romance, and was ready to wait about, in such rendezvous as were alone available—museums, public galleries, and the remoter walks in the Park—whilst Lilias and Christopher told one another, in words and phrases that varied but slightly, exactly what they had told one another at their previous meetings.

There was one thing, however, that Christopher said only once; and that was when he asked Lilias if she would agree to a runaway marriage. But she was so much shocked and distressed, that he never made the suggestion again. Neither, however, did he ever again ask her to marry him—but it was enough for Lilias to hear him say that he loved her. She scarcely looked beyond that, even in her inmost thought, but lived and moved in a dream of stifled self-reproach and ecstatic happiness.

Abruptly and violently, the dream was shattered.

The London season was drawing to a close: Lilias was to be taken to a house party in Scotland, and then a series of country-house visits. She and Christopher said goodbye—said it, indeed, again and again—at six o'clock on a July evening, beneath the august bulk of the Albert Memorial.

It was agreed that they should write to one another.

'Is it safe?' Christopher asked doubtfully.

'Oh, quite. Mother gave up opening my letters after I left school,' Lilias replied.

She had forgotten that at home, in Cavendish Square, her letters were brought up to her bedroom when she was called in the morning, and that now, they would lie upon strange breakfast-tables, or perhaps even be handed to her mother with Mrs Ivey's own correspondence. And she did not realize that all over her young, expressive face, and in every tone of her dream-laden voice, she was betraying herself to the vigilant and suspicious eyes of her mother, the experienced ones of her grandmother, and the sharply critical one's of her mother's friends.

The crisis came a month after the Iveys had left London. The younger girls and Henry had been sent with the governess to Felixstowe, as usual, Edward was paying a visit to a school-friend, and Edward Ivey was dutifully keeping grandmama—who was always immovable—company at home.

Mrs Ivey and Lilias were staying with a great-aunt whom Lilias did not very much like—her mother's aunt, a childless widow with a house in Yorkshire, far away in an isolated country district.

One morning, when Lilias had hoped for a letter from her lover and had not received one, Aunt Gertrude called her into the library.

Unsuspiciously, Lilias went in.

The very next moment, *she knew*.

Her mother, white-faced and in tears, was sitting on the sofa, the many pages of a letter, in handwriting that Lilias knew well, strewn around her.

Aunt Gertrude, with stern, shocked face, and compressed lips, stood over her.

The scene that followed was dreadful.

Lilias, in floods of tears, admitted everything. Indeed she could hardly do otherwise, for her mother told her that Christopher's letter put it beyond question that they had been meeting, and corresponding, clandestinely for some time. She said, again and again, that she could never have believed it of a daughter of hers—and added at the same time that she had had her suspicions, for weeks. So, it seemed, had Aunt Gertrude, and it was in fact she who had plainly told Lilias's mother that it was her duty to find out from whom it was that her child received all those letters. . . .

They would not give Lilias the letter. Aunt Gertrude, indeed, put it into the fire, declaring that it was a most disgraceful thing that anyone calling himself a gentleman should have written it.

Mrs Ivey, after working herself into a state that required bed and a visit from the doctor, sent a telegram to her husband, asking him to come at once.

Aunt Gertrude, until he arrived, never let Lilias out of her sight for five minutes together, and all writing materials were taken away from her.

Lilias who by this time felt that she had behaved with a wickedness, immodesty, deceit, and ingratitude such as no decently brought-up girl had ever been guilty of before, wondered wildly for a day or two if Christopher would perhaps come up to Yorkshire, and insist upon seeing her. But he did not come, nor, so far as she knew, did he ever write to her again, nor answer the scathing letter that she was told had been sent by her father.

Thus, although none of them realized it at the time, ended, at the age of nineteen, the youth of Lilias Ivey.

Aunt Gertrude offered to keep Lilias with her for a long visit.

'Quite out of the way of *anything* here,' she significantly observed.

Mr and Mrs Ivey agreed the more thankfully because it was evident, when they returned to London, that the scandalous behaviour of their daughter had not passed unobserved. One or two people even ventured to question Mrs Ivey, but she quickly turned the subject. She made Adelaide put her hair up, and come everywhere with her, and fortunately Adelaide was a great success, and became engaged to the eldest son of a Baronet almost at once.

Lilias came home for the wedding, and was bridesmaid to her younger sister.

The night before the wedding, her mother sobbed and cried, and said to Lilias: 'Oh, my darling, if *only* you hadn't ruined all your chances, *you* might be as happy as Adelaide, now. How could you do it—how could you? It's broken my heart—and your father's.'

The younger ones looked at their eldest sister with a certain awe. They knew that she was under a terrible cloud, but they did not know what she had done.

Grandmama, however, was not told anything at all, except that Lilias had been ill and was to stay in the country, because Mrs Ivey said that the truth would probably kill her.

After the marriage, Lilias went back to Yorkshire.

There, she gradually fell into a routine of parish visiting, tête-à-tête meals with Aunt Gertrude—when they talked about the dogs, the parrot, and the headlines in the newspapers—aimless little walks, desultory gardening, novel-reading, and bed at ten o'clock.

Some years later, when Aunt Gertrude had a stroke, it seemed right to the point of inevitability that the niece to whom she had given a refuge should share, with a trained nurse, the care of her.

By this time the other children of Edward and Mary Ivey were all grown-up. Indeed Edward and Constance were both married, and Isabel was 'the daughter at home'.

It was felt by all of them that Lilias belonged to great-aunt Gertrude, and the life in remote Yorkshire. She felt it to be so herself.

At the end of another ten years, Lilias looked very much older than she was—which might have been owing to the responsibility of the household, which had now entirely devolved upon her, for Aunt Gertrude, although she lingered, semi-insensible, to an incredible age, never again recovered the use of her faculties.

When, eventually, she died, Lilias had already lost her father, killed in a riding accident, her mother was living in Brighton with Isabel, close to Henry and his young wife, and the house in Cavendish Square belonged to Edward and his family.

Aunt Gertrude had, naturally, left the house, her furniture, and most of her money, to a male, rather than to a female relative. Her jewellery, and some personal effects, went to her nieces and great-nieces, and there was a legacy for Lilias. It was not a very large one, and it was so arranged that Lilias should receive the income of the sum invested during her lifetime, but might not touch the capital. For she had never regained her aunt's confidence.

Lilias, who had grown accustomed to living under a cloud, and had moreover lost touch with people and things unconnected with the York-shire village, decided to remain there, and moved into a small house near the church.

There she continued, on a smaller scale, the life that she had led for so many years with Aunt Gertrude. The clergyman found in her a most useful parish-worker.

In only one respect did she ever disappoint the good man: never would she extend more than the barest charity to any girl guilty of a lapse from virtue.

He once ventured to remonstrate with her, but Lilias Ivey, now grey-haired and inclined to gauntness, replied only that, in her opinion, there should, in such cases, be no such thing as a second chance.

SARA MAITLAND

An Edwardian Tableau

ʚɞ

True to their word, the Suffragists marched on the House of Commons yesterday, and the scenes witnessed exceeded in violence the utmost excesses of which even these militant women had previously been guilty.

It was an unending picture of shameful recklessness. Never before have otherwise sensible women gone so far in forgetting their womanhood.

Daily Sketch. Saturday, 19 November 1910

Dinner seemed interminable, and yet Caroline was not sure that she wanted it to end. Afterwards there were two things to be faced; she was so tired that they seemed the same, equally important, equally unimportant, it did not seem to matter. Richard would propose to her, and her mother would lecture her about coming down to dinner without stays. She knew the first from her father's heartiness; Richard and he had been in the library together before the other guests had arrived; also her mother at the last, the very last, moment had changed the seating so that Caroline and Richard were sitting next to each other. And would she accept him? His face moved backwards and forwards, in and out of focus, she was so tired that she did not know what she would do, what she wanted. He would be a bishop one day they all said, he was a canon already, he was too old for her, she was too young for him, she would be a bishop's wife perhaps, perhaps not; how could she not know? How could she not care? She had known about the lecture from her mother from the very moment she had come downstairs. How could she have thought that her mother, who noticed everything, would not notice? Her mother's standards were liberal but fixed, like her father's politics, and Caroline knew every shade of them. She was allowed to smoke when there were no guests in the house: she was allowed to hunt escorted by only the groom if Graham was away, but not if he was home and did not want to go out himself; and she could leave off stays in the daytime, but not in town and not for dinner. The lecture would cover these and other points and would include her mother's favourite little joke, 'Impropriety is one thing, Indecency is another.' She should not have risked it, she could not face it, the pain would have been better; no it wouldn't; even without the stays she could feel the pain, the bruise where only yesterday one of her whalebones had been snapped and driven into her side. Hunting falls never hurt like this, but people were gentle over hunting falls and they were your own fault or bad luck, not inflicted, deliberately laughingly inflicted, the way Graham had hurt her when they had both been very small—run to nurse and she would make it better—but now there was no nurse and no one she could tell. She was going to fall asleep, during dinner, at the table, no please not, please not God. They, they out there, her mother, her father, stout Lady Corson, the They outside her pain and tiredness were talking, listen to them, don't fall asleep, not Here, not Here.

They were discussing some minor corruption, some political corruption, something mildly bad, mildly important, Caroline could not remember the details. Sir George Corson kept saying how dreadful it was, how very dreadful, how it just went to show, how monstrous it was. Caroline's mother laughed her silvery laugh—and had her laugh always been like that or had she read somewhere of a silvery laugh and set out to procure one, just as she procured good cooks and beautiful dresses?—she laughed her beautiful, silvery laugh and said, 'Of course, Sir George, these things wouldn't happen if you gave women the Vote, that would purify politics,' because of course Caroline's mother believed in the vote in her beautiful, decorous way. Sir George responded, 'Come now Mrs Allenby, women purify in the Home, you make the politicians of the future and it's far too important a job for us to give you time off to go running in and out of polling booths. You wouldn't like it if you had to do it, you wouldn't be in a position to purify anything then, you know. No, no, women don't need the vote, they have the sons of England to look after, and they have husbands to do the sordid things like voting for them.' 'What about the unmarried women?' asked someone down the other end of the table; the conversation was going to become general, it always did when the Vote came up: there were subjects, the Vote, the Hysterical Militants, the Impossible Irish, the Ridiculous Workers, the Poor Peers, subjects that no one could resist. Sir George looked swiftly round the table, all the women except Caroline were married and he could guess what was meant by the odd seating easily enough, so he laughed and said, 'The unmarried women? Dear Madam there shouldn't be any, and in any case we don't want to be ruled by the failures, that's not democracy to my way of thinking. Remember a *Saturday Review* article that hit the nail on the head, said that a woman who failed to marry had failed in business and nothing can be done about that. I agree; it seems a little harsh at first, but think about it, think about it.' There was a little silence and then Richard raised his head and started speaking slowly and gently. Dear Richard, Caroline thought as his profile swum into focus and, yes, they would make him a bishop and she would be a bishop's wife. 'It seems to me,' he was saying, 'that all this unrest is a symptom of a massive breakdown in trust. Everyone seems to be frightened, frightened and too proud, women don't trust their men any more and the working people don't trust us. But it does seem to me that it must in some way be our fault, and if they can't trust us then we can't be worthy of the trust and must allow them something they do trust, the Vote or Unions or Home Rule or whatever it is. I think they're wrong, I think they would do better to trust people than institutions, but they don't and they must somehow be freed from their fear. There's too much fear and not enough trust and love.'

Caroline's father laughed, 'Come now Souesby, where there are separate interests there's going to be distrust. We don't trust the workers come to that, and I for one don't trust the Irish, neither lot of them, and I don't trust those screaming women, and I don't see my way to doing so. Universal love indeed, you sound like one of those Russian Anarchist fellows.'

But Richard was not daunted; he's brave, she thought, gentle and brave, just as he was out hunting. He went on, 'That's not fair, Sir, and you won't scare me from my truth with an anarchist bogey. I'm not an anarchist, as you know perfectly well, but I will say that I have read a lot of their literature and I think they have some pretty sound ideas. Just building up more and more institutions is not going to help any of us; we must have more trust in each other, more of a common interest, and stop pinning our faith on all these organizations, and institutions, or at any rate look at them more closely and see if they deserve to continue.'

Someone said, 'That's a fine way for a good churchman to talk,' but he smiled and replied in his politest voice, 'If you knew my record, Madam, as a lunatic ritualist, practically an idolatrous Papist, I assure you, you probably wouldn't think of me as a good churchman at all. I don't think I care so much about being a good churchman as I do about being a good man, and I still say that we all, all of us, on every side, need more love and more trust. Speaking as a churchman I could say simply that Perfect love casteth out fear.' And even as she thought how superb he was, how her mother herself could not have done it better, could not have been more firm, more silencing, more politely rude, Caroline heard her own voice, in the distance, far away, out there, say, 'So does hate,' and even then it might have been all right. She had said it very quietly, almost to herself, but Richard, attentive and loving, turned round and asked quite clearly what she had said and there was no escape. For a timeless moment her eyes seemed fixed on her mother's beautiful chest, her pure white shoulders rising up from the exquisitely ruched chiffon and the line of her neck running up past her pearls and into her lovely, lovely hair, and why thought Caroline, or half thought into the endless gap in time, why don't I look like that so that I could say things like this and no one would mind?

Then she said rather loudly, 'I said, "so does hate". Perfect hate casteth out fear.' And in the silence, the astonished silence, that followed she could hear her head tapping out thoughts. That will teach them, that will teach them to sit here, so pompous and liberal and benign and intelligent and talk about purity and trust and love, when outside there is anger and meanness and hate, beautiful hate which made you feel six foot tall, which made you feel as you felt when you knew that your mare was going to take a fence that other people were refusing at, was going to take it perfectly, take it in her stride. That will teach you Canon Richard Souesby to keep your white hands clean and turn the other cheek and trust them all while they beat you and throw you about and laugh in your face. And then Lady Corson, kind, well meaning, stout Lady Corson, said loudly and carefully, 'That reminds me of the most peculiar book I was reading, young women are so much more imaginative I think than we ever were. I wonder if you've heard of it, Mr Allenby? It's called *Dreams*, by an Olive Schreiner. It was lent to me by . . .' And they went on talking and gradually everyone else joined in, but not Richard; he sat silent beside her, did not look at her, looked at his food, and Caroline was afraid, afraid for herself, afraid of herself, and now he would not marry her and what would she do? How would she manage without him? How could she have thought that she did not care, that it did not matter whether she accepted him? And now he would not ask her and she loved him, she loved him, she loved him. But he did not turn round, did not smile, just sat looking at his food, eating his food. And her mother's chilly white shoulders were waiting, waiting till afterwards, till all the people had gone, and the fact that she was not wearing stays had ceased to matter compared to what she had done, she had silenced a whole dinner party, she had embarrassed everybody. The white, cold shoulders and the tiredness and the dreadful, dreadful pain in her side all became one cold blur and Richard would not ask her to marry him and she was getting colder and colder and further away and then Emma was beside her and 'Would Miss Caroline like a drink of water?' Emma, so quiet and discreet, pouring water quietly, discreetly, and she drinking it and feeling better and gradually the dining-room coming back towards her and she back into it, all so quietly and discreetly that no one noticed; no one except Richard and he turned towards her looking concerned and asked almost soundlessly if she were all right. He smiled sweetly, lovingly, and she thought that after all he would ask her to marry him and she would accept and he would look after her always, and she felt well and strong again and started listening to the conversation.

By now it had moved on to the awful Incident the day before, when hundreds of women had fought with the police for six hours, trying to get to the House of Commons. Well, she thought, it was bound to come up, it was bound to; she felt strong enough for it, she need not say anything, they need never know, she need only listen, even the pain seemed bearable. Sir George Corson was speaking again. 'I don't often find myself agreeing with that dreadful *Mirror*. This time I did, they hit the nail on the head, *The Times* was far too soft on them. The *Mirror* said those women were the disgrace of the Empire and a source of shame to all womanhood. I couldn't agree more. Glad they left the word "ladies" out. I wouldn't even call them "women", female, that's what they are, females. Disgusting.' And no, she couldn't keep quiet, could not listen and say nothing, could not hear her friends spoken of like that, because they were her friends, her real friends, so she said, 'I was there.' She saw the rucheing of her mother's dress move as her mother's shoulders tightened a little, disapproving. Then one of the ladies down the table said, 'Caroline dear, I didn't know you were a militant,' and there was perhaps a hint, a slight tone, of admiration, envy. It was not clear but it was enough for Caroline to go on, 'Oh no, I wasn't. I mean I'm not a member of the Union or anything. I was there by mistake, but I got involved, because of the crowd and separated from Emma, there was this enormous crowd, watching them you see.' How could she be so cool? Her mother's shoulders had relaxed again, she could go on, there would be no trouble, no trouble so long as she kept calm. 'I didn't understand the newspapers this morning, it didn't seem like that then, the police were very brutal.' Sir George interrupted, 'Now, now, Miss Allenby, they were only doing their duty.' 'I thought,' she said as carefully as possible, 'that it was their duty to arrest anyone who assaulted them. They wouldn't arrest us.' The shoulders tightened again. Was her whole life to be governed by the rise and fall of a pair of perfect, beautiful shoulders? 'You see,' she hurried on, 'I got involved.' Involved. There must be a better word: committed, converted. She had been standing, pushed about by the crowd, trying to see Emma, when suddenly a funny old, no not old, middle-aged lady had fallen to the ground at her feet. She had bent down to help and asked, 'Are you all right?' But the woman was hysterical, she lay on the ground and sobbed, 'They won't arrest us, they won't arrest us, they won't arrest us' over and over again. How could she explain to these safe people how nothing had made sense, how the police were refusing to arrest them; them, us, me? 'Sir George, you don't understand, the crowd was all around pushing in; if one tried to get out, and at first I tried very hard, I did not want to be there, I don't believe that militancy will work, I didn't see the point of it, I wanted to get out, but if you tried then the horrible men in the crowd pushed you back in again, back to the police and they would not arrest you whatever you did. There was an old lady there in a wheelchair, perhaps she was as mad as anything, perhaps she should have stayed at home, but she was there, and the police pulled her out of the chair and threw her on the ground and then they shoved the chair away; I saw them do that. It was very frightening, if the ladies there did foolish things it was because they were so frightened.' The panic had been the worst thing, she had been so frightened, so lost, so confused, turning in circles, pushing against other women, pushing them down, knocking them over herself in her desperate efforts to escape. Finally she had run into a policeman and had grabbed him thinking that here was safety, that he would help her, 'Get me out of here, please get me out of here.' He had seized her in his arms, crushing her so tightly that she could hardly breath, tearing her blouse on his buttons and he had tried to kiss her and when she had protested he had laughed and said, 'That's what you really want; that's what you're here for, isn't it, sweetie?' She had started to struggle, to kick, even to bite. The policeman, suddenly angry, no longer smiling, had literally thrown her to the ground and as she landed she had felt one of her stays snap and ram itself up under her ribs. The pain and the shock had been too much and she had lain there with a red film running over her eyes for a moment, and then she had opened her eyes and seen the policeman standing there smiling, pointing her out to another of himself who was also smiling, but who looked almost frightened. A great wave of hatred, the sort she had not felt since she had been a little girl, filled her up, lifted her to her feet and she had realized that she was not frightened any more. She was a fighting force, she was Deborah, and Joan of Arc, and Boadicea and there was no fear but only waves of beautiful hatred which made her feel six foot tall and insuperable. Hitting and shoving and insulting policemen had felt like Mafeking Night, only the bonfires were all inside her and hotter and brighter and better. But these things she could not explain, and she said, still quite calmly, to the dinner party, 'You cannot imagine how horrible it was, how frightening; some of the members of Parliament came out on the steps to watch, they were smiling and laughing. One of them had a little child with him, she cannot have been more than ten, and he kept pointing us out to her and trying to make her laugh and she just stood there and looked amazed. He is probably a good honest man who would not go himself, let alone take his daughter to a fight, a match, whatever it is that men go to, Father, what's the word? I know, a "mill", but he still thought that seeing a thousand ladies abused by the police, English ladies by their own police, was a suitable amusement for her. And all those ladies were trying to do was to present a petition, asking for what they believe to be their rights. Apart from the vote, surely they have the right to petition Parliament? I hated that member of Parliament so much at that moment that even if they had been fighting for something that I thought bad, thought totally wrong, I would not have left those ladies then, I would not have wanted to, even if I'd been able to.' She was getting excited, she knew it was a mistake, that it might spoil everything, that it would do the suffrage cause no good; but her excitement was not for the cause, it was for herself, because she had discovered that she need not be afraid, that she could be strong, that she need not be tied down in awe of her mother, that beauty was unimportant compared to the strength of her feelings, that militancy might not do much good for its cause but it did wonderful things for the militants. They knew what she knew, how good it was to be angry, to be really angry and show it, that when you were really angry nothing else mattered, that there was no pain, no fear, no restraint, no anything but an enormous space that you could fill up with yourself and see how huge and strong you really were. She knew how good it was to have an enemy and know that he hated and feared you, because the police had been frightened of the women and of what they had found in themselves, but that you only hated him and were not frightened so that you would win really even when he appeared to have won. And so she finished up almost panting. 'Sir George, I have told you how dreadful it was, how humiliating and disgusting, how shameful to the government that let it happen. I haven't told you, I cannot tell you, how fine it was, how good I felt hating and fighting with the police, how good it was to abuse members of Parliament at the top of my voice, how fine and beautiful and lovely those muddy women on the ground were, how much I loved those "Females", as you call them when we helped each other. They have been waiting, all women have been waiting for fifty years for the Vote, waiting patiently to accept it as a pretty present from the men who laugh at them, who abuse them mentally and physically. I tell you that after yesterday I am beginning to believe that after thirty years of patience and waiting and teaching calmly, Our Lord must really have enjoyed hurling over the tables of the money lenders in the temple.'

'Caroline? That is quite enough!' That was her father, his good natured face red with embarrassment. Her mother was calmer and far colder, 'I think we have heard quite enough about the hysterical conduct of some unhappy, unbalanced women for one evening. Emma, please offer Mrs Lettering some more of the fruit shape. Tell me, Sir George, have you seen the Martins since they got back from Dresden? So strange to come back at this time of year, one can't help wondering why they've returned.'

Caroline sat at her place and the warmth she had felt died away, but she wasn't sorry, she could not be sorry, neither for what she'd done nor for what she'd said. She would be sorry in the morning when she had to listen to her mother and watch the beautiful neck take on its curve of disdain; she would be sorry if Richard did not ask her to marry him, and sorrier still when he married someone else, but it would not be the right kind of sorry, not the kind they would expect. That kind was out of the question now she knew how strong she could be, how it felt to be free of fear, how it felt to be totally herself. Then she looked at Richard and he was smiling, not pityingly, not even kindly, but with open admiration. For a moment she was tempted into humility, into wondering what she had done to deserve this wonderful man, but her courage was high and with a final rush of bravery she thought, 'Of course I have deserved it, of course I deserve this man, of course.'

LIST OF STORIES

In chronological order of publication

NOTES AND SOURCES

'The World Without Man' by [Sir] Charles G[eorge] D[ouglas] Roberts (1860–1943). *London Magazine* (May 1912); reprinted in *In the Morning of Time* (1919).

Poet, professor, journalist, editor, and novelist, Charles G. D. Roberts was regarded during the latter part of his life as the dean of Canadian writers. Though a prolific novelist, he was celebrated for his many books on the natural world and the creatures of the wild. Books such as *The Kindred of the Wild* (1902), *The Watchers of the Trails* (1904), and *Haunters of the Silences* (1907) were bestsellers and created a new genre. *In the Morning of Time* was perhaps Roberts's most original venture. Utilizing state-of-the-art knowledge of prehistory (though perhaps prompted by Conan Doyle's more populist, but less accurate, *The Lost World* (1912)), he created a convincing scenario portraying the growth of primitive man.

'Overture' by Eugene O'Sullivan (pseudonym of Owen Dudley Edwards, b. 1938). *Edinburgh Review* (July 1978).

Born in Dublin, Owen Dudley Edwards is Reader in Commonwealth and American History at the University of Edinburgh, where he has been on the teaching staff since 1968. His publications include biographical studies of P. G. Wodehouse, T. B. Macaulay, and Arthur Conan Doyle, and he is the General Editor of The Oxford Sherlock Holmes.

'A Trooper of the Thessalians' by Arthur D[ouglas] Howden Smith (1887–1945). *Adventure* (23 Aug. 1926); reprinted in *Grey Maiden: The Story of a Sword Through the Ages* (1929).

Journalist, war correspondent (1914–18), and biographer, Smith was born in New York and travelled extensively from an early age. He produced a mass of superior popular fiction, specializing in adventure and historical stories and writing mainly for Arthur Sullivant Hoffman's magazine *Adventure*. A great sequelizer and Stevenson enthusiast, Smith concocted effective sequels to both *Treasure Island* and *Kidnapped* in, respectively, *Porto Bello Gold* (1924) and *Alan Breck Again* (1934). His novels of American frontier life in the eighteenth century (*The Doom Trail*, *Beyond the Sunset*, etc.) were both praised by the critics and bought by the reading public.

Setting: the campaigns of Alexander the Great, *c*.330 BC.

'Greek Meets Greek' by Jack Lindsay (1900–90). From *Come Home At Last* (1936).

Jack Lindsay was born in Melbourne and educated at Queensland University. He settled in England, dying in Cambridge. Poet, dramatist, novelist, art critic, and biographer, he wrote extensively on classical life and literature, his novels including *Cressida's First Lover* (1931), *Rome for Sale* (1934), and *Caesar is Dead* (1934). He described the setting of 'Greek Meets Greek' as 'Boundary of Macedonia and Moesia, 3rd century BC'.

'Cottia Went to Bibracte' by Naomi [Mary Margaret] Mitchison (b. 1897). From *When the Bough Breaks and other stories* (1924).

Naomi Mitchison is the daughter of John Scott Haldane (1860–1936), the philosopher and physiologist, and brother of the scientist J. B. S. Haldane (1892–1964). In a long and distinguished writing career she has produced novels, short stories, plays, travel books, biographies, and children's books. Her historical novels include *The Conquered* (1923), *Cloud Cuckoo Land* (1925), *The Blood of the Martyrs* (1939), *The Bull Calves* (1947), and *Sea-Green Ribbons* (1991).

Setting: Gallic Wars, *c*.52 BC.

'A Judgement of Tiberius' by W[illiam] Outram Tristram (b. 1859). *The Graphic* (20 July 1907).

Outram Tristram's chief work was *Coaching Days and Coaching Ways*, a heavily nostalgic evocation of pre-steam train Britain which, with its stunning illustrations by the young Hugh Thomson, was first serialized in the *English Illustrated Magazine* and then issued in book form by Macmillan in 1888. It was an instant success and was reprinting well into the twentieth century. Outram Tristram had already published his first novel, *Julian Trevor* (1883), and a melodrama with a touch of the weird, *A Strange Affair* (1887). His last book, *Moated Houses*, was published in 1910.

'The Cupbearer' by Richard Garnett (1835–1906). From *Twilight of the Gods* (1888).

The son of the linguist and philologist of the same name (1789–1850), Garnett, like his father, held a post in the British Museum. He was superintendent of the Reading Room from 1875 to 1884, chief editor of the library's first printed catalogue 1881–90, keeper of printed books 1890–9. His erudition was legendary and he produced several volumes of translated verse and biography. He also wrote learnedly allusive poetry and critical introductions to innumerable standard works, as well as being a prolific contributor to periodicals such as the *Saturday Review*, *Macmillan's*, *Temple Bar*, and *Fraser's*. He is perhaps best remembered for his *Relics of Shelley* (1862) and for his collection of pagan tales, *Twilight of the Gods*, from which this story is taken. *Twilight of the Gods* was reissued in 1903 with several additional stories. Garnett was father of the publisher Edward Garnett (1868–1936) and grandfather of the novelist and autobiographer David Garnett (1892–1981). The story is set during the reign of the Byzantine Emperor Basil II (958–1025).

'A Fragment of Medieval Life' by [Richard Horatio] Edgar Wallace (1875–1932). *Louis Wain's Annual: 1911/12* (1911).

Illegitimate and largely self-educated, Wallace dragged the late-Victorian 'shocker' into the twentieth century, creating—virtually singlehanded—the modern, fast-paced 'thriller'. For the last six years of his life he was the most popular writer in the Western world, his novels outselling everything but the Bible. He was fascinated by history and an admirer of such historical novelists as Quiller-Couch, Maurice Hewlett, and Dumas *père*. His own contribution to the historical genre—apart from the story reprinted here—was *The Black Avons*, a four-book series for younger readers that followed the fortunes of a family, the Avons, from the reign of Elizabeth I to the aftermath of the First World War. The series was published in 1925 and was sparked off by the critic Augustus Hare's rueful comment, 'Histories used often to be stories. The fashion now is to leave out the story.'

'The Death of Lully' by Aldous [Leonard] Huxley (1894–1963). From *Limbo* (1920); reprinted in *Twice Seven. Fourteen Selected Stories* (1944).

Aldous Huxley was the third son of Leonard Huxley, an assistant master at Charterhouse and later editor of the *Cornhill Magazine*. His mother was the former Julia Arnold, granddaughter of Thomas Arnold of Rugby and niece of Matthew Arnold; his eldest brother was the scientist (Sir) Julian Huxley. Huxley began writing poetry and short stories whilst an undergraduate at Oxford. His first novel, *Chrome Yellow*, appeared in 1921 and was followed by *Antic Hay* (1923) and *Those Barren Leaves* (1925). His most ambitious novel, *Point Counter Point* (1928), was a best-seller. In the 1950s his avant-garde reputation was increased by his well-publicized experiments with hypnosis and with mescalin and other psychedelic drugs, as described in *The Doors of Perception* (1954) and *Heaven and Hell* (1956).

Lully was the common English rendering of Raimon Lull (*c*.1235–1315), a Marjorcan-born Catalan, Franciscan mystic, and author of several controversial treatises. He died of wounds sustained during a missionary crusade to North Africa in his 80th year.

'Pretty Prickly English Rose' by Fred Urquhart (b. 1912). From *Proud Lady in a Cage* (Edinburgh, 1980).

Fred Urquhart left school at the age of 15 and went to work in an Edinburgh bookshop. His first novel, *Time Will Knit*, appeared in 1938, by which time he had published his first short stories. He became literary editor of *Tribune* in 1946 and then worked as a literary agent. He subsequently became a reader for MGM, London scout for Walt Disney, and publisher's reader. As well as novels and short-story collections, Fred Urquhart has also edited anthologies and is the co-compiler of the *Everyman Dictionary of Fictional Characters*. One critic has remarked that Urquhart's specialities are 'good dialogue and bad women'.

Joan (or, more properly, Jane or Johanna) Beaufort (d. 1445) was the daughter of John Beaufort, earl of Somerset. James I of Scotland, whilst a prisoner of the English at Windsor, saw her walking in a garden and fell instantly in love with her (his passion is recorded in the 'King's Quair'). They were eventually married, at St Mary Overy Church, Southwark, in February 1424 and were crowned King and Queen of Scotland at Scone on 21 May. James was assassinated by Sir Robert Graham and his followers at Perth in February 1437 (during which Catherine Douglas, afterwards called 'Bar-lass', suffered a broken arm after trying to keep the assassins at bay). Johanna later married Sir James Stewart (the 'Knight of Lorne').

'The Sacrist of Saint Radegund' by Arthur Gray (1852–1940). *Chanticlere* [the Jesus College, Cambridge, magazine] (No. 56, May 1911); reprinted in the *Cambridge Review* (4 June 1919) and in *Tedious Brief Tales of Granta and Gramarye* (Cambridge, 1919).

Arthur Gray was educated at a school in Blackheath and then at Jesus College, Cambridge, where he remained—as Fellow and, from 1912, as Master—for the rest of his life. His published works included *A Chapter in the Early Life of Shakespeare* (1926) and *Shakespeare's Son-in-Law* (1939), as well as several books on Cambridge history. He is now remembered by supernatural-fiction enthusiasts for the extremely rare *Tedious Brief Tales of Granta and Gramarye* (Cambridge, 1919)—nine stories published separately (mostly in the *Cambridge Review*) under the pseudonym 'Ingulphus' and accompanied by charming drawings by E. Joyce Shillington Scales. Gray's choice of pseudonym was an academic joke: the History of Crowland Abbey attributed to Ingulphus (d. 1109, secretary to William the Conqueror, and Abbot of Crowland) was a late medieval forgery. A tenth Ingulphus story, 'Suggestion', appeared in the *Cambridge Review* in October 1925.

'A Lodging for the Night' by Robert Louis [originally Lewis] [Balfour] Stevenson (1850–94). *Temple Bar* (Oct. 1877); reprinted in *New Arabian Nights* (1882), vol. ii.

Born in Edinburgh, Stevenson abandoned the study of engineering to read law at Edinburgh University, being admitted advocate in 1875. But by then he had determined on becoming a writer and his work was already appearing in periodicals. His life was dogged by chronic ill health, but despite this he travelled widely. His name was made by *Treasure Island* (1883) and he consolidated his reputation as a consummate storyteller with *The Strange Case of Dr Jekyll and Mr Hyde* (1886) and a sequence of Scottish romances that included *Kidnapped* (1886), *The Master of Ballantrae* (1889), and *Catriona* (1893), all turning on the '45 rebellion and the fascination exerted by an outlaw culture. Stevenson also wrote many travel books, short stories, and poems. He died suddenly at Valima in Samoa of a brain haemorrhage whilst working on *Weir of Hermiston* (published in 1896).

The French poet François Villon (1431–?) was a Master of Arts of the University of Paris. He led what is usually described as a 'colourful' life and spent much of his time avoiding arrest. In 1455, following a quarrel, he killed an ecclesiastic, Philippe Sermoise, and on Christmas Eve the following year, with some companions, he broke into the Collège de Navarre and stole 500 gold pieces. He was imprisoned in 1461, on the orders of the Bishop of Orléans and after his return to Paris in 1462 was twice more placed under arrest. On the second occasion, for his involvement with an affray in which a papal notary was wounded, he was sentenced to be hanged: this was commuted to ten years' exile from Paris 'in view of his evil life'. The temporary death sentence inspired his poem *Ballade des pendus*. The date of his actual death is unknown.

'Twilight' by Marjorie Bowen [Gabrielle Margaret Vere (Campbell) Long] (1886–1952). From *God's Playthings* (1912).

Marjorie Bowen's bestselling first novel, *The Viper of Milan* (1906), was written when the author was only 16. In a little under fifty years she went on to produce over 150 books, including 104 novels under five pseudonyms, twenty volumes of short stories, three plays, and biographies of such figures as Cobbett, Hogarth, William III, and Lady Hamilton. Today she is best remembered for her supernatural stories, which present a powerful evocation of gloom and decay, forlorn landscapes and tarnished grandeur.

The story is set in the early sixteenth century. Lucrezia Borgia (1480–1519) was the daughter of Rodrigo Borgia (Pope Alexander VI, a Spaniard by birth) and the sister of Cesare Borgia (1476–1507), whose career is said to have inspired Machiavelli's *Il Principe*. Lucrezia's marriage to Don Gasparo de Procida was annulled by her father and she became betrothed to Giovanni Sforza; but Rodrigo Borgia cancelled the engagement and she married Alfonso of Aragon. After her husband's murder by Cesare Borgia, Lucrezia (at the age of 22) married Alfonso d'Este, heir to the Duke of Ferrara.

'The Scapulary' by Rafael Sabatini (1875–1950). *Nash's Illustrated Weekly* (16 Dec. 1919); reprinted in *Turbulent Tales* (1946).

Born in Italy but brought up in Liverpool, Sabatini was a romantic with unusually strong leanings towards realism and accuracy in his fiction. In this he was influenced by Stanley J. Weyman and was an apt pupil, subsequently outgrowing his teacher both in narrative technique and in the material rewards his fiction brought him. A champion of the Borgias (especially Cesare, about whom he wrote a massive although by no means hagiographic biography, 1911), he had a weakness for rogues, charlatans, quacks, and heroic confidence tricksters (some of his best tales concern the shameless chicaneries of Casanova and 'Count' Cagliostro). During the 1920s his brand of historical fiction—swashbuckling adventure but always built on a firm bedrock of accurate historical fact—came into its own and Sabatini became a world bestseller, with books such as *Scaramouche* (1921) and *Captain Blood: His Odyssey* (1922) selling in their hundreds of thousands. His novels glowed with life, but he was also a master of the short story.

'The King's Stratagem' by Stanley J[ohn] Weyman (1855–1928). *Strand Magazine* (Apr. 1891); reprinted in *In King's Byways* (1902).

Weyman was a barrister who earned far more from writing for the *Cornhill Magazine* and the *English Illustrated Magazine* than at the bar. A novel in the Trollopean mode, *The New Rector* (1891) was a disaster; but *A Gentleman of France* (1893), set in the closing years of the last Valois, Henri III, was such a roaring success that it brought about a reinvigoration of the whole historical genre. Weyman proceeded to write bestseller after bestseller (e.g. *Under the Red Robe*, *The Red Cockade*, *Shrewsbury*, *Count Hannibal*, *The Long Night*), his favourite periods being sixteenth- and seventeenth-century Europe and early nineteenth-century England. In 1908, fearing that success was turning him stale, he took the remarkable step of ceasing to write and retired to North Wales. A decade later he picked up his pen again, and launched into a creative Indian summer, producing some of his best work, including *The Great House* (1919), *Ovington's Bank* (1922), and *The Traveller in the Fur Cloak* (1924), a superb period thriller.

'The Masque at Ludlow' by Anne Manning (1807–79). From *The Masque at Ludlow and Other Romanesques* (1866).

Manning was a writer who specialized in 'domestic-historical' stories and novels, often in diary or epistolary form and usually written from the woman's point of view—although *Tasso and Leonora: the Commentaries of Ser Pantaleone degli Gambacorti* (1856) is related by the gentleman-usher of Leonora d'Este (sister of the Duke of Ferrara). Her popular *The Maiden and Married Life of Mary Powell, afterwards Mistress Milton* was published in 1849 and a sequel, *Deborah's Diary*, the supposed autobiography of Milton's daughter, in 1858. The same pastiche technique (Manning's model was the 1847 bestseller *The Diary of Lady Willoughby* by Hannah M. Rathbone) was used in the equally popular *The Household of Sir Thomas More* (1851), written in the form of a diary by More's daughter Margaret. Manning had a sure narrative touch and a sensitive ear for period nuances. Her other works include *Cherry and Violet: A Tale of the Great Plague* (1853) and *The Lincolnshire Tragedy: Passages in the Life of the Faire Gospeller, Mistress Anne Askew* (1866), based on the life of the sixteenth-century martyr.

Comus by John Milton (1608–74) was written at the suggestion of Henry Lawes (1596–1662), composer and song-writer. *Comus*, presented on Michaelmas night 1634 at Ludlow Castle to celebrate the Earl of Bridgewater's appointment as Lord President of Wales, was first printed, anonymously and untitled, in 1637. Bridgewater's daughter Alice was aged 15 at the time of the performance.

A translation of Letter VIII is as follows:

JOHN MILTON TO HIS FATHER

A few days ago the Countess of Bridgewater asked of me a play, or so to speak a satyric drama; the idea was suggested to her by that most skilled musician, my friend Henry Lawes. The plan of the work is that I provide the poetry, which he will set to music, and the scene be set at Ludlow Castle, where if I may have your permission, I am to go in order to embark on the business. Please therefore inform me of your wishes, my father, and be persuaded of this, that you have a son most full as always of love and duty towards you.

'Lord Jerningham' by [Newton] Booth Tarkington (1869–1946). *Lady's Realm* (Sept. 1906).

Tarkington is chiefly remembered as the author of *The Magnificent Ambersons* (1918), for which he won his first Pulitzer Prize (his second was for his 1921 novel *Alice Adams*) and which was later made into a film by Orson Welles. Tarkington also wrote numerous short stories, many about Penrod Schofield, a piratical 11-year-old argued by some to be an unacknowledged influence on Richmal Crompton's William Brown. Tarkington's early novels included the costume drama *Monsieur Beaucaire* (1900), a failure in the USA but a bestseller when published in Great Britain. In 1902 he was elected to the Indiana legislature on the Republican ticket, and in his final years suffered from partial blindness.

'Iconoclasts' by Norah [Robinson] Lofts [Mrs Robert Jorisch] (1904–83). From *I Met a Gypsy* (1935).

Norah Lofts (who also wrote as Juliet Astley and Peter Curtis) was one of the undisputed Queens of historical romance for nearly fifty years. She combined historical accuracy with storytelling capabilities of a high order. She won the Georgette Heyer prize for historical fiction in 1978.

'Mr Codesby's Behaviour' by A[lfred] M[cLelland] Burrage (1889–1956). *Chambers's Journal* (May 1936).

On the sudden death of his father, A. M. Burrage was driven to writing for a living at the early age of 16 to provide for his mother, sister, and aunt. The experience coloured his later life, and his fiction. Many of his heroes are part of the public school world to which he (ripped untimely from Douai College) never truly belonged, and often in his tales there is a harking back to a Golden Age of careless youth which he never really experienced. The desperate need to earn money turned him into one of the most prolific British writers of popular short stories of the twentieth century, and at the same time one of the most skilful. His work appeared in every important monthly fiction and general interest magazine of the inter-war years, as well as scores of weeklies. Although he wrote, inevitably, to a formula, his work often transcended its setting, distinguished as it was by Burrage's humour and fascination for character. Almost as a sideline he wrote ghost stories that were admired by Conan Doyle and M. R. James. A collection of these, *Someone in the Room* (1931), was published under the pseudonym 'Ex-Private X', as was his autobiographical memoir of the 1914–18 conflict, *War is War* (1930), one of the classics of Great War literature.

'Mr Mitchelbourne's Last Escapade' by A[lfred] E[dward] W[oodley] Mason (1865–1948). *Black and White* (Christmas Number, 13 Dec. 1900); reprinted in *Ensign Knightley and other stories* (1901).

Mason is perhaps best remembered for his novel *The Four Feathers* (1902), which, in the fifty years after its publication, sold over a million copies. He became Liberal MP for Coventry in 1906. A man of action, he sailed, climbed mountains (Mont Blanc five times, each time by a different route), and travelled all over the world; he also worked for British Intelligence during the First World War (among other experiences, he foiled a plan to infect British troops with anthrax and was present when Mata Hari was executed by firing squad). An accomplished and proficient writer, he published over thirty novels and many short stories.

'Martin's Close' by M[ontague] R[hodes] James (1862–1936). From *More Ghost Stories of an Antiquary* (1911).

James was the son of an Evangelical cleric, the rector of Great Livermere in Suffolk, and was educated at Eton and King's College, Cambridge, where he became a Fellow and eventually, in 1905, Provost. After the First World War he returned to take up the provostship of his old school. James was one of the outstanding scholars of his generation with an international reputation, principally in the fields of palaeography and apocryphal studies, though he is now generally remembered for his ghost stories, the first collection of which, *Ghost Stories of an Antiquary*, appeared in 1904. These and their successors—*More Ghost Stories of an Antiquary* (1911), *A Thin Ghost* (1919), and *A Warning to the Curious* (1925)— have remained in print ever since and set in train a still fertile antiquarian strain of English supernatural fiction. Only a handful of his stories had historical settings, though everything he wrote was imbued with a deep sense of how the past impinges on the present. 'Martin's Close' contains a brilliant demonstration of

James's mimetic powers and draws on his familiarity with the State Trials of the late seventeenth century, of which he was particularly fond. The setting, Sampford Courtenay in Devon, was a King's College property and was visited by James, as Dean of King's, in 1893.

'The Emancipation of Mrs Morley' by Clemence Dane [Winifred Ashton] (1888– 1965). *The Passing Show* (4 May 1935); reprinted in *Fate Cries Out* (1935).

Before the age of 30, Clemence Dane (her pseudonym was taken from the famous church at the head of London's Fleet Street, St Clement Danes) had a peripatetic career, teaching French in Geneva, studying art at the Slade School, London, and then in Dresden, more teaching in Ireland; then, abandoning both teaching and a promising career as a portrait painter, she took to the stage. Her first novel, *Regiment of Women* (1917), was a lesbian tragedy, and most of her work displays a strong, though never strident, feminist bias. Her stage experience led her to write a number of theatrical novels, of which *Broome Stages* (1931), a saga following the fortunes of a theatrical family through several generations, is the most famous. Dane also wrote furiously for the theatre itself: over thirty plays, including translations and adaptations. She also wrote a number of historical novels, though her best work in the genre is to be found in her short stories. At the age of 70 she took on the job of general editor for the pioneering 'Novels of Tomorrow' series published by Michael Joseph.

'The Squire's Story' by Elizabeth [Cleghorn] Gaskell (1810–65). *Household Words* (Christmas Number, 1853); reprinted in *Lizzie Leigh and other tales* (1855).

Born at Chelsea, the daughter of a Unitarian minister and sometime Keeper of the Treasury Records, Gaskell was brought up by an aunt in Knutsford, Cheshire— the original of Cranford. In 1832 she married William Gaskell, also a Unitarian minister. Her first book, *Mary Barton*, was published in 1848 and brought her into contact with Dickens, to whose periodicals *Household Words* and *All the Year Round* she contributed regularly thereafter. *Ruth* and *Cranford* both appeared in 1853. Though she wrote powerfully on themes of contemporary concern (as in her great novel describing industrial conditions, *North and South*, first serialized in *Household Words* in 1854–5) she often turned to history for her settings, notably in her novel of the eighteenth century *Sylvia's Lovers* (1863), as well as in several of her short stories.

'The Missing Shakespeare Manuscript' by Lillian de la Torre [Lillian Bueno McCue] (1902–93). *Ellery Queen's Mystery Magazine* (July 1947); reprinted in *The Detections of Dr Sam: Johnson* (1960).

Lillian de la Torre added a new twist to the 'Mystorical'—the mystery set in historical times. Fascinated by the past, and particularly by the era covered by the 'Newgate Calendar' (the rich, gamey, often scurrilous contemporary record of malefactors hanged at Tyburn during the period 1680–1820), she brought an en- quiring and imaginative mind to the subject and a determination to solve ancient mysteries and old crimes. Her solutions to various historical puzzles (what really happened to the serving-wench Elizabeth Canning in 1753?; was Archibald Douglas the true heir to the vast fortune of the Duke of Douglas in 1769 or a French glass-blower's son?) are both ingenious and plausible. She also wrote cookbooks, novels, poetry, and plays. The work for which she will be best remembered is the series of short detective stories featuring Dr Johnson as an eighteenth-century Sherlock Holmes, with James Boswell as his Watson. This success (from 1946 onwards) sparked off an entire sub-genre of famous historical figures acting the sleuth. The story is Lillian de la Torre's version of the notorious William Henry Ireland's forgery of *Vortigern* (1796), transferred to Garricks 'Shakespeare Jubilee', which Boswell attended (dressed as described), although Johnson did not.

'My Dear Clarissa' by Brinsley Moore (*fl.* 1910–30). *London Magazine* (Aug. 1928).

Brinsley Moore was an industrious contributor to the fiction magazines of the first third of the twentieth century, though he appears never to have published in book form (at least under this name). His speciality was the comic tale told in epistolary form, usually, in the early part of his career, with East End or dockland settings, following in the footsteps of such 'Cockney' or 'suburban' chroniclers as Barry Pain, Keble Howard, W. W. Jacobs, and W. Pett Ridge. Invited to contribute to the 1916 Christmas issue of the *Premier Magazine* he turned to the eighteenth century for inspiration, altering his approach to take in the historical background. Where before his stories had consisted of a succession of letters building up to a comic climax, he now (as narrator) took on the persona of a serendipitist, forever stumbling across caches of old manuscripts and bundles of faded letters and extracting a story from these. An artful touch was to add footnotes to his tales, references to apparently well-known eighteenth-century diaries or journals (a favourite source was 'The Diary of Anne Somers') that were in fact all concocted by Moore. This technique was successful with magazine editors and enormously popular with readers. His comic Regency tales ran for some years in the *Premier* before transferring to the prestigious *London Magazine*.

'The Company of the Marjolaine' by John Buchan, first Baron Tweedsmuir (1875–1940). From *The Moon Endureth: Tales and Fancies* (1912).

Born in Perth, the son of a minister, Buchan's first work—including a novel, *Sir Quixote of the Moors*—was published whilst he was still an undergraduate at Glasgow University. In 1901 he went out to South Africa with the High Commission charged with the reconstruction of the country following the Boer War; thereafter he combined writing with an active public life, his career culminating in the governor-generalship of Canada in 1935. *The Thirty-Nine Steps* (1915) inaugurated an era of heroic spy fiction populated by clean-living, patriotic heroes and ruthless foreign opponents and was followed by other novels featuring Richard Hannay, including *Greenmantle* (1916) and *Mr Standfast* (1919). Buchan also wrote many non-fiction works, including biographies of Montrose (1913, 1928) and Scott (1932).

The story's last line indicates a setting of 1785–6, well after the Peace of Paris (1783) but before the Constitutional Convention (1787): the reference to a 'Convention' is to the Continental Congress, which governed the United States

1776–89. Charles Edward Stuart (1720–88) had separated from his wife in 1780. The narrator is fictional but is evidently supposed to be a nephew of 'Champagne' Charles Townshend (1725–67), who as Chancellor of the Exchequer introduced duties that helped propel the American colonists towards rebellion.

'Words for Things' by Emma Donoghue (b. 1969). First published in *The Penguin Book of Lesbian Short Stories*, ed. Margaret Reynolds (1993).

Born in Dublin, Emma Donoghue published *Passions Between Women: British Lesbian Culture 1668–1801* in 1993. She has written two modern novels, *Stir-Fry* (1994) and *Hood* (1995). Her historical plays are *I Know My Own Heart* (1993, based on Anne Lister's Regency diaries) and *Ladies and Gentlemen* (1995, set in 1880s American vaudeville). Donoghue lives in Cambridge, where she is completing a Ph.D. on friendship between men and women in eighteenth-century fiction.

In October 1786 Mary Wollstonecraft took up the position of governess to the children of the Kingsborough family of Cork. Her employment lasted for nearly a year, ending in August 1787. Her letters to her family and friends present a graphic account of her time in Cork.

'Two Good Patriots' by Baroness [Emmuska Magdalena Rosalia Maria Joseph Barbara] Orczy [Mrs Montagu Barstow] (1865–1947). *Printer's Pie* (July 1912); reprinted in *The League of the Scarlet Pimpernel* (1919).

Born in Hungary, the daughter of Baron Felix Orczy, a musician and experimental agriculturalist, Emmuska Orczy studied painting at the West London School of Art and later at the Heatherley School of Art, where she met her husband, the artist Henry Montagu Barstow. As well as illustrating children's stories, Baroness Orczy also published translations and short stories, her series of 'Old Man in the Corner' detective stories appearing in the *Royal Magazine* from 1901. The novel that made her name, *The Scarlet Pimpernel*, was published in 1905, although it had been written three years earlier and rejected by several publishers before being turned into a play. The book achieved huge popularity and was made into a film by fellow-Hungarian Alexander Korda in 1933. Orczy went on to write several sequels to *The Scarlet Pimpernel*, as well as many other historical romances of indifferent quality. She and her husband settled in Monte Carlo after the First World War, though the Baroness returned to London when Montagu Barstow died in 1943.

'A Committee Man of "The Terror"' by Thomas Hardy (1840–1928). *Illustrated London News* (Christmas Number, 23 Nov. 1896); reprinted in *A Changed Man* (1913).

Hardy was born at Upper Bockhampton, Dorset, the son of a builder. He was apprenticed to an ecclesiastical architect in Dorchester from 1856 to 1861 before moving to London to study under Sir Arthur Bloomfield. His first published novel, *Desperate Remedies*, appeared (at his own expense) in 1871 and was followed by *Under the Greenwood Tree* (1872) and *A Pair of Blue Eyes* (1873). The following year the publication of *Far From the Madding Crowd* allowed him to give up his architectural career and to marry Emma Gifford, whom he had met some years earlier. Over the next twenty years he published a dozen novels, including *Tess of the D'Urbervilles* (1891) and *Jude the Obscure* (1895). Both attracted a good deal of hostile criticism—especially *Jude*—and their reception was instrumental in Hardy's giving up fiction to concentrate on poetry. Emma Hardy died in 1912, and two years later Hardy married Florence Dugdale. His popular historical novel *The Trumpet Major* (1880) was set during the Napoleonic Wars, a period to which he often returned in his short stories, as well as in his great blank-verse epic *The Dynasts* (1904–8).

'Runaway Match' by Georgette Heyer (1902–74). *Woman's Journal* (Apr. 1936).

Despite her name being inextricably linked to the historical (or, more properly, the Regency) romance—all forsooth and furbelows—Georgette Heyer was a deliberately, and seriously, comic novelist. Even the Elizabethan *Beauvallet* (1929)—which begins 'The deck was a shambles. Men lay dead and dying . . .'—has its frivolous moments. Her first novel, *The Black Moth* (1921), a frothy confection of highwaymen, lost brothers, mettlesome heroines (a particularly Heyerian type), and a fortune going begging, was written at the age of 17 and published two years later. She began to write for the fiction monthlies during the early 1920s and went on to write over fifty novels, including a dozen detective stories. Her historical plots were complicated, her characters compelling and often very funny, her dialogue extraordinary.

'The Singular Adventure of a Small Free-Trader' by [Sir] Arthur [Thomas] Quiller-Couch (1863–1944). *Woman at Home* (Feb. 1897); reprinted in *Old Fires and Profitable Ghosts* (1900).

Scholar, novelist, short-story writer, and anthologist, Arthur Quiller-Couch was born at Bodmin, Cornwall, and educated at Newton Abbot College, Clifton College, and Trinity College, Oxford. He was a lecturer in Classics at Trinity in 1886–7 before moving to London, where he was assistant editor of the *Speaker* 1890–9. He was knighted in 1910 and two years later succeeded A. W. Verrall as King Edward VII Professor of English Literature at Cambridge. Under the pseudonym 'Q' he had by this time been publishing novels for over twenty years —including *Troy Town* (1888), *The Splendid Spur* (1889), and *The Ship of Stars* (1899). He also wrote poetry and criticism and was a brilliant lecturer. But perhaps his finest work is to be found amongst his dozens of well-crafted short stories, many of which have historical settings. As an anthologist he has an enduring monument in *The Oxford Book of English Verse*, published in 1900 and still going strong. In 1937 he was elected Mayor of Fowey—the Troy Town of his fiction.

'How the Brigadier Played For a Kingdom' by [Sir] A[rthur] Conan Doyle (1859–1930). *Strand Magazine* (Dec. 1895); reprinted in *The Exploits of Brigadier Gerard* (1896).

Born in Edinburgh and educated by the Jesuits at Stonyhurst, Conan Doyle studied medicine at Edinburgh University, graduating in 1881. His most famous character, Sherlock Holmes, first appeared in *A Study in Scarlet* in 1887, but only came to mass prominence after the first Holmes short stories, beginning with 'A

Scandal in Bohemia', were published in the *Strand Magazine* (from July 1891). When Holmes was 'killed off' in 1893, Conan Doyle replaced his great detective with Brigadier Étienne Gerard, a cavalry officer in Napoleon's army, writing a sequence of individual stories based round this character that were published in the *Strand* and later collected as *The Exploits of Brigadier Gerard* (1896) and *Adventures of Gerard* (1903). Conan Doyle also wrote several novels in the historical vein, beginning with his first major work (on the Monmouth revolt of 1685), *Micah Clarke: His Statement* (1889), and including *The White Company* (1891), *Rodney Stone* (1896), and *Uncle Bernac* (1897).

The setting of the story is March 1813, when Prussia had concluded a secret alliance with Russia against Napoleon.

'Catherine Carr' by Mary [Eleanor] Wilkins [Freeman] (1852–1930). From *The Love of Parson Lord and other stories* (1900).

Born in Randolph, Massachusetts, where she continued to live until her marriage in 1902, Mary E. Wilkins set most of her fiction in rural New England. Her first collection of stories, *A Humble Romance and other stories*, was published in 1887, but her reputation was established by *A New England Nun and other stories* (1891). Her other novels include *Jane Field* (1892), *Pembroke* (1894), and *Jerome, A Poor Man* (1897); she was also the author of a play about the Salem witch trials and a historical novel, *The Heart's Highway* (1900). She was at her best, however, in the short story and is perhaps best remembered for the supernatural tales gathered together in *The Wind in the Rose Bush* (1903).

'Wash' by William [Harrison] Faulkner (1897–1962). *Harper's Magazine* (Feb. 1934); reprinted in *Dr Martino and other stories* (1934).

Faulkner (originally Falkner) grew up in Oxford, Mississippi, in a family not unlike that of the Sartoris clan in his novels. Failing US height requirements, he joined the British Royal Air Force in Canada, but the First World War ended before he completed his training. He published a volume of pastoral poems, *The Marble Faun*, in 1924, his first novel, *Soldiers' Pay* appearing in 1926 with the help of Sherwood Anderson. He found his full voice in *Sartoris* (1929), the first novel in his long, loosely connected Yoknapatawpha saga charting the decline of the Sartoris, Benbow, and McCaslin families (representing the Old South) and the rise of the unscrupulous Snopes. Amongst his many novels are *The Sound and the Fury* (1929), *As I Lay Dying* (1930), *Absalom, Absalom!* (1936), *The Unvanquished* (1938), and *A Fable* (1954), which earned him the Pulitzer Prize. Many of the characters in Faulkner's novels reappear in his short-story collections (e.g. *These 13* (1931), *Idyll in the Desert* (1931), *Miss Zilphia Grant* (1932), and *Dr Martino* (1934)). During the 1930s and 1940s Faulkner worked in Hollywood as a scriptwriter and was awarded the Nobel Prize for Literature in 1950.

'The Way We Lived Then' by E. M. Delafield [Edmée Elizabeth Monica Dashwood, *née* de la Pasture] (1890–1943). *Good Housekeeping* (Apr. 1932).

E. M. Delafield (her pseudonym was supplied by her younger sister Yolande) spent most of her early career trying to escape the enfolding influence of her mother, the society novelist Mrs Henry de la Pasture (in Delafield's work there is a preponderance of tyrannical matriarchs). Her first book, *Zella Sees Herself* (a wry essay in self-deception), was rescued from the Heinemann slush pile by the novelist Fryn Tennyson Jesse and published in 1917. From then until her early death in 1943 she published over thirty novels, five collections of short stories, a couple of plays, and four volumes chronicling the exploits of her 'Provincial Lady' (who, like her creator, writes novels and has a mildly humdrum husband and two exasperating children). These were written in diary form, were a huge success during the 1930s, and still retain their freshness, humour, and charm today. Most of the rest of Delafield's output, which mainly concerns women struggling against adversity, was seriously comic—witty and astringent, and distinctly darker in tone than the 'Provincial Lady' books. She had a fascination for the Victorian period and edited for Leonard and Virginia Woolf *Ladies and Gentlemen in Victorian Fiction* (1937) as well as a book on the Brontës as seen by their contemporaries (1935).

'An Edwardian Tableau' by Sara Maitland (b. 1950). From *Introduction 5: Stories by New Writers* (1974).

Sara Maitland was educated privately and at St Anne's College, Oxford, where she read English. She won the Somerset Maugham Award in 1979 for her first novel, *Daughter of Jerusalem*. Her other works include *Virgin Territory* (1984), a study of the music-hall artist Vesta Tilley (1986), and a collection of short stories, *Telling Tales* (1983).

PUBLISHER'S ACKNOWLEDGEMENTS

The editors and publishers gratefully acknowledge permission to reprint copyright material in this book as follows:

Marjorie Bowen, 'Twilight'. Copyright the Estate of Marjorie Bowen.

A. M. Burrage, 'Mr Codesby's Behaviour'. Copyright 1936 A. M. Burrage.

Clemence Dane, 'The Emancipation of Mrs Morley', first published in *The Passing Show* (1935). Reprinted by permission of Laurence Pollinger Ltd., on behalf of the Estate of Clemence Dane.

Emma Donaghue, 'Words for Things', © 1993 Emma Donaghue. Reprinted by permission of the Carol Davidson and Robert Lucas Literary Agency on behalf of the author.

William Faulkner, 'The Wish'. Copyright 1934 William Faulkner, from *Dr Martino and Other Stories*. Reprinted by permission of Curtis Brown Ltd.

Georgette Heyer, 'Runaway Match'. Reproduced by permission of Sir Richard Rougier: copyright 1936 by Georgette Heyer.

Aldous Huxley, 'The Death of Lully' from *Limbo*. Reprinted by permission of Random House UK Ltd., and HarperCollins, USA on behalf of Mrs Laura Huxley.

Jack Lindsay, 'When Greek Meets Greek' from *Come Home At Last* (Nicholson & Watson). Reprinted by permission of Murray Pollinger Literary Agent.

Norah Lofts, 'Iconoclasts'. Copyright 1935 Norah Lofts.

Sara Maitland, 'An Edwardian Tableau', first published in *Introduction 5: Stories by New Writers*. Reprinted by permission of the author.

A. E. W. Mason, 'Mr Mitchelbourne's Last Escapade', from *Black and White* by A. E. W. Mason. Reprinted by permission of A. P. Watt Ltd. on behalf of Trinity College, Oxford.

Naomi Mitchison, 'Cottia Went to Bibracte' from *When the Bough Breaks and Other Stories*. Reprinted by permission of David Higham Associates Ltd.

Baroness Orczy, 'Two Good Patriots' from *The League of the Scarlet Pimpernel* by Baroness Orczy. Reprinted by permission of A. P. Watt Ltd. on behalf of Sara Orczy-Barstow Brown.

Eugene O'Sullivan, 'Overture', first published in *New Edinburgh Review*, © 1978 Eugene O'Sullivan. Reprinted by permission of the author.

Rafael Sabatini, 'The Scapulary', reprinted in *Turbulent Tales*. Reprinted by permission of A. P. Watt Ltd. on behalf of The Royal Institute for the Blind, the Imperial Cancer Research Fund and Action Research.

Arthur D. Howden Smith, 'A Trooper of the Thessalians', from *Grey Maiden: The Story of a Sword Through the Ages* (1929). Copyright the Estate of Arthur D. Howden Smith.

Booth Tarkington, 'Lord Jerningham'. First published in *Lady's Realm*, September 1906. Copyright the Estate of Booth Tarkington.

Lillian de la Torre, 'The Black Stone of Dr Dee', from *Ellery Queen's Mystery Magazine*. Reprinted by permission of David Higham Associates Ltd.

Fred Urquhart, 'Pretty Prickly English Rose', from *Proud Lady In A Cage*. Reprinted by permission of the author.

Any errors or omissions in the above list are entirely unintentional. If contacted the publisher will be pleased to make corrections at the earliest opportunity.

OXFORD

MORE OXFORD PAPERBACKS

This book is just one of nearly 1000 Oxford Paperbacks currently in print. If you would like details of other Oxford Paperbacks, including titles in the World's Classics, Oxford Reference, Oxford Books, OPUS, Past Masters, Oxford Authors, and Oxford Shakespeare series, please write to:

UK and Europe: Oxford Paperbacks Publicity Manager, Arts and Reference Publicity Department, Oxford University Press, Walton Street, Oxford OX2 6DP.

Customers in UK and Europe will find Oxford Paperbacks available in all good bookshops. But in case of difficulty please send orders to the Cash-with-Order Department, Oxford University Press Distribution Services, Saxon Way West, Corby, Northants NN18 9ES. Tel: 01536 741519; Fax: 01536 746337. Please send a cheque for the total cost of the books, plus £1.75 postage and packing for orders under £20; £2.75 for orders over £20. Customers outside the UK should add 10% of the cost of the books for postage and packing.

USA: Oxford Paperbacks Marketing Manager, Oxford University Press, Inc., 200 Madison Avenue, New York, N.Y. 10016.

Canada: Trade Department, Oxford University Press, 70 Wynford Drive, Don Mills, Ontario M3C 1J9.

Australia: Trade Marketing Manager, Oxford University Press, G.P.O. Box 2784Y, Melbourne 3001, Victoria.

South Africa: Oxford University Press, P.O. Box 1141, Cape Town 8000.

OXFORD REFERENCE

THE CONCISE OXFORD COMPANION TO ENGLISH LITERATURE

Edited by Margaret Drabble and Jenny Stringer

Based on the immensely popular fifth edition of the *Oxford Companion to English Literature* this is an indispensable, compact guide to the central matter of English literature.

There are more than 5,000 entries on the lives and works of authors, poets, playwrights, essayists, philosophers, and historians; plot summaries of novels and plays; literary movements; fictional characters; legends; theatres; periodicals; and much more.

The book's sharpened focus on the English literature of the British Isles makes it especially convenient to use, but there is still generous coverage of the literature of other countries and of other disciplines which have influenced or been influenced by English literature.

From reviews of *The Oxford Companion to English Literature*:

'a book which one turns to with constant pleasure . . . a book with much style and little prejudice' Iain Gilchrist, *TLS*

'it is quite difficult to imagine, in this genre, a more useful publication' Frank Kermode, *London Review of Books*

'incarnates a living sense of tradition . . . sensitive not to fashion merely but to the spirit of the age' Christopher Ricks, *Sunday Times*